Engineering Society

Also by Kerstin Brückweh

THE VOICE OF THE CITIZEN CONSUMER
A History of Market Research, Consumer Movements, and the Political Public Sphere (*editor*)

Also by Dirk Schumann

RAISING CITIZENS IN THE 'CENTURY OF THE CHILD'
The United States and German Central Europe in Comparative Perspective (*editor*)

Also by Richard F. Wetzell

INVENTING THE CRIMINAL
A History of German Criminology, 1880–1945

Also by Benjamin Ziemann

CONTESTED COMMEMORATIONS
Republican War Veterans and Weimar Political Culture

Engineering Society

The Role of the Human and Social Sciences in Modern Societies, 1880–1980

Edited by

Kerstin Brückweh
Research Fellow, German Historical Institute London

Dirk Schumann
Professor of Modern and Contemporary History, Georg-August-Universität Göttingen

Richard F. Wetzell
Research Fellow, German Historical Institute Washington DC

and

Benjamin Ziemann
Professor of Modern German History, University of Sheffield

palgrave
macmillan

Softcover reprint of the hardcover 1st edition 2012 978-0-230-27907-0

First published 2012 by
PALGRAVE MACMILLAN

Palgrave Macmillan in the UK is an imprint of Macmillan Publishers Limited, registered in England, company number 785998, of Houndmills, Basingstoke, Hampshire RG21 6XS.

Palgrave Macmillan in the US is a division of St Martin's Press LLC, 175 Fifth Avenue, New York, NY 10010.

Palgrave Macmillan is the global academic imprint of the above companies and has companies and representatives throughout the world.

Palgrave® and Macmillan® are registered trademarks in the United States, the United Kingdom, Europe and other countries.

ISBN 978-1-349-32680-8 ISBN 978-1-137-28450-1 (eBook)
DOI 10.1057/9781137284501

This book is printed on paper suitable for recycling and made from fully managed and sustained forest sources. Logging, pulping and manufacturing processes are expected to conform to the environmental regulations of the country of origin.

A catalogue record for this book is available from the British Library.

A catalog record for this book is available from the Library of Congress.

10 9 8 7 6 5 4 3 2 1
21 20 19 18 17 16 15 14 13 12

Contents

List of Figures and Tables

Figures

Tables

Notes on Contributors

Peter Becker is Professor of History at the University of Vienna. He researches the cultural history of public administration, the history of criminology and policing, and the history of biological reasoning on the social during the twentieth century. His main publications include *Verderbnis und Entartung. Zur Geschichte der Kriminologie des 19. Jahrhunderts als Diskurs und Praxis* (2002); *Sprachvollzug im Amt. Kommunikation und Verwaltung im Europa des 19. und 20. Jahrhunderts* (editor, 2011); *The Criminals and their Scientists. The History of Criminology in International Perspective* (editor together with Richard Wetzell, 2006); and *Little Tools of Knowledge: Historical Essays on Academic and Bureaucratic Practices* (editor together with William Clark, 2001).

Kerstin Brückweh is a Research Fellow at the German Historical Institute London. She is the author of *Mordlust: Serienmorde, Gewalt und Emotionen im 20. Jahrhundert* [Mordlust. Serial Killings, Violence and Emotions in 20th-Century Germany] (2006) and the editor of *The Voice of the Citizen Consumer: A History of Market Research, Consumer Movements, and the Political Public Sphere* (2011). She is now working on a history of British self-observations in survey research and censuses in the nineteenth and twentieth centuries.

Chris Dols is a cultural historian who works as a junior researcher at the Radboud University Nijmegen. His recent publications include two books on the nineteenth- and twentieth-century Catholic temperance movement. In his current (Ph.D.) project, he focuses on the scientization of the social in the Catholic community in the Netherlands (1921–72).

Thomas Etzemüller is Adjunct Professor in Modern History at the University of Oldenburg. His recent publications include *Ein ewigwährender Untergang. Der apokalyptische Bevölkerungsdiskurs im 20. Jahrhundert* (2007); *Die Romantik der Rationalität. Alva & Gunnar Myrdal – Social Engineering in Schweden* (2010); and *Die Ordnung der Moderne. Social Engineering im 20. Jahrhundert* (editor, 2009). His current research focuses on the invention of 'Population' as a means to check modernity, imaginary landscapes, and couples in modernity: negotiating gender relations.

Sarah E. Igo is Associate Professor of History, Political Science, and Sociology at Vanderbilt University in Nashville, Tennessee. She is the author of *The Averaged American: Surveys, Citizens and the Making of a Mass Public* (2007) and is currently working on a cultural history of privacy in the twentieth century.

Till Kössler is Professor of the History of Education, Childhood, and Youth at the Ruhr-Universität Bochum. He received his Ph.D. in 2003 on the basis of a study on the social and cultural history of communism in West Germany after 1945. Between 2003 and 2011 he was assistant professor for modern European history at the Ludwig-Maximilians-Universität in Munich, and a visiting scholar at the Universidad Complutense de Madrid in 2007–8. He has written extensively on the history of communism and the German workers' movement, the history of political parties, Spanish history, and the history of childhood. He recently finished a book-length study on Catholicism and childhood in Spain between 1890 and 1936, which will be published in 2012. He is also the co-editor of a volume on miracles and wonders in twentieth century culture, titled *Wunder. Poetik und Politik des Staunens im 20. Jahrhundert* (2011).

Anja Kruke is Head of Department of the Archiv der sozialen Demokratie at the Friedrich-Ebert-Stiftung in Bonn. She has held lectureships at the universities of Hamburg, Bochum, and Cologne. Her most recent books include *Demoskopie in der Bundesrepublik Deutschland. Meinungsforschung, Parteien und Medien 1949–1990* [Public Opinion Polling in West Germany: Polls, Parties and Media 1949–1990] (2007; second edition, 2011) and *Solidargemeinschaft und Erinnerungskultur im 20. Jahrhundert. Beiträge zu Gewerkschaften, Nationalsozialismus und Geschichtspolitik* [Communities of Solidarity and Memory Culture in the 20th century: Essays on Trade Unions, National Socialism and the Politics of History] (co-editor, 2009).

Martin Lengwiler is Professor of Modern History at the University of Basel, Switzerland. His most recent books include *Praxisbuch Geschichte. Einführung in die historischen Methoden* [Practical Manual History: Introduction to Historical Methodology] (2011); *Das Präventive Selbst. Eine Kulturgeschichte moderner Gesundheitspolitik* [The Preventive Self: A Cultural History of Modern Public Health Policy] (co-editor, 2010); *Umbruch an der „inneren Front“: Krieg und Sozialpolitik in der Schweiz 1938–1948* [Change at the 'Inner Front': War and Social Policies in Switzerland 1938–1948] (co-editor, 2009).

Elizabeth Lunbeck is Nelson Tyrone, Jr. Professor of History, and Professor of Psychiatry, at Vanderbilt University. She is the author of *The Psychiatric Persuasion: Knowledge, Gender and Power in Modern America*, and *Family Romance, Family Secrets: Case Notes from an American Psychoanalysis* (with Simon Bennett, 2003). She has edited several volumes, the most recent being *Histories of Scientific Observation* (with Lorraine Daston, 2011). She is currently finishing *The Americanization of Narcissism* (Harvard University Press).

Julia Moses is Lecturer in Modern History at the University of Sheffield. She is currently completing her first book, titled *The First Modern Risk: Workplace Accidents and the Origins of European Welfare States*, based on her Cambridge Ph.D. thesis, and has co-edited, with Michael Lobban, *Comparative Studies in the Development of the Law of Torts in Europe: The Impact of Ideas* (Cambridge University Press, forthcoming).

Harry Oosterhuis teaches history at the Faculty of Arts and Social Sciences of Maastricht University and is affiliated with the Huizinga Research School for Cultural History in Amsterdam. His research focuses on the cultural and social history of psychiatry and mental health care, of sexuality and gender, of health and citizenship, and of bicycling. He is the author of *Stepchildren of Nature: Krafft-Ebing,*

Psychiatry, and the Making of Sexual Identity (University of Chicago Press, 2000), co-author of *Verward van geest en ander ongerief. Psychiatrie en geestelijke gezondheidszorg in Nederland (1870–2005)* (2008), and co-editor of *Psychiatric Cultures Compared. Psychiatry and Mental Health Care in the Twentieth Century: Comparisons and Approaches* (Amsterdam University Press, 2005).

Lutz Raphael is Professor of Contemporary History at the University of Trier. He was visiting professor at the EHESS Paris (in 1999, 2003, and 2009), and during 2010–11 was fellow of the international research centre 'Work and Human Lifecycle in Global History' at the Humboldt-Universität zu Berlin. His publications include *Recht und Ordnung. Herrschaft durch Verwaltung im Europa des 19. Jahrhunderts* (2000); *Geschichtswissenschaft im Zeitalter der Extreme* (2003); *Nach dem Boom. Perspektiven auf die Zeitgeschichte seit 1970* (with Anselm Doering-Manteuffel, 2008); and *The Atlas of European Historiography: The Making of a Profession 1800–2005* (co-editor with Ilaria Porciani, 2010).

Dirk Schumann is Professor of Modern and Contemporary History at Georg-August-Universität Göttingen. His most recent books include *Raising Citizens in the 'Century of the Child': The United States and German Central Europe in Comparative Perspective* (editor, 2010), *Political Violence in the Weimar Republic, 1918–1933: Fight for the Streets and Fear of Civil War* (2009), and *Between Mass Death and Individual Loss: The Place of the Dead in Twentieth-Century Germany* (co-editor with Alon Confino and Paul Betts, 2008).

Stefan Schwarzkopf is Associate Professor of Business History at Copenhagen Business School and his work on the history of advertising, marketing, market research, and political propaganda in Europe and North America has appeared in journals such as *Contemporary British History*, the *Journal of Macromarketing, Management & Organizational History*, *Journal of Cultural Economy*, and *Theory, Culture & Society*. Together with Rainer Gries, he is the editor of *Ernest Dichter and Motivation Research: New Perspectives on the Making of Post-War Consumer Culture* (Palgrave Macmillan, 2010).

Mathew Thomson is Reader in the Department of History at the University of Warwick. He is author of *The Problem of Mental Deficiency: Eugenics, Democracy and Social Policy in Britain, c. 1870–1959* (Oxford University Press, 1998) and *Psychological Subjects: Identity, Culture, and Health in Twentieth-Century Britain* (Oxford University Press, 2006).

Emil Walter-Busch is Senior Lecturer Emeritus of Applied Social Research and Social Psychology at the Universities of St Gall and Zurich. His most recent books include *Burckhardt und Nietzsche. Im Revolutionszeitalter* (2012); *Geschichte der Frankfurter Schule. Kritische Theorie und Politik* [History of the Frankfurt School: Critical Theory and Politics] (2010); *Arbeits- und Organisationspsychologie im Überblick* [Introducing Organizational Psychology] (2008); and *Faktor Mensch. Formen angewandter Sozialforschung der Wirtschaft in Europa und den USA, 1890–1950* [The Human Factor: Applied Social Research in Europe and the USA, 1890–1950] (2006).

Richard F. Wetzell is a Research Fellow at the German Historical Institute in Washington, DC. He was a Postdoctoral Fellow at Harvard University and has taught at the University of Maryland, Georgetown University, and the Catholic University of America. His publications include *Inventing the Criminal: A History of German Criminology, 1880–1945* (University of North Carolina Press, 2000) as well as two edited collections – *Criminals and their Scientists: The History of Criminology in International Perspective* (co-edited with Peter Becker, Cambridge University Press, 2006) and *Crime and Criminal Justice in Modern Germany* (Berghahn Books, 2012). He is currently writing a legal and political history of penal reform in Germany from 1880 to 1970.

Benjamin Ziemann is Professor of Modern German History at the University of Sheffield. His most recent books include *Contested Commemorations: Republican War Veterans and Weimar Political Culture* (2013); *Sozialgeschichte der Religion. Von der Reformation bis zur Gegenwart* [Social History of Religion: From the Reformation to the Present] (2009); *Katholische Kirche und Sozialwissenschaften 1945–1975* [The Catholic Church and the Social Sciences 1945–1975] (co-editor, 2007); *Reading Primary Sources: The Interpretation of Texts from 19th and 20th Century History* (2008); and *War Experiences in Rural Germany, 1914–1923* (2007).

Acknowledgements

This volume pursues three major aims. First, we seek to analyse the impact of the human and social sciences in various fields of application, taking a comparative perspective that includes case studies from Great Britain, Italy, Germany, the Netherlands, Sweden, Switzerland, Spain, and the United States. Second, we employ the notion of the 'scientization of the social' in order to integrate the history of the human and social sciences into the mainstream of the historiography on Western societies in the twentieth century. The endeavours of pollsters, psychotherapists, statisticians, and other experts who tried to bring scientific knowledge to bear on social problems were part and parcel of the many social reform projects that have characterized Western societies since 1880. Finally, we aim to reveal the ambivalences and unintended side effects of the scientization of the social in order to contribute to a reflexive social history of the twentieth century.

Most of the chapters in this volume were presented in preliminary form at an international conference that took place at the Humanities Research Institute, University of Sheffield. During the conference, Sabine Maasen and Felix Keller offered comments on two panels and stimulated our discussion with their sociological expertise. Both the conference and the work on this volume were generously supported by the German Historical Institute in London and the German Historical Institute in Washington, DC. We are indebted to their directors, Andreas Gestrich and Hartmut Berghoff, for this support. During the preparation for the conference and this volume, we received intellectual and practical support from a number of colleagues. In particular, we would like to thank Moritz Föllmer, Murray Goot, Lynda Hodge-Mannion, Holger Nehring, Theodore Porter, Dan Scroop, and Anja Kruke and Gabriele Lutterbeck at the Archiv der sozialen Demokratie in Bonn, which provided the cover image. Jon Ashby did a superb job copy-editing some of the chapters. The kind permission of Riksbyggen Press in Stockholm, The History of Advertising Trust in Norwich, and the Katholiek Documentatie Centrum in Nijmegen to print copyrighted material is gratefully acknowledged. Last but not least, we wish to thank Michael Strang, Ruth Ireland, Jenny McCall, and Clare Mence at Palgrave Macmillan for their interest in our project and the anonymous reviewers for their helpful suggestions.

The editors
February 2012

1
Introduction: The Scientization of the Social in Comparative Perspective

Benjamin Ziemann, Richard F. Wetzell, Dirk Schumann, and Kerstin Brückweh

In a memorandum submitted to Germany's Catholic bishops in 1905, the Jesuit priest Hermann Krose called for the establishment of a central statistical office in the Catholic Church. Krose was strongly influenced by the ideas of the Belgian astronomer and statistician Alphonse Quetelet (1796–1874). According to Quetelet, the behaviour of individuals in a given social context displayed a certain probability to commit crime, suicide or other social actions. Since differences between individuals disappeared in larger aggregates, societies as a whole showed characteristic patterns that could be compared by statistical means.[1] A central statistical office for Germany's Catholic Church, Krose proposed, should gather data on the number of Catholics and priests in each diocese, the numbers of conversions and lapses, the percentage of Easter communicants, and the average number of church-goers on Sunday. Such data, he argued, would be a crucial weapon in the confessional conflict with the Protestants and also demonstrate the modern administrative standards of the church. Krose began publishing his own annual compilation of data in 1908 and set up a provisional office for church statistics in 1909. A full-fledged statistical office of the Catholic Church was finally set up in 1915. Well into the post-Vatican II period, statistical data on church-goers and other statistical parameters of practised piety provided a crucial basis for internal debates on pastoral shortcomings and reform strategies in the German Catholic Church.[2]

Krose is just one vivid example of the many practitioners of the social sciences who have endeavoured to analyse and reform social relations in various fields of society. They all contributed to what this volume calls the 'scientization of the social'. In the first section of this Introduction, we discuss this concept, issues of periodization, and theoretical frameworks and implications. Reflecting the organization of the chapters in this volume, the following three sections of the Introduction examine the dynamics of scientization in three fields of applied knowledge: first, social and penal policy; secondly, therapeutical approaches; and thirdly, the application of opinion polling, marketing techniques, and organizational research by business firms, political parties, and churches.

This volume uses the notion of the 'scientization of the social', coined by the historian Lutz Raphael, to analyse the intended and unintended consequences that the 'continuing presence of experts from the human sciences, their arguments, and the results of their research had in administrative bodies and in industrial firms, in parties and parliaments'. Where possible, it also seeks to trace the impact of these social science experts on the discursive construction of meaning in the context of 'social groups, classes and milieux' and their everyday lives.[3] An analysis of the scientizing of the social allows us to chart the ways in which the practical application of a broad range of human sciences conceptualized social problems and those of individuals, and offered to solve them by means of statistical calculation, testing, surveying, counselling or other forms of therapy.

The term 'scientization' should not be confused with 'scientism', a term introduced into the English language in the 1940s by the economist Friedrich von Hayek, who meant to criticize, as methodological 'scientism', the view that the methods of the natural sciences are the most 'appropriate way of understanding social phenomena'.[4] While such a tendency to ape the epistemology of the natural sciences did indeed pervade American social science,[5] our concept of the scientization of the social simply refers to the application of social scientific knowledge in various social fields. Furthermore, our use of the term 'scientization' should be distinguished from that of Jürgen Habermas, who used the German term *Verwissenschaftlichung* in an article on the 'scientization of politics and public opinion', first published in 1964.[6] A key proponent of the Frankfurt School of 'Critical Theory', Habermas bemoaned the colonizing effects that the modern social sciences were having on the life-worlds of ordinary people. For Habermas, 'scientization' was thus shorthand for a pervasive and dangerous tendency of the social sciences to replace free political deliberation among equal and enlightened citizens with a technocratic 'calculation' of political decisions.[7] While we also stress the ambivalent results of the scientization of the social, we, unlike Habermas, use 'scientization' as a value-neutral analytical concept.

The scientization of the social shaped not only social relations (between welfare state clients and their case workers, for instance) and the ways in which complex organizations such as firms, churches or parties operated. The application of social scientific knowledge also changed the self-descriptions of society. In the late twentieth-century America, for instance, 'narcissism', a term originally coined to probe clinical phenomena, was transformed into a category of social critique to highlight deficiencies in American mass culture.[8] In another transfer of meaning, the scientization of market research techniques such as panel surveys and consumer interviews contributed to the idea that the marketplace is a 'democracy of goods', and thus helped to establish semantic links between mass consumerism and Western democracy.[9] A classic example of the power of sociological self-descriptions is, of course, the Marxist terminology of 'classes' with its statistical underpinnings during the early period of the Soviet Union.[10] For a historical perspective on the scientization of the social it is therefore helpful to distinguish between the social structures of society and the semantic forms of social self-description, and to bear in mind that the human sciences can transform both.[11]

In the first essay in this volume, Lutz Raphael argues that the historical analysis of the embedding of the human and social sciences in Western societies must overcome the limits of discourse analysis by paying attention not just to ideas, discourses or metaphors but also to four other factors: the role of experts as the hidden protagonists in the implementation of science-based programmes; the clients and users of the social sciences, from the administrative officials who choose to use social science methods to the psychiatric patients who become the objects of social engineering; the specific histories of the different techniques of the human and social sciences, such as sampling, polling, interviewing or classifying; and, finally, the history of the institutions that provide the infrastructure that both enables and limits the spread of the human sciences throughout society.

Raphael proposes a periodization of the scientization of Western societies that distinguishes four configurations. The configuration of 'social reform' (1880–1910) started from a general critique of liberal laissez-faire policies and used techniques such as statistics and social surveys to promote gradual social reform in order to improve the lot of the labouring poor. After the First World War, the configuration of 'social engineering' legitimized social science-based interventions by reference to national recovery in the broadest sense. While earlier fears of degeneration gained new plausibility, the metaphor of 'community' also acquired increasing importance as an antidote to the atomizing effects of capitalism. After the Second World War, the configuration of 'planned modernization' (1945–70), which first took shape in the United States and Great Britain, responded to communist ideology during the Cold War by conceiving social engineering as a technology of planned and peaceful modernization. Relying on modernization theory, behaviourism, and instrumental positivism, this configuration profited from new technologies such as sampling and polling as well as new clinical therapies. Starting in the 1970s, Raphael argues, 'The Age of Therapy' began to supplant planned modernization. Born from new uncertainties and disputes about the future of welfare state, this final configuration has placed new emphasis on therapy as a way of asserting the individual self.

Periodization

Taking its cue from Raphael's periodization, this volume's chronological focus begins in 1880. The 1880s serve as a useful starting point because key tenets of nineteenth-century liberalism were transformed during this period of economic crisis and growing uncertainty about the stability of bourgeois society. Moral statistics, for example, provided reasons to believe that increasing rates of suicide and divorce were symptoms of a larger problem: immoral behaviour by individuals was held to reflect more general anomic tendencies in society and the gradual disintegration of social bonds.[12] Similar problems were identified in the field of social policy, where more systematic forms of enquiry and expertise emerged, starting in the 1880s. In academic and political debates on welfare legislation, the specific hazards of industrial society, such as unemployment or workplace accidents, were no longer attributed to the contingent misbehaviour of individuals, but regarded as the symptoms of general risks embedded in the workings of industrial society, and therefore in need of systemic solutions.[13]

Another key tenet of progressive liberalism was the figure of the law-abiding citizen who could distinguish between right and wrong and thus be held legally responsible for his deeds. From the 1880s, however, a moral panic about rising crime rates gripped large sections of the public. In judicial practice, criminal defendants' soundness of mind was questioned and probed in an increasing number of cases. In this situation, the emerging science of criminology advanced sociological and biological explanations of criminal behaviour that began to inform debates on penal reform.[14] Psychiatric expert opinions were increasingly called for to help judges to determine the soundness of mind of those who stood in the dock for offences against criminal or military law.[15] In these and in other policy fields, the 1880s marked the starting point for intensive debates on the systematic application of social scientific knowledge. In all these fields, the tension between individual moral responsibility and the social determination of individual behaviour became a prominent issue. Changing patterns of social differentiation caused anomic situations, and highlighted the risks of individualization, a process that Emile Durkheim described as the 'great disease of our time'.[16]

While the reasons why the 1880s are an appropriate starting point for a historical analysis of the scientization of the social are clear, limiting the scope of our inquiry to 1980 at the other end has been more arbitrary. In fact, a few chapters also cover the following one or two decades, and the social sciences have certainly continued to influence Western societies beyond 1980. Nonetheless, there are good reasons for treating the 1980s as a caesura. First, in many respects this decade marks the beginning of what can be called a period of secondary scientization.[17] In the century from 1880 to 1980, a succession of different disciplines and methodological approaches were applied in various policy fields and organizations. By about 1980, however, any further application of methods from the human sciences was confronted with the conceptual framework and the empirical results of earlier interventions at all levels of enquiry. This period of 'secondary scientization' thus required a new reflexivity from social scientists and the advocates of their solutions in the respective policy fields. In social science-led political planning, another facet of this reflexivity can be found in attempts to anticipate unintended side-effects of planning and to incorporate possible outcomes into the design of survey methods and communication strategies.[18] But not only those who promoted the engineering of social change through planning and surveying entered a period of more reflexive, secondary scientization since about 1980. Social scientists themselves began to scrutinize more closely the contexts in which their empirical knowledge was employed.[19]

The social context of scientization also changed from the late 1970s and 1980s onwards. Historians seem to be in broad agreement that economic crises, transformations in the regime of industrial production, and the exhaustion of Keynesian optimism fundamentally transformed capitalism and perhaps even modernity. These developments forced left liberals and Social Democrats on the European continent to begin a fundamental rethinking of their agenda for progressive social reform.[20] The transformation of capitalist society since the late 1970s also had immediate effects on the scientization of the social. A period of prolonged economic crisis triggered widespread debates and initiatives for a reform of welfare state policies, which had been a crucial field for the application of social scientific knowledge since the 1880s. In social and economic policy, the optimistic agendas for steering developments through science-based state intervention were now abandoned. Instead, social science experts now recommended gradual changes and incremental forms of intervention, thus bringing to the fore the (already established) notion of a 'science of muddling through'.[21]

Dorothy Ross and Theodore Porter have written that '[t]he history of the social sciences invites reflection on the ways in which historians and social scientists are mutually implicated in each other's work',[22] a formulation that perfectly captures one of the threads that bind together the chapters in this volume. A historical analysis of the scientization of the social should change the ways in which historians conceive of and write general histories of particular countries and periods. To date, the practical interventions of the *sciences humaines* in various fields of society are virtually absent even from the most sophisticated and comprehensive general accounts of twentieth-century European and North American history.[23] While religion, schools, the arts, and the institutional structures of higher education are usually covered at least briefly in most accounts, the practical relevance of social scientific expertise is hardly ever mentioned.[24] Only when a historian is keen to discuss structures of social inequality is a more detailed discussion of different sociological surveys and their implications provided.[25]

When historians treat the history of the social sciences as a separate specialist field, they are bound to miss important implications of expert knowledge for topics that figure prominently in their general histories. Writing about the welfare state without taking into account the technical underpinnings of social insurance schemes provided by the statistical calculations of actuaries neglects a crucial element.[26] As welfare agencies sought to rationalize and control female sexuality and to treat individuals with mental health problems, demographers, psychologists, and physicians assumed positions that allowed them to define the boundaries of acceptable behaviour. Thus the discourses that regulated the inclusionary and exclusionary tendencies of welfare state intervention were framed by experts. Starting already in the interwar period, the welfare state provided powerful tools for the definition and practical implementation of citizenship rights. Social science knowledge played a major part in the formulation of these rights and of concomitant duties assigned to the individual.[27] Likewise, it would be difficult to write the political history of postwar Western Europe and the United States without considering the fundamental reshaping of the public sphere through the widespread use of opinion polls.[28] Finally, it is impossible to write the history of the darkest chapter in twentieth-century history, the Holocaust, without a close analytical look at the social sciences. In the framework of a *völkisch* sociology that took racial inequality for granted, a large number of geographers, population experts, area planners, and human biologists provided statistical calculations and other data on social structures in Nazi-occupied Poland and the Soviet Union for those who planned and implemented the 'final solution' and thus provided crucial blueprints for the genocidal policies of the Nazi regime.[29]

Theoretical frameworks: Michel Foucault and Niklas Luhmann

These examples must suffice to demonstrate why general histories for the period from 1880 onwards must address the scientization of the social. But the history of the social sciences is also relevant to the work of historians in a second respect, namely the methodological issues it raises. Looking at the historiography on the scientizing of the social, two approaches seem to be promising for a historical perspective on this topic. The first is the work of Michel Foucault, particularly his approach to discourse analysis as an 'archaeology of knowledge'. Anyone who has studied the emergence of social scientific discourses on particular subjects has been able to analyse the process of a 'formation of objects', by which discrete phenomena such as degeneration, neurosis, stress, deviance, accident risks or consumer choice are delineated in the framework of specific disciplinary approaches. Once the formation of these objects has been situated in its historical context, the next step of historical analysis is to reconstruct the (institutional, professional, legal, social) position of those who are legitimately speaking about the elements of this discourse.[30]

Historians working on the scientization of the social have made effective use of the basic toolkit of Foucauldian discourse analysis.[31] More complicated, however, is the notion of 'governmentality' that Foucault developed in the late 1970s in order to describe forms of self-regulation and self-management. As a more indirect and flexible form of exercising power, governmentality is a complex of technologies for influencing the behaviour of other human beings and one's self.[32] In a series of influential books, sociologist Nikolas Rose has suggested interpreting the rise of the psy-sciences (psychiatry, psychotherapy, psychoanalysis) in Britain as the emergence of new forms of governmentality. In this perspective, the scientization of the social results from the need to develop more flexible, individualized forms of exercising power over subjects by exploring the depths of their 'selves'.[33] To be sure, the massive increase in the number of psychiatrists and psychologists in many (but not in all) Western societies in the postwar decades points to the fact that more flexible, therapeutic forms of intervention into the problems of individuals were deemed appropriate.[34] Nonetheless, important doubts about the notion of governmentality remain in the context of the scientizing of the social. In particular, the extent to which the psy-sciences actually exercise power over individuals remains dubious. Moreover, an empirical investigation of the psy-sciences in the British context has shown that the practitioners of the psy-sciences did not marshal enough therapeutic resources to substantially shape fields of intervention such as education or the workplace.[35]

While Foucault's theory of discourse formation remains an important reference point for historical work on the scientization of the social, a number of historians and sociologists have taken a cue from the work of the German sociologist Niklas Luhmann (1927–1998), one of the most prolific and thought-provoking sociologists of the postwar period, whose main project was to develop a theory that would reflect the complexity of modern society.[36] The key notion for this endeavour was that of functional differentiation. In different ways, Emile Durkheim's work on the division of labour, Georg Simmel's notion of an 'intersection between social circles', Max Weber's *Protestant Ethic* and his theory of different value spheres, and last but not least, Talcott Parsons's AGIL pattern of systemic differentiation all contributed to this strand of theory.[37] To this state of the art, Luhmann added two core ideas. First, he rejected the rigid framework of Parsons' analysis and conceptualized differentiation as an open process that could lead, for instance, to the emergence of sports as a new functional sub-system of society. Secondly, Luhmann developed the notion of communicative codes that delineate and regulate the workings of functionally differentiated, self-referential social systems. Scientific communication, for instance, operates on the true/false distinction, while the legal system is based on the legal/illegal distinction. The extension of social systems is therefore not confined to specialized institutional settings. Rather, they operate wherever their specific code is used in social communication.[38]

Luhmann's version of the theory of functional differentiation is helpful because it allows us to conceptualize the scientization of the social as a process of translation and adaptation, in which specific knowledge is transferred from one functional sub-system, science, to others such as the economy (consumer research), politics (statistics and polling), religion (pastoral counselling, sociography) or the legal system (criminology and criminal justice). In the realm of social policy at the workplace, scientization even required a two-step translation process, as all forms of social science expertise had first to be framed in the legal language of compensation law before political decisions could be taken.[39] Seeing scientization as a translation between systems is a useful guard against Whiggish notions of scientization as a success story in which social scientific knowledge easily invades different fields and effortlessly occupies a dominant place based on claims of universalist truth. Quite to the contrary, the translation of knowledge from the social sciences into different fields of application is usually, like every translation, much more complicated, and often fraught with problems.

Conceptualizing scientization as a translation of knowledge between functionally differentiated sub-systems has two important methodological implications. First, it points to the significance of metaphors in this translation process. Historians of science have devoted substantial attention to metaphors and have demonstrated that scientific language, like every other linguistic domain, is rich and varied in its use of metaphorical language. In our context, the focus is on 'metaphors as messengers of meaning'.[40] In the difficult transfer of scientific knowledge, metaphors work according to the literal meaning of the term, carrying meaning from one system to another and thus facilitating the scientization of the latter. Sometimes these metaphors had only an illustrative function. But metaphors could also have a constitutive function, opening up fundamentally new perspectives on a given problem. Biological metaphors of 'disease' and 'degeneration' have often worked as constitutive elements in discourses that called for and then practised radical exclusion: of inmates in mental asylums, for instance, or of Jews or other minorities who were deemed 'inferior' by a social science based on racial categories.[41]

Although Luhmann's systems theory takes communication as the core operation and is less interested in actors, his notion of functional differentiation can also be used to identify historical constellations in which actors facilitate the building of bridges across the gaps between differentiated fields. Peter Wagner has proposed the concept of 'discourse coalitions' to describe a constellation in which social scientists develop descriptions that strengthen the arguments of a group of actors in the political system, whose policies might, in turn, support the standing of these scientists in academia.[42] Forging these informal alliances depends on the social capital both sides have at their disposal and on elective affinities with regard to the cognitive structure of the social scientific knowledge. The post-1945 era of expanding welfare state provision and social and organizational reform in many fields of society provides ample evidence for discourse coalitions between reform-minded progressive practitioners and both functionalist and critical sociologists.[43]

Some historians of science have raised doubts about the notion of functional differentiation as a premise for historical studies on the scientization of the social. Looking at the many political fields in which scientific knowledge was applied during the twentieth century, they point to the overwhelming presence of 'cross-linking' and interconnectedness between scientific actors (and institutions) and the state.[44] It goes without saying that these links at the institutional level existed, and that many scientific institutions were in fact dependent on state funding in any event. But these institutional links should not distract from the fact that paying salaries to academics is basically an economic operation and does not guarantee that they will produce assertions that can deemed to be 'true'. The fundamental difference between science and politics lies in the self-referential procedures that the science system uses to distinguish between true and false. Nevertheless, scientists can never fully control the uses of their knowledge in different systemic contexts and are thus not in control of a process of scientization. In that sense, 'scientific knowledge is', as Luhmann has formulated, always 'a construction of the end-user'.[45] For a successful translation of knowledge from the social sciences into contexts of application, 'common sense' and 'tacit knowledge' are as important as any refined methodologies or theories.[46] As Theodore Porter has demonstrated with regard to a number of fields of statistical expertise, the 'push for quantitative rigor was in the first instance an adaptation to public exposure, not the achievement of a well-insulated community of researchers'.[47] Bureaucratic routines that can absorb and work with scientific expertise are often more important than the cutting-edge nature of the social scientific knowledge they use.

The ambivalent effects of scientization

A focus on functional differentiation makes clear that scientization can never be construed as a one-way street on which the scientists provide crucial resources while their own position remains unchanged. Quite to the contrary, as sociologist Peter Weingart argued already 30 years ago, the scientization of society and the 'politicization of science' are two complementary developments. The ever-increasing presence of social scientists on advisory boards or panels set up by governments or government agencies meant that these experts were increasingly involved in a different set of selection criteria, where the distinction true/false was eclipsed by arbitrary decision making.[48] The many corporatist management boards of Swiss and German welfare state agencies, such as accident insurance or social insurance bodies, in which academic experts were part of a tripartite structure together with representatives for employers and trade unions, provide ample evidence for this trend.[49] Even in systemic contexts that seem, at least at first glance, remote from political conflicts, the politicization of expert knowledge was part and parcel of the scientizing of the social.

Scientization led to highly ambivalent and problematic effects. In a seminal formulation, Max Weber regarded 'disenchantment' as a key trait of the modern world, characterized by rationalization and secularization. Precisely those social sciences that were meant to contribute to this process, however, turn out to be themselves 'disenchanted', locked into a cycle of experts and counter-experts frequently making contradictory claims, and thus undermining the notion that the social sciences can deliver reliable, objective knowledge.[50] For a historical analysis of these developments, it might be helpful to distinguish between the diagnostic and performative functions of social science experts. While the objective knowledge of social scientists is crucial for the former, their 'credibility' and reputation, sometimes even merely their 'prominence', is decisive in the latter context, when press conferences can be used to reassure the public that a political body or organization is about to implement new procedures or policies based on expert advice. In these cases, expert advice is certainly more than a 'transfer of knowledge' into a different field. Rather, it is staged as a crucial part of a decision-making process.[51] The symbolic effects of expert knowledge are therefore often just as important as their material effects.[52]

Ambivalence is thus the keyword for an understanding of the scientization of the social, both from a historical and a sociological perspective. Scientization is not a secular, inevitable process that rode roughshod over Western societies from the 1880s. Depending on context, country, and the systemic field in which social scientific knowledge was applied, this process had limits or suffered repeated setbacks. In addition, there were deliberate attempts to roll back the institutional and discursive positions that progressive social scientists had managed to occupy since the 1960s. The scientization of society was also fraught with intrinsic problems emerging from some of the disciplines that were involved. This was often the case if the epistemological profile of a discipline was low, that is, if internal benchmarks for the assessment of truth were not clearly developed, the disciplinary field was highly fragmented (most often through competing approaches and schools), and the professionals were thus not able to exert full control over the production of knowledge in their own field. In many respects, psychology is a classic example for this particular ambivalence of scientization. Although the postwar decades saw a boom in bottom-up demand for therapeutic cultivation of the self, psychologists, in Britain and in other countries, were not really able to exercise full professional authority over this field of applied science.[53]

Ambivalence is also the keyword for an assessment of the results of scientization. The scientizing of the social did most certainly not consist in a progressive accumulation of empirical knowledge that promoted enlightenment and reform. Nevertheless, on many occasions the social sciences did inform attempts to foster participation, develop notions of citizenship, or debate the consequences of changing moral values. Even then, these achievements were never uncontested, as the history of opinion polling, for instance, demonstrates.[54] It would be equally misleading, however, to assume that scientization contributed to a 'totalizing logic' of disciplinary control over (or self-control by) individuals, as some pessimistic interpretations of governmentality in the wake of Foucault tend to assume.[55] Such an interpretation underestimates the fact that the application of social scientific knowledge does not simply reduce complexity and create certainty, but rather tends to increase the amount of uncertainty about the future trajectory of complex systems. In more general terms, this reflects the difficulties in the attribution of causes and effects, a difficulty that is well known among practitioners of the social sciences. At the same time, we see here one of the reasons why the neurosciences have gained new currency in recent decades, for example, with regard to the control of criminality. For an important part of their appeal is the promise to establish direct causal links between violent behaviour and specific pathologies of the brain.[56] But such claims notwithstanding, indeterminate conclusions are the staple of the scientizing of the social. In that sense, scientization is part and parcel of a 'dangerous modernity', in which societies are always confronted with unforeseeable side-effects of reform strategies and decisions.[57]

Social and penal policy

The four chapters in this volume's first section, on social and penal policy, share several major themes. The first of these is the shift from individual responsibility to risk management in both social and penal policy during the last third of the nineteenth century.[58] This theme is most prominent in Julia Moses's chapter on the emergence of workplace accident insurance in Germany, Britain, and Italy in the period 1870–1920. Early nineteenth-century court cases brought by workers demanding compensation for injuries caused by workplace accidents regularly resulted in verdicts rejecting such claims based on the freedom of contract. As free agents, the courts argued, workers had accepted all risks associated with their work when they entered into their contracts of employment. As the number of industrial workplace accidents increased with the spread of industrialization, all three countries began to consider the adoption of employer liability laws to address the problem. Liability laws passed in Germany (1871) and Britain (1880) required employers to compensate workers for injuries from workplace accidents if the employer had caused the accident, which, however, was extremely difficult to prove in most cases. Since such court battles between workers and employers threatened to inflame further already volatile class relations, employers and legal experts began to press for compulsory workplace accident insurance that would eliminate the question of who was responsible for workplace accidents. Such compulsory insurance laws were first passed in Germany in 1884, and by Britain and Italy in 1897 and 1898 respectively.

These laws represented a momentous shift from a system based on assigning individual legal responsibility to collective risk management. This shift resulted from the rise of statistics as a new form of knowledge.[59] By the late 1860s, the British, German, and Italian governments began to systematically collect statistics, including statistics about workplace accidents. This new statistical knowledge transformed the perception of workplace accidents, which began to be seen as a work-related risk displaying statistical regularities, rather than the result of individual fault. Although the risk-management approach involved some prevention efforts, its main consequence was a new consensus that the costs of workplace accidents had to be borne collectively instead of holding individual workers or employers legally responsible. The seemingly neutral information provided by statistics allowed the state to defuse the worker–employer conflicts arising from accident litigation under the liability laws. As Moses shows, new kinds of specialized knowledge continued to play an important role after compulsory workplace accident insurance was adopted: statisticians monitored accident and compensation rates, jurists interpreted the relevant legal provisions, and medical experts assessed compensation claims. The continuing role of jurists shows that the shift from individual legal responsibility to collective risk management was not a process of de-juridification. Although litigation decreased, the insurance schemes were established by laws that had to be interpreted by jurists. Thus the intrusion of statistical and medical expert knowledge did not displace but was mediated by legal expertise.

The shift from individual responsibility to collective risk management also characterized the expansion of social insurance schemes into health, disability, and unemployment insurance. Adverse events such as illness, disabling accidents, and unemployment, which had previously been considered the responsibility of individuals, were now treated as risks that were statistically predictable on the level of populations and should be managed collectively. Perhaps somewhat surprisingly, the same shift from individual responsibility to collective risk management also occurred in late nineteenth-century penal policy. Penal reformers, too, argued that criminal justice should no longer focus on the individual legal responsibility of criminal offenders but on managing the risks associated with the danger of their committing further crimes in the future. The primary purpose of criminal justice, the reformers insisted, should not be exacting retribution for past offences, but protecting society against future crimes. In penal policy, too, this shift from responsibility to risk was associated with the impact of new forms of specialized knowledge, especially statistics and medicine.

In a 1978 article, Michel Foucault suggested that there was not just an analogy but a direct connection between changes in criminal law and changes in the civil law treatment of industrial accidents. The penal reformers' demand that punishments should match an offender's dangerousness (rather than degree of guilt), he argued, was related to the change from fault-based to no-fault liability for industrial accidents. Under no-fault liability, an enterprise's liability for an accident no longer depended on proof of criminal responsibility but merely on establishing a causal link. Just as no-fault liability dispensed with fault, the penal reformers' new concept of 'dangerousness' as the basis for punishment dispensed with the notions of guilt and responsibility.[60] Ironically, the only example of no-fault liability that Foucault mentioned was the German *Causalhaftung* (causal liability), which played a very limited role in German law,[61] rather than the more influential French legal concept of *risque professionnel*, introduced by the 1898 French law on workplace accidents.[62] Foucault's thesis was picked up and further developed by François Ewald in his 1986 book *L'État Providence*, in which he argued that turn-of-the-century jurists such as Raymond Saleilles and Adolphe Prins[63] used the civil law notion of risk in order to refocus criminal justice on the dangerousness of the individual criminal.[64] In the 1990s, the German literary scholar Stefan Andriopoulos used Foucault's thesis as the point of departure for a discourse analysis that revealed numerous connections between 'accident' and 'crime' in turn-of-the-century legal and literary discourses. He also noted that it was French, Belgian, and Swiss penal reformers who explicitly connected the new criterion of dangerousness to no-fault liability in civil law, whereas German penal reformers did not. This was most likely because no-fault liability and *risque professionnel* played an important role in French and Swiss civil law, whereas the German Accident Insurance Law of 1884 had removed the issue of industrial accidents from private civil law into the realm of *Sozialrecht* (social law).[65]

If direct references to no-fault accident liability as a model for the concept of 'dangerousness' in criminal law appear only in a small number of sources, the evidence that late nineteenth-century social and penal policy underwent analogous shifts from individual responsibility to collective risk management is overwhelming. The adoption of compulsory accident insurance reflected the transformation of laissez-faire liberalism into a more interventionist liberalism. By the last third of the nineteenth century, the classic liberal argument that workers had assumed all risks associated with their work under 'freedom of contract' had lost credibility. More generally, the recognition of social inequalities had undermined the liberal belief that a society composed of rational and autonomous individuals freely entering into contracts with one another could regulate itself without government intervention. An analogous development took place in penal policy. If the development of the welfare state resulted from a fraying of the belief in 'freedom of contract', late nineteenth-century penal reform arose from an erosion of the concepts of 'free will' and individual responsibility. This erosion was due to the emergence of criminology – not as a professional discipline (which criminology did not become until the second half of the twentieth century) but as an intellectual enterprise: the search for a scientific understanding of the causes of crime. While academics and practitioners from a wide variety of professions were engaged in criminological research, the most decisive impact came from statistics and medicine.[66]

Already in the 1840s, Adolphe Quetelet's pioneering studies of French crime statistics had revealed astonishing regularities, which led him to conclude that 'the crimes which are annually committed seem to be a necessary result of our social organization ... society prepares crime, and the guilty are only instruments by which it is executed'.[67] Other statisticians studied the correlation between crime rates and particular social factors (such as grain prices) to reveal the social causes of crime. At the same time, free will and individual responsibility were also undermined by medical research on crime. The key development here was not that medical doctors were increasingly consulted by the courts when a defendant's mental state was in doubt (which they were), but that the medical profession, psychiatrists in particular, began to argue that a large number, perhaps even a majority, of criminals were mentally abnormal ('degenerated', 'psychopathic', 'mentally defective', 'feeble-minded'). This medical challenge to criminal justice resulted from the development of degeneration theory,[68] a new psychiatric focus on 'borderline states' and 'mental abnormalities' falling short of full-fledged mental illness,[69] the medical profession's ambition to address social problems through 'social hygiene',[70] and the impact of Cesare Lombroso's theory of the 'born criminal', which, although it was mostly rejected outside of Italy, spurred medical research into the connection of crime and mental abnormality.[71]

Starting in the 1880s, these statistical and medical studies were seized upon by a new generation of penal reformers who interpreted them as evidence that criminal behaviour was determined by individual and social factors, and that the notion of individual responsibility had therefore become untenable. As a result, they argued, the idea that criminal justice should serve the purpose of retribution (through prison sentences calibrated to match an offender's individual degree of guilt) must also be discarded. Instead of exacting retribution for past offences, criminal justice should protect society by taking the measures necessary to prevent future ones. Although contemporary reformers did not use the term 'risk management', their term 'social defence' designated exactly this purpose. In order to protect society, an offender's punishment would be based on his or her 'dangerousness' rather than the degree of guilt. To be sure, being 'guilty' of a criminal offence would still be a condition for conviction, but the purpose, length, and content of punishment would depend on the offender's dangerousness. Because dangerousness varied among individuals, punishment would have to be individualized. More importantly, the penal sanction was to be transformed from a punitive measure into a preventive one. What did this mean? Most penal reformers distinguished five categories of offenders and five corresponding sanctions: occasional offenders should have their prison sentences suspended on probation; juvenile offenders were to be educated in homes for wayward youth; mentally abnormal offenders should receive medical treatment; corrigible habitual offenders were to be rehabilitated in detention; and incorrigible habitual criminals should be incapacitated through indefinite detention. Some reformers also advocated the sterilization of mentally abnormal offenders whose abnormalities were considered hereditary.[72] In short, the penal sanction was to be transformed into a spectrum of preventive measures many of which were identical to other types of state intervention, such as juvenile welfare, medical treatment, or detention in a workhouse. Of course, the rate at which these reforms were implemented in different countries varied; but it is fair to say that even though criminal justice retained important retributive elements, between 1880 and 1930 most Western European countries implemented a majority of these measures, thus shifting the focus of the criminal justice system, especially in the punishment phase, from a retributive focus on individual guilt towards 'social defence' and risk management.[73]

Just as in the case of workplace accident insurance, the role of the new experts was not limited to toppling the old system, but became even more prominent under the new regime of risk management: forensic psychiatrists assessed offenders' 'dangerousness' in court;[74] prison doctors determined incoming prisoners' 'social prognosis' through 'criminal-biological examinations';[75] welfare workers assisted the courts in gathering information on defendants' backgrounds;[76] statisticians tracked recidivism rates; and, generally, penal reform remained closely connected to the development of criminological research.

Despite the increased participation of these experts, the nature of the relationship of the new criminological knowledge to criminal justice is a difficult and contested issue. First, just as criminology influenced penal policy, criminological research was shaped by legal and moral categories: criminologists' concept of 'crime' was a legal category derived from the penal code, and new psychiatric diagnoses such as 'antisocial personality' were little more than moral judgments dressed up in medical terminology. Secondly, key parts of the penal reformers' categorization of offenders, especially the distinction between occasional and habitual criminals, preceded any empirical criminological findings. While some historians have therefore concluded that one cannot speak of the 'scientization' or 'medicalization' of criminal justice, others have argued that criminal justice underwent a significant but juridically framed process of medicalization: although the jurists who led penal reform were never going to abandon juridical control over criminal justice by letting medical doctors (or welfare workers) take over, they incorporated criminological (and especially psychiatric) knowledge and experts into a reformed criminal justice system.[77]

Peter Becker's chapter examines the relationship of the human sciences and penal policy during the 'second biological wave' in explanations of crime, which started in the 1980s, following several decades during which sociological explanations of crime prevailed. If the first biological wave (from the 1880s to the 1940s) was dominated by psychiatrists, the second wave has been shaped by neuroscientists. While both seek to predict the future dangerousness of individuals, they differ in their view of the human brain: whereas the psychiatrists of the first wave considered the brain to be a fixed entity (subject to genetic and environmental influences only during its formation) and therefore envisioned eugenic measures to remove criminogenic genetic factors from the gene pool, contemporary neuroscientists view the human brain as malleable, thereby opening up the possibility of therapeutic intervention through neuroceutical drugs. The neuroscientists' key claim is that violent behaviour is the result of genetic polymorphisms interacting with negative environmental impulses.

Whereas some neuroscientists concede that predictions of dangerousness based on neurological malfunctioning are only possible for populations (and thus probabilistic), others claim to have discovered a direct causal link between specific forms of neurological malfunctioning and violent behaviour. In order to understand why neuroscientists have been so readily recognized as experts in matters of social and penal policy, Becker examines the 'demand side', the penal system's need for classifying offenders as well as recent shifts in social and penal policy from sociological approaches to approaches that focus on 'subjects themselves and their performance in markets'.[78]

The second important theme among the first section's chapters is the transnational aspect of the scientization of social and penal policy. The past decade has witnessed a burgeoning interest in transnational history, the history of nineteenth-century 'internationalism', and the role of transnationalism in social reform and social policy. In a pioneering volume, Martin Geyer and Johannes Paulmann proposed the term 'internationalism' to describe two interrelated processes that gained critical momentum in the mid nineteenth century: (1) the internationalization of cultural, political, and economic practices, and (2) the effort to reform society by way of transnational cooperation. They argued that, from the 1840s to 1914, the second type of internationalism unfolded in three overlapping stages: first, 'voluntary internationalism', in which 'voluntary activists' crossed national boundaries to build strategic transnational alliances to promote their reform projects; secondly, 'professional internationalism', the stage at which self-selected 'experts' became involved and set up international meetings with a view to lobbying for their reform schemes; and finally, the stage of 'institutionalized internationalism', during which governments became officially involved in the system of periodic congresses, permanent bureaux, and exchanges of information.[79]

While this three-stage model was developed to apply to internationalism in many spheres, it applies especially well to social and penal policy, which was strongly represented among the international congresses held with increasing frequency from the mid-nineteenth century onwards.[80] It must be borne in mind, however, that not only were the three stages overlapping, but the model's distinction between 'activists', 'experts', and government representatives describes ideal types; in reality, the boundaries between these categories were fluid, and hybrid combinations abounded. The international penal and prison reform movement, for instance, included prison wardens, judges, criminal law professors, government bureaucrats, statisticians, medical doctors, as well as laypeople and clergy active in prison societies. Even if we classified the latter as 'activists', it is clear that the remaining groups all regarded themselves as 'experts'. This great variety of different types of experts gradually led to the diversification of international congresses and organizations. In the realm of penal reform, the last third of the nineteenth century witnessed the emergence of three different transnational networks: prison congresses (attended mainly by prison officials), congresses of criminal anthropology (dominated by medical doctors), and the International Union of Criminal Law (comprised mostly of jurists).[81] Furthermore, some participants attended international congresses not only as experts but also as government representatives; at the prison congresses, for instance, the top officials in charge of the administration of prisons in each country usually participated in the dual role of individual expert and government representative. This hybridity of participants and the resulting hybrid nature of many international congresses (as both gatherings of experts and intergovernmental meetings) only strengthened the influence of the human and social sciences on the development of social and penal policy.

More recent research on transnational history has not taken up Geyer and Paulmann's use of the term 'internationalism'. Instead, most historians have adopted the term 'transnationalism' to designate cross-national interactions whose participants (or at least a majority of them) are non-governmental actors, and have reserved the terms 'internationalism' and 'international relations' to refer to intergovernmental interactions.[82] Patricia Clavin has refined this distinction by suggesting that transnational and international relations differ not just in the status of their participants but also in the nature of the transnational connection that is created. In contrast to international relations, she argues, transnationalism creates a 'transnational community' that is not 'an enmeshed network' but a 'honeycomb' whose 'boundaries remain open, porous, revisable, and interactive'.[83] Madeleine Herren has stressed the importance of transnational perspectives for the history of social policy. In her analysis, the First World War was a watershed in the transnational history of social policy: whereas the prewar period was characterized by the 'multi-lateralization of social policy as a result of a modernizing bourgeois internationalism', after 1918 this internationalism was replaced by a 'corporatist arrangement', best embodied by the tripartite structure of the International Labour Office, in which unions, employers, and governments were represented.[84] Pierre-Yves Saunier meanwhile has proposed a model of three sequential but overlapping transnational 'circulatory regimes' in the realm of social policy between 1800 and 1940. Under Saunier's first regime (similar to Geyer/Paulmann's 'voluntary internationalism') the participants in transnational discourse are scholars, political activists, clergy, and entrepreneurs who focus on various causes related to the 'social question', including prison reform, temperance, abolitionism, and poor relief. Under the second regime, starting in the late nineteenth century, transnational exchanges become more institutionalized, increasingly becoming intergovernmental, whether at the level of national, regional or local governments. Under the third regime, which begins after the First World War, transnational exchanges and organizations begin to be shaped by universalist aspirations to create international norms and standards in policy, scholarship, and the social domain.[85]

Julia Moses's analysis of the development of accident insurance legislation in Germany, Britain, and Italy fits well with Saunier's 'second regime' because governmental actors are beginning to play a crucial role in transnational exchanges. Her chapter is also an excellent example of 'entangled history' because it combines comparative history (comparing accident insurance legislation in the three countries) with the study of transnational transfers (especially British and Italian references to the German model of accident insurance) and the history of transnational networks (international congresses on workplace accidents and industrial hygiene).[86] As Moses demonstrates, these transnational exchanges, which were facilitated by the 'universal language of statistics', helped to create 'an international public sphere' focused on workplace risk that helped to spur domestic accident insurance legislation.

The transnational aspect of social policy is even more central to Martin Lengwiler's chapter, which explores the role of expert knowledge in the shaping of social policy by international organizations and transnational networks during the twentieth century. Like Herren, Lengwiler regards the First World War as a watershed: many congresses did not survive the national antagonisms of the war, although more technical associations (such as the International Association of Actuaries) outlasted the war; in the postwar period the role of the prewar congresses was filled by the emergence of international organizations such as the League of Nations and the International Labour Organization (ILO), which began to pursue the universalist aspirations characteristic of Saunier's 'third circulatory regime'. The ILO's tripartite structure (representing unions, employers, and governments) reflected the legacy of political internationalism, while its strong reliance on expert knowledge reflected the legacy of scientific internationalism. The significance of expert knowledge within the ILO is demonstrated by the establishment of the Social Insurance Section (1921) as an advisory expert group and, in 1947, the foundation of the International Social Security Association (ISSA) as a think tank on social security systems. When the ILO became embroiled in Cold War conflicts in the 1970s, the Organization for Economic Cooperation and Development (OECD) assumed much of the ILO's former role in international social and economic policy.

Lengwiler's central question reflects another theme among this section's chapters: where did the demand for scientific expertise in solving social problems come from? Lengwiler explains the spread of expert knowledge in the international organizations and networks he examines by reference to three main factors: first, as the expansion of welfare states increased their complexity, expert knowledge was in demand for technical reasons; secondly, expert knowledge 'had the symbolic effect of enhancing the public's trust in the stability of social insurance'; and finally, the persuasiveness of expert knowledge functioned as compensation for the international organizations' lack of political authority – put differently, scientization gave international organizations a kind of 'soft power'.

Julia Moses, too, poses the question of why expert knowledge was in demand. Her main answer is that experts played a crucial role in resolving a politically contentious issue by providing 'seemingly objective' information that 'appeared to transcend politics'. And yet, as Moses points out, specialist knowledge could also become repoliticized in periods of crisis, as illustrated by the concern about 'pension addiction' after the First World War. Once compensation schemes were adopted, the processing of individual cases and the task of risk management required the participation of legal, medical, and statistical experts.

Peter Becker also raises the question of why policymakers have been willing to regard neuroscientists as experts qualified to offer solutions to the crime problem. Becker sees the answer in four interrelated developments: the penal system's need to classify offenders creates a demand for technologies of risk assessment; dissatisfaction with sociological approaches to crime has made policymakers receptive to experts who promise new solutions, especially solutions focused on the offender rather than social reform; research in neuroscience has been prominent in the media because it answers the media's demand for spectacular discoveries; and, finally, neuroscientists are able to use new visualization methods, especially functional magnetic resonance imaging (fMRI), as an extremely effective tool for popularizing their (actually quite complicated) findings. In sum, the most salient factor in explaining society's demand for expert knowledge in these case studies seems to be the promise of the human and social sciences to transcend politics and social conflict through their apparent 'objectivity' and 'neutrality'.

The section's other chapter, by Thomas Etzemüller, shares some of the themes discussed earlier: its examination of Alva and Gunnar Myrdal's role as social engineers has a transnational aspect because the Myrdals took formative study trips to Britain and the United States; and the Swedish approach to social engineering had an important element of risk management not only in the form of eugenics but through a general concern with deviations from social norms. Nevertheless, Etzemüller's chapter differs from the others in two major ways: whereas the other chapters are transnational and concerned with *communities* of experts, Etzemüller presents a case study of one country and, in the chapter's second half, focuses on two individual experts, the Myrdals.[87] One of the advantages of focusing on one country is that Etzemüller is able to sketch a broad panorama of social engineering. The Myrdals' investigation of the 'population question' led them to conclude that many women remained childless because they did not want to give up their jobs or could not afford children. This single diagnosis led them to develop a wide-ranging programme of social reform, including the socialization of production, a policy of redistribution, rationalized housing construction, the collectivization of child-rearing, and the socialization of consumption.

While these reform proposals may, Etzemüller fears, 'seem shocking today', he argues that the Myrdals' goal was not 'controlling people' but 'emancipation through self-normalization'. With a comparative glance at the Nazi regime, Etzemüller describes Sweden from the 1930s the 1960s as a 'normalizing society'.[88] While both Nazi Germany and Sweden were shaped by the 'utopia of a socially harmonious national community' and both endorsed eugenics and passed sterilization laws, he rightly insists that the two societies were very different. Whereas Nazi Germany was a 'totalitarian society' focused on disciplining through coercion, he argues, Sweden was a 'normalizing society' focused on 'the systematic production of knowledge in order to differentiate between normality and problem areas' and to develop 'rational techniques of steering society'. While one might argue that the Nazi regime, too, was engaged in 'the systematic production of knowledge' on the distinction between normal and abnormal, Etzemüller's study of social engineering in a democratic society convincingly demonstrates that it is misguided to draw a teleological connection from biopolitical social engineering to Nazism and the Holocaust, as has been done by scholars who have argued that the 'final solution' was the 'logical outcome' of a 'fatal racist dynamism' in the human sciences.[89] The history of the human sciences and their biopolitical approach to social and penal policy suggests that they contain both repressive and emancipatory potentials; they are not inherently more compatible with totalitarian than with democratic politics. Which potentials are realized depends on the larger social and political forces that determine a country's political system.[90]

Diagnosis and therapy

The chapters in the volume's second section examine the ascent and deployment of the 'psy-disciplines', that is, psychiatry, psychoanalysis, psychology, and psychotherapy, for diagnostic and therapeutic purposes since the late nineteenth century. Overall, the rise of these disciplines profited from the fact that they were never completely integrated into the academy, which favoured a multitude of approaches and facilitated their adaptation to the needs of a wide variety of fields, such as education and the business world, as well as the impregnation of the vernacular with psychological terminology.[91]

The four chapters in this section approach their topics from different theoretical and conceptual perspectives. Mathew Thomson, focusing on Britain, seeks to modify Nikolas Rose's Foucauldian interpretation of the influence of the psy-disciplines by distinguishing between 'scientization' and '(social) engineering', using the latter term to refer more specifically to direct interventions in society. This differentiation aims at capturing more precisely the unfolding of scientization on different levels of society and, not least, the limits of the process. In the following chapter, Harry Oosterhuis, discussing the Netherlands, takes the concept of 'citizenship' as his point of departure, defining it as the ability of the self to take autonomous decisions as the precondition for allegiance to and participation in a democratic political system. The psy-sciences are then seen as bringing about and guaranteeing individual autonomy by creating 'mental health', supported by public institutions established under the framework of the welfare state.[92] While 'citizenship' is thus conceived as a normative category, it implies ambiguities and conflicts about its proper definition and change over time. Focusing on the influence of the new 'child sciences' on social discourses and practices, Till Kössler's chapter reconceptualizes the history of Spain in the early twentieth century. In contrast to previous studies, he describes its course not as a conflict between the backwardness of Catholic nationalists and the modernity of liberal-progressives but as a struggle between the two camps about Spain's appropriate path to modernity in the wake of the lost war of 1898. While emphasizing how childhood and education became politicized as a result of the reception of the 'child sciences', he also notes that calls for their depolitizication drew on the same science-based logic. Finally Elizabeth Lunbeck's chapter is a piece of historical discourse analysis. Discussing how, during the 1970s, the concept of 'narcissism' inscribed itself into a long-standing tradition of social critique in the United States, she identifies the malleability of the concept and its moral framing as reasons why this psychoanalytical discourse was readily drawn upon in public debate, even though key elements of the concept were recast in this process of translation. Tracing back the concept to its Freudian origins, she also explains why its Americanized version focused on the female consumer and explores the potential for a change of this focus towards the powerful male.

In his contribution to this volume (Chapter 2), Lutz Raphael identifies a first 'configuration' of scientization, which saw its heyday between 1880 and 1910 and centred on the concept of 'social reform', aiming at the working poor. As the optimistic belief in liberal individualism waned, a broad coalition of experts devised new tools of intervention, most notably social insurance, and oversaw their application. The psy-disciplines emerged in the shadow of this configuration, with significant national differences. Psychiatry, which was connected to the academy through its link to medicine but remained on its margins, turned from the asylum to the new university clinic, including its outpatient facilities, and shifted its focus away from the specific symptoms that defined the mentally ill to the wide range of symptoms on the borderline of 'normality', thus defining a 'psychopathic personality'. Freud's psychoanalysis, which became known in America following his visit in 1909, came to be embraced there by physicians and the general public alike, serving not only as a novel tool for psychotherapeutic intervention but also as a 'general theory of human behaviour'.[93] This overall broadening of interest of psychiatry, however, did not imply a swift training of a wide range of specialists prior to 1914, as Mathew Thomson notes in his essay for Britain.

Psychology became institutionalized as a discipline in a more complicated process. While the natural sciences with their focus on the laboratory emerged as a potential model by the end of the nineteenth century, psychologists in Britain were divided between those adhering to statistical methods and others taking a more theoretical and systematizing approach. In France, the discipline remained linked to both medicine and philosophy. Intelligence testing, a new tool of intervention created by Alfred Binet, met with harsh resistance by schoolteachers in the very field for which it had been devised. In France as in Britain, academic institutionalization of the discipline remained limited.[94] By contrast, psychology was firmly institutionalized in the academy in Germany and the United States, but it became an applied discipline on a broad scale much earlier in America than in Germany. In Germany, a strong experimental orientation, for which Wilhelm Wundt's laboratory was but one example, remained embedded in a tradition of middle-class self-inspection and kept its links to university philosophy. When the German laboratory came to the America via Wundt's students during the 1880s and 1890s, it took on a utilitarian bent and, given the expansion of the American university system at the time, became a crucial element of institutionalized academic psychology. Influenced by Darwin and Spencer, American psychology came to focus on developmental issues, with a specific interest in children and education. Joining the progressive reform movement, psychologists, more so than in Germany, claimed the status of important academic experts who translated 'objective' knowledge into solutions of social problems.[95] A new wave of immigration, the closing of the 'frontier' in the West, and rapid industrialization and urbanization had created these problems and triggered a 'search for order'[96] to which the social and the psy-sciences were to make their contributions. A first case in point was the juvenile court, meant to replace punishment by education, but ultimately failing to reach its goal on account of limited resources and adequate staff training. When an Indiana superintendent described the methods used in his reformatories 'applied psychology', he referred to wielding the paddle, not to psychotherapy, while proving by the terminology he used how important psychology as a discipline had become.[97]

The First World War had a major impact on the 'scientization of the social' via the psy-disciplines. They now became harnessed by the state, albeit to different degrees, to help sustain the war effort in the belligerent countries and to contribute to stabilization and regeneration in the interwar years. Hence, they were a key element of the second configuration, that of 'social engineering', which Lutz Raphael describes in his essay. While the American psy-sciences emphasized the malleability of the human psyche, their continental European counterparts underlined constitutional factors and dismissed the existence of trauma when it came to dealing with the masses of veterans who had returned from the war with some form of mental illness.[98] The influence of the psy-disciplines was particularly pronounced in the United States, even though the country entered the war late and did not suffer casualties on a scale comparable to that of the European belligerents. However, as the United States introduced military conscription for the first time in its history, against the backdrop of a highly diverse population, thoroughly examining the aptitude of recruits seemed indispensable. For this diagnostic purpose, intelligence testing, based on Binet's approach and simplified for speedy use by Stanford professor Lewis Terman, seemed the best method. Its hardly uplifting results – one in three recruits was rejected as unfit – had profound consequences. It reinforced ethnic prejudice and helped bring about legislation in 1920 that imposed restrictions on immigration. Furthermore, mass testing was adopted after the war by a greatly expanding school system as it seemed an appropriate tool for assigning children to the different tracks of the system. More than simply a technical process, it left an imprint on language: terms such as 'imbecile' or 'moron' became popular on account of the tests and remained fixtures long after psychology had abandoned them as meaningless, bolstering not least racial prejudice.[99] The idea of testing was taken to yet another level by the mental hygiene movement of the interwar years. Following a simplified Freudian and behaviourist line of argument, this movement advocated a therapeutic model of schooling and changed the terminology of the public discourse on education.[100]

In Europe, diagnostic psychology was given the most visible boost in the wake of the First World War in Britain, where Binet-style testing became a fixture of research and practical work in many fields, fitting in well with the British tradition of sociological work.[101] As Mathew Thomson argues in his contribution, in the interwar period British psy-sciences and the therapies they offered moved away from a focus on the mentally ill and emphasized prevention and general mental health. As the scientific status of psychology in particular continued to be academically insecure and torn between that of a science and a social science, this was a chance for popularizers and encouraged broad claims on social issues, driven also by the competition of a number of schools of thought. In the Netherlands, as Harry Oosterhuis shows in his essay, the introduction of universal suffrage in the wake of the First World War heightened fears about the potential irrationality of the newly enfranchised lower classes and gave a boost to the psycho-hygienic movement. Largely without state control, a network of outpatient mental health services emerged that focused on prevention and addressed a wide range of issues, such as providing counselling centres for alcoholics and child-guidance clinics. In Spain, as Till Kössler shows, it was the defeat in the war of 1898 against the United States that gave an enormous boost to the reception of the child-related psy-sciences. As a psychobiological conceptualization of childhood replaced a moral one, child protection was institutionalized, initially with a narrow focus on health and the prevention of criminality, and then, over the course of the 1920s, given additional tasks following the new science-based model of the energetic, carefree child. While this new scientific conceptualization of childhood came to be shared by the two main political camps, Catholic nationalists and liberal-progressives, it led them to decry each other's allegedly corrupting influence on children and greatly increased the pressure on the Spanish state to modernize and create new institutions for childcare and education.

These developments reflected a certain pessimism about the dangers arising from modern mass society as well as optimism about the chances of science-based interventions in the 'national body'. Germany would be the most extreme case, turning eventually to deadly forms of intervention after the Nazi seizure of power, with the well-established discipline of psychiatry playing its part in the euthanasia programme.[102] German psychotherapy was able to assert its professional autonomy vis-à-vis psychiatry, helping to improve the combat efficiency of Luftwaffe pilots during the Second World War. Likewise, psychology was given a boost as it developed techniques for the selection of officers and saw its status as an applied discipline strengthened by the introduction of the diploma in 1941.[103] Subservience to a totalitarian regime proved perfectly compatible with ongoing professionalization, not least during the war.

For American psychiatrists and, even more so, psychologists, the Second World War was a defining moment, ushering in an era of 'planned modernization', the third configuration identified by Lutz Raphael. Following an enormous increase in numbers – membership in the American Psychological Association grew more than elevenfold to over 4400 in the years from 1920 to 1946 – psychologists drew upon their wartime expertise in sustaining morale at home and on the front-line to claim a prominent role in giving advice on domestic social problems and on foreign policy in a Cold War framework and to remain committed to providing for the mental health of the population at large. While the psy-disciplines thus continued their wartime alliance with the state, which was attractive for the latter as it seemed to help depoliticize problems such as racial conflict, the feminist movement, along with the anti-psychiatry movement, subsequently developed its own critical and democratizing reading of psychological expertise, calling into question the experts' status as omniscient problem-solvers.[104] In the postwar era,

American psychology, together with its British counterpart, became dominant in the Western World, even though clinical psychology came to be introduced in West Germany and other parts of Europe only ten years later than in the United States.[105]

In the Netherlands, as Harry Oosterhuis points out, the approach to mental health lost its defensive stance towards modernization over the course of the 1950s and saw another rupture in the late 1960s, when criticism of the former patronizing approach mounted, inspired by leftist political ideas and the self-organization of patients, and 'spontaneous self-development' became the guiding principle. Under this framework, health services expanded substantially, now integrated in the Dutch welfare state. This is a case in point for Raphael's argument about the transition to an 'age of therapy', his fourth configuration, in which, following the failure of the grand visions of 'planned modernization', the self became a prominent focus of science-based intervention and advice, and the psy-disciplines saw the apogee of their influence. While the expansion of services based on the psy-disciplines and a more uncertain relationship between experts and their clients were developments along general Western European lines, Dutch secularization in the 1960s was particularly pronounced, as Oosterhuis emphasizes, making for a sharp break with the past and propelling female emancipation. From the mid-1980s onwards, neoliberalism set the mental health system on a different course, however, attempting to reduce costs and narrow the focus of the system to the severely ill.

For Britain, Mathew Thomson offers a more cautious assessment of the influence of the psy-sciences in the post-1945 period. While British mental health experts assumed a prominent role on the international stage, for example, in the World Health Organization (WHO), domestically they failed to set the agenda for social reform and never provided a figure comparable in prominence to Christopher Lasch in the United States, losing out against sociologists, whose scientific status was more secure. Psychological experts continued to play their role in the schools, the workplace, and the expanding health system, but resources remained limited and psychological therapy largely confined to the severely ill. The psy-sciences were most influential, Thomson argues, when embraced in a process of popularization from below. While in the interwar years and in the two postwar decades in Britain and elsewhere, most demand for the psy-sciences had been created by state agencies, not least the military, it now came mainly from private individuals.

How influential the psy-disciplines had become on the other side of the Atlantic was apparent when in 1978 Christopher Lasch published his scathing criticism of American society, *The Culture of Narcissism*. As Elizabeth Lunbeck shows in her essay, Lasch's argument was not original but made a big impression as the 'dark double' of therapeutic culture in a society shaken by the Vietnam War, Watergate, and economic troubles. Lasch, taking up a criticism that sociologist David Riesman and others had articulated earlier, described Americans as wedded to a superficial consumer culture in which adherence to material objects had taken the place of human empathy. Borrowing from psychoanalytic ideas of his time, Lasch popularized the Freudian concept of narcissism, while replacing the asceticism at its core with outward affluence.

As the thrust of his argument was directed against consumerism, the narcissist was cast as a female or homosexual figure, marginalizing the strong male leader. Lasch's critique would not remain the last of its kind. The alleged 'triumph of the therapeutic' has continued to generate fears that in a pervasive 'therapeutic culture' individuals have dissociated themselves from moral obligations and communal bonds and embarked on a permanent quest for emotional gratification and personal happiness, taking to its seemingly logical conclusion the disenchantment and rationalization of the world Max Weber had described.[106]

This criticism brings up the question of the range and the limits of the impact the psy-sciences have had since the 1880s and, by implication, raises the issue of theoretical approaches to conceptualize this impact. Kössler argues that the influence of the psy-disciplines on politics in the field of childcare and education was pervasive, both in terms of defining new objects of political activities and of creating experts as a new group of political actors. Thomson clearly stresses the limits of their influence, pointing out, apart from the continuous lack of resources, the predominance of genuine political and economic considerations with regard to schooling and the workplace, and underlining how the 'common sense' of the British public determined the fate of broader concepts. While 'Bowlbyism', which emphasized the importance of close bonds between mother and child and suggested a return of mothers to the domestic sphere in the postwar era, was widely accepted, a critique of capitalism informed by the psy-sciences met with rejection. Oosterhuis, in contrast, emphasizes the scope of Dutch mental health services and their transition to a system providing for increased autonomy of its clients. Thomson calls into question the notion of a comprehensive governmentality, as put forward by Nikolas Rose, while Oosterhuis prefers an Eliasian approach to a Foucauldian one, pointing to the underlying processes of individualization and informalization as prime causes for the long-term development of the Dutch mental health system. Lunbeck's argument cannot be neatly categorized in this respect. Her point, however, that a stinging critique of society in an age of therapy that was couched in psychoanalytic language met with widespread approval seems to indicate rather limited effects of such therapy, while proving the appeal of such scientized language. This reading would be in line with Peter Stearns' notion of the profound ambiguity of expert advice – the more advice is provided, the more insecure its recipient becomes – and evidence from recent studies that have emphasized how American clients drew their own conclusions from the expert advice they received on matters of child-rearing and psychiatric treatment.[107]

Polling, marketing, and organizations

'The story begins in 1946.'[108] While British market researchers believe that their story began with the first informal get-together of what later became the influential Market Research Society, historians of the social sciences usually take a longer perspective. They trace the history of the disciplines that became recognized in the twentieth century as the social sciences back to older branches of knowledge.[109] Peter Wagner, for example, relates the idea of developing social knowledge for the purpose of social betterment to the Enlightenment, and regards the American and French Revolutions as the first large-scale applications of modern social and political theory.[110] The following decades witnessed different (pre-)histories for the various applied social sciences, but there seems to be a general agreement that by the late nineteenth and early twentieth century the various strands coalesced into fields to which specialists devoted their principal efforts of research, reflection, and training.[111] Dorothy Ross, for example, calls this era the 'disciplinary formation',[112] and Peter Wagner speaks of the 'classical period of the social sciences'.[113]

However, opinion polling, marketing techniques, and organizational research, which are at the heart of the last section of this volume, needed a bit longer to develop. Susan Herbst describes the time between 1930 and 1950 as the 'vital period' for polling in politics and industry in several national settings.[114] This is also the period on which Emil Walter-Busch, Sarah Igo, and Stefan Schwarzkopf concentrate in their contributions to this volume. Walter-Busch focuses on business organizations, foundations, and the state as promoters of applied social sciences in the United States and Switzerland. In particular he looks at industrial psychology, industrial relations research, and market and political opinion research. Sarah Igo focuses on the beginnings of opinion polling in the United States from the 1930s to the 1960s. She analyses how the origins of polling intersected with larger political and economic developments, and looks at some of the implications for modern public spheres. She also examines public reactions to opinion polling in order to trace the ways that this new technology helped to remake not just politics and public life but the public itself. Looking at the United States and the United Kingdom in a similar period (1920–60), Schwarzkopf discusses the role of political rhetoric and the social sciences in the engineering of British and American consumer societies. All three essays concentrate on the implementation and development of different social science methods, their funding, and their relationship to 'the public', the clients, and the state.

While the decades from 1920 up to the 1960s can be seen as the vital period for these applied social sciences, their techniques have a prehistory that dates back into the nineteenth century.[115] Straw polls are one of the techniques that gained influence especially in the United States from the 1820s, when they were first used to predict the outcome of presidential elections.[116] They made it possible to sound out public opinion about a candidate on various occasions (e.g. festivals) or from various sources (e.g. mailing lists of car owners) and were usually featured in newspapers. The best-known protagonist was the *Literary Digest*, which had great success in predicting Herbert Hoover's election in 1928. In the vital period from 1920 to 1960, however, failures seem to have been as important as successes for the development of polling. Less than ten years later, the *Literary Digest*'s 1936 straw poll got it completely wrong when it predicted Alfred Landon as the next American president with 57 per cent of the vote, when in fact he received only 38.5 percent.[117] The *Literary Digest*'s failure was even more significant because George Gallup's American Institute of Public Opinion (founded in 1935) correctly predicted Franklin D. Roosevelt's victory. Gallup did not use straw polls but the new technique of statistical sampling. The 1936 election was therefore also a competition between two technologies, which was won by Gallup and marked a milestone for modern opinion polling.

The different techniques were unthinkable without the international developments in statistics in the early nineteenth century as well as the increasing importance of national censuses, local surveys on poverty, and surveying in the colonies.[118] Statistical sampling was the most important invention in this field because it helped to make the expensive counting of every single citizen or targeted group member redundant.[119] Instead, only a small number of people who were considered representative were asked about their opinions, tastes, desires, sentiments, etc. To be sure, sampling-based opinion polling had its setbacks. In particular the erroneous prediction of a Thomas Dewey victory in the 1948 US presidential election showed the limits of this technique – a technique that was based on probability rather than 100 per cent accuracy. Pollsters were also wrong about British elections in 1970 and in 1992. However, failures like these had remarkably little impact in the long term. Among other things this was because once a method like statistical sampling was introduced it was not set in stone. Different forms of sampling, such as random or quota sampling, were introduced and further developed. In the same way polling was not a static methodology that followed the same approach over decades but it changed over time. Felix Keller points out that statistical sampling was promoted as a technique that continuously sought to improve and to adapt to new circumstances; it promised to be a self-reflexive method that responded to ever-changing needs.[120] In making this promise pollsters sold failures as part of the progress of the technique.

While the decades from the 1920s to the 1960s were a vital period, this was not at all a homogeneous era. These years tell the success story of opinion polling, marketing techniques, and organizational research, which were applied to an increasing number of areas. Two fields of application are the focus of two other essays of this section, both of which start in 1945. In the first case study, Anja Kruke and Benjamin Ziemann (Chapter 12) analyse the impact of opinion polling on the West German political system. According to Lutz Raphael's definition,[121] this is a classic example of the scientization of the social; the case study shows the increasing presence of opinion-polling experts in political parties, government institutions, and the mass media as well as the reciprocities of the technique, concepts, and strategies the pollsters had on offer. The second case study, by Benjamin Ziemann and Chris Dols (Chapter 15), seems less obvious but convincing: the authors demonstrate that in the Catholic Churches in the Netherlands and West Germany the advance of organizational research was dependent on a reform coalition between social scientists, theologians, and progressive bishops. Together these two chapters show that the end of the Second World War was a significant caesura. Postwar reconstruction offered opportunities to apply and enhance techniques, and planning was high on the agenda of many states and other organizations in the years after 1945.[122]

The Catholic Church is one example of numerous organizations – such as companies, administrations, schools or political parties – all of which are part of our everyday (working) lives.[123] Organizational theories about them have become increasingly important since the late nineteenth century.[124] One of the first names associated with organizational research was that of Frederick Taylor and his principles of scientific management.[125] The shift from Taylorism to human relations occurred in the Hawthorne experiments of the 1920s and especially the 1930s.[126] However, 1945 also marked a significant caesura for organizational research. Peter Wagner related this to developments since the late nineteenth century when bureaucracies in state, business, and political parties rose to ever-increasing importance and it became clear that there would be no withering away of the state and no self-organization of society: 'Such observations were at the root of a political sociology of organizations and bureaucracy, which later turned into an organizational theory that became the main paradigm in management studies and the new discipline of political science after the Second World War.'[127] From the middle of the twentieth century functionalism, associated with names like Talcott Parsons and Amitai Etzioni, became one of the major paradigms in the social sciences, but the field was far from monolithic; especially Herbert Simon and his theory of 'bounded rationality' became one of the major alternatives in organizational research during the 1940s and the 1950s.[128] In the United States, Samuel Stouffer's studies in battlefield sociology, published under the title 'American Soldier' in 1949, marked the breakthrough for a thorough investigation of battle performance within a functionalist paradigm.[129] In a somewhat ironic contrast, the West German Armed Forces, the Bundeswehr, relied on the services of Max Horkheimer and the 'Critical Theory' of his Frankfurt Institute for Social Research when the Federal Republic started recruiting new officers in the 1950s and wanted to test both aptitude and political attitude.[130]

Historians of polling, marketing techniques, and organizational research as well as practitioners in these fields seem to agree that 1945 was a significant caesura. In setting 1946 as the starting date of their profession, British market research and polling professionals claimed that 'market research in Britain is a postwar success story' that lasts until today.[131] As proof, market researchers refer to growing numbers of institutes, areas of applications, and turnover. However, from a critical historical perspective another important caesura can be seen in the 1970s when economic development entered a more decisive period, and planning became less predictable. In addition, criticism has arisen within the social science communities since the 1970s. In response to Theodor Adorno's analysis of the rise of 'administered society' with its accompanying form of social knowledge, developments such as the imbalance of demand-driven knowledge production and academic research were increasingly criticized within the social sciences.[132] Human resources departments and industrial psychologists in industrial firms began to place the internal democratization of the workplace at the top of their agenda during the 1970s.[133] The 1970s and their impact on postwar history are currently a major field of research.[134] While there seems to be agreement on the influences of global developments, such as the oil crises, national chronologies differed. It therefore seems relevant to take a closer look at the similarities and differences of the case studies provided in the last section of this volume.

To explain the dynamic growth of the American social sciences since the 1920s Walter-Busch (Chapter 14) points to two factors: competitive managerial capitalism and, as a side-effect, wealthy private foundations. In contrast to the United States, Switzerland had a regime of cooperative managerial capitalism. The political and business elites of the Swiss cantons preferred to promote 'practically useful or academically proven sub-disciplines of traditional sciences of state and the new social sciences' (Walter-Busch). On the federal level, support for social scientific research only started in the 1950s. The belated Swiss engagement in polling also resulted from the country's specific form of direct democracy: with dozens of referenda each year, public opinion researchers found it particularly difficult to convince politicians and citizens of the usefulness of polling. Switzerland also differed from its European neighbours in its anti-elitism and populist anti-intellectualism, which tended to prevent the establishment of those social scientific disciplines that were considered to be academically or practically useless. Especially in the first half of the twentieth century, the Swiss political elite and businessmen were satisfied with how the dense network of private interest associations and public institutions kept them informed.

The United States is often described as the mother country of polling, marketing techniques, and organizational research. But some authors also stress the importance of other nations in the development of these techniques, for example, the United Kingdom and its tradition of poverty studies,[135] and the jury is still out on this. However, the myth of the United States as the forerunner makes them a steady point of reference. And indeed, companies such as Gallup did start in America and then established polling affiliates abroad. To be sure, other countries, such as Britain, also started consumer research and audience research before the Second World War, but the war significantly shaped their research environment.[136] One important result of Nazism was the forced emigration of Jewish social scientists from Germany and Austria. It is also an example of entangled history because the work of these practitioners later on influenced polling, marketing, and organizational research in Europe – with different success, as the example of Ernest Dichter demonstrates.[137]

In postwar Germany, the introduction of sample-based opinion polls was closely connected to the reconstruction of the democratic system by the three Western Allies and their survey research. Transnational developments such as these also led to a less nationally focused approach to applying the social sciences. However, as Dorothy Ross has stated, the 'contents and borders of the disciplines that resulted ... were the product as much of national cultures, local circumstances, and accidental opportunities as of intellectual logic'.[138] In France, for example, students and left-wing intellectuals such as Pierre Bourdieu criticized polls, while in West Germany criticism mainly came from liberals and conservatives.[139] Whereas in Australia opinion polls were the monopoly of a conservative media conglomerate until 1971, West German polls were characterized by a pluralism of pollsters, methodological approaches, and media outlets, as outlined by Anja Kruke and Benjamin Ziemann in Chapter 12. The specific fields to which these technologies were applied also had a significant influence, as can be seen in the example of the Catholic Church in different nations. Other denominations used sociological research as well – the Board of Deputies of British Jews, for example, established their own research department. The decreasing significance of sociological research in the Catholic Churches since the mid-1970s should therefore not be seen as a general decrease in all fields but as specific to the Catholic Church and as the beginning of what can be described as secondary scientization.[140]

Apart from national peculiarities, the national case studies also show significant similarities. One such similarity in the field of opinion polling, marketing techniques, and organizational research is the constant reference to democracy. On the one hand, polling claimed to be a new technology of mass feedback that represented all citizens and not just a specific powerful group, that is, the upper classes.[141] Polls also made it possible not to equate newspapers' and journalists' opinions with public opinion. Public opinion polls were therefore presented as a democratic instrument that was indispensable to modern societies. On the other hand, marketing rhetoric equated citizens with consumers. In Chapter 13 Stefan Schwarzkopf follows this equation back to the interwar period. At that time the metaphor of the marketplace as a democracy of goods and as a consumer democracy in which consumer choices functioned like votes that decided the fate of products and companies became prominent. Schwarzkopf characterizes the 1930s as the key decade in the United States and Great Britain, during which the intellectual construction of terms like free choice and consumer sovereignty and the defence of marketing communication in the public sphere went hand in hand with innovations in actual marketing and market research practices. Instrumental in this shift away from the phenomenon of production to the political economy of consumer decision-making was a group of liberal economists, especially the members of the Austrian School of Economics who had to flee from Nazi repression. Kruke and Ziemann observe a similar equation when in the mid-1950s both West German Social Democrats and Christian Democrats began to conceptualize the electorate in terms of consumers in an open marketplace. In the following decade focus groups and group discussions were used in addition to traditional sampling methods, and by the 1970s a wide range of psychological approaches – so far used only in marketing and advertising – were applied to political opinion polling, which made it similar to a form of market research.[142] Polling had to be sold rhetorically and literally to different audiences. For the American case, Sarah Igo makes this point with regard to the beginnings of opinion research in the 1930s, when social science methods as a scientific but also democratic technology had to be advertised to corporate sponsors and to the public. Organization researchers also referred to democracy because the theories and strategies they developed were meant to increase democratic structures within organizations (such as political parties).

What does polling do to democracy? This is one of the major underlying questions, which also lies at the heart of critiques of these techniques. Susan Herbst stresses that surveys inevitably narrow public debate by defining public problems in specific ways.[143] Other forms of expressing public opinion, such as letters to public officials, pressure groups, arts, etc., seem to have become less newsworthy than polls.[144] Indeed, the cooperation of opinion pollsters and newspaper journalists has a long history dating back to the straw polls. Another problem is rooted in the assumption of representative polls that everybody is equally important. This also means that polling ignores 'the complexity of social structure and power dynamics by overlooking social inequality and missing key aspects of policy formation'.[145] If we take into account that opinion pollsters also often were market researchers we arrive an even more complex picture. As Daniel J. Robinson has argued with regard to Canada, many of the sponsors and beneficiaries of opinion polling were in fact

> the very 'Interests' – pols, state propagandists, PR men, advertisers – whose power, according to Gallup and CIPO [Canadian Institute of Public Opinion] officials, was meant to be tempered by the sample survey. The polling-generated 'People', itself a distorted facsimile of the citizenry, constituted more a malleable entity to be acted on or manipulated by others than a virtuous source of democratic authority.[146]

This observation leads to another similarity, which is the struggle between objectivization and the politicization of experts. The establishment of a group of professional experts is a major indication of a process of scientization. While experts promise to objectivize controversial topics, the increasing use of expert advice also leads to a politicization of the social sciences.[147] When social science experts are in increasing demand for a variety of purposes and actors, the ensuing political controversies also affect their disciplinary fields and academic institutions. As the example of the applied fields of polling, marketing techniques, and organizational research shows, these fields were highly competitive. The institutes who carried out work in the different areas were dependent on the commissioning of research projects, be it through the market or through institutional links. As Peter Wagner has shown for organizational research, these sponsors were organizations (such as public agencies, big business, the media or political parties) that were large enough to be able to afford the production of knowledge on demand.[148] If we concentrate on experts, we can observe that opinion research in 1930s America was not a well-demarcated field and that its practitioners were not a defined group (see Igo's chapter). Likewise, Emil Walter-Busch shows that after 1900 in the United States and Switzerland multiple participants were engaged in a continuous struggle to occupy and define a sharply contested but never clearly demarcated discursive and practical field.[149]

It is important to note that the self-proclaimed experts were not only driven by objective interests but also by economic and political reasons. They had to sell their services to different clients (such as companies, government departments or political parties) to maintain their business. In many cases this led to a double strategy: from their foundation in the 1940s up to the present, West German polling institutes, for example, always performed market research for companies in addition to their political polling for government and political parties. Moreover, since the 1990s it has not been sociologists but consulting firms like McKinsey who have been employed by the Catholic Church, which leads Ziemann and Dols to the conclusion that finance was the hidden agenda of church reform in West Germany. Regarding America and Britain, Schwarzkopf points out that knowing and testing consumers and equating them with voters is 'big business' for those who benefit from this illusion. The idea that commercial advertising and market research are a necessary part of modern democracy is mainly propagated by those who benefit financially from testing, surveying, measuring, etc.

Experts in polling, marketing, and organizational research sometimes had a difficult standing because they seemed to follow obvious political interests. Schwarzkopf observes that during the 1920s advertising and marketing practitioners were eager to use political-philosophical constructs in order to legitimize their role in economic and social life. He also stresses that during the 1940s and 1950s Paul Lazarsfeld and other market and opinion researchers made a deeply political – not scientific – decision to equate consumer choices and political choices as equivalent or even identical. Likewise, Kruke and Ziemann come to the conclusion that in the postwar period polling experts were politicized, and Ziemann and Dols argue that, far from being flag-bearers of objectivity, experts in the Catholic Church contributed to a substantial politicization of the struggles about the implementation of Vatican II. Walter-Busch's chapter on Switzerland deplores the chronically underdeveloped and politically dependent status of Switzerland's social sciences.[150]

All the chapters in this section share one significant finding: scientization does not need to be thoroughly persuasive to do its work. Although it was criticized, contested or rejected by parts of the public, scientists, social critics, and government officials, it still had an effect in various fields of modern societies. However, while the discussion about the validity of the applied social sciences continues, the application of these sciences is not contested. Thus, market research practitioners at the end of the twentieth century are able to present their history with reference to turnover, appearance in the media, etc. as an overall success story – despite the more critical histories presented by historians of the social sciences.

Conclusion

Through a historical analysis of the scientization of the social, the history of the social sciences can be embedded in the broader trajectory of Western societies since the late nineteenth century. Starting in the 1880s, statistics, the psy-sciences, survey research and other forms of social research were applied in a variety of different settings in Western Europe and the United States. While Eastern European countries also developed practical applications of the social sciences, especially the state socialist societies during the Cold War,[151] the chronology and the overall context of scientization differed substantially from those in Western societies. At a general level of comparison, the existence of university-based research was important, but it was not the primary driving factor in these developments in the West. To be sure, countries with highly developed university systems had a head start in some fields of applied research, mainly the United States and Germany, which pioneered 'big science' – scientists working in large-scale laboratories – and were the trailblazers in the development of research-driven university teaching in the two decades around 1900. Other countries, such as Switzerland, France, Great Britain, and Italy, lagged behind in that respect until the post-1945 expansion of their university systems took place.

But the proliferation of university-taught social science disciplines was neither a necessary nor a sufficient factor in the scientization of the social. At least until 1945, three other factors had greater potency: first, the existence of state bureaucracies that were keen to use the social sciences for the purposes of nation-building, welfare state intervention or education; secondly, an acute sense of crisis or moral panic among the elites, a factor that was crucial both for the statistics-based welfare state intervention in favour of the labouring poor and for the Spanish interest in psychology-based education; and, thirdly, contingent factors such as wars: it is hard to overestimate the extent to which military recruitment for the First World War and medical care for its mental casualties gave a boost to the application of the psy-sciences, while the Second World War kick-started a massive upsurge in organization research and social science-driven state planning.

In the absence of major wars since 1945, the first two factors remained crucial in the postwar period. In addition, however, another key driving force of the scientization of the social came to the fore, starting in the United States and the Great Britain during the 1930s, but quickly advancing in other Western European societies in the post-1945 period: firms that were governed by competitive managerial capitalism and rapidly expanding in size experienced increasing demand for social scientific knowledge that would allow them to optimize their internal structures (industrial psychology and organizational research) and for information on the anonymous multitudes that constituted the external environment for their managerial strategies. These were, in the first instance, the consumers and their preferences. But as other formal organizations such as parties, media corporations and – last but not least – the churches adopted similar strategies, they also applied both forms of social science knowledge, the latter mostly in the form of opinion polling.

In that sense, a comparative analysis of the scientization of the social requires more than comparisons between different countries. Especially for the period since 1945, a functionalist form of comparison, that is, charting the different ways in which various functional sub-systems of Western societies (economy, politics, education, mass media, religion) employed social scientific knowledge is also relevant. The chronologies of these encounters show characteristic differences. Such a differentiated approach also has important conceptual implications. It demonstrates that the notion of an all-pervasive, secular process of scientization, which is advanced by some Foucauldian interpretations, does not match the historical record. The scientization of the social did have political implications, not least for the scientists themselves. But these implications were contingent on specific historical contexts and did not represent an inherent feature of 'modernity'. Moreover, the scientization of the social was not only directed from above, by political and other elites and scientists who offered their expertise. It was also embraced from below, as clients and consumers, most notably after the Second World War, made their own uses of a popularized scientific knowledge that conveyed an ever-increasing variety of messages.

Notes

1. See B. Ziemann (2005), 'Krose, Hermann A.', in *Biographisch-Bibliographisches Kirchenlexikon*, vol. 24 (Herzberg: Bautz), pp. 983–6.
2. B. Ziemann (2007), *Katholische Kirche und Sozialwissenschaften, 1945–1975* (Göttingen: Vandenhoeck and Ruprecht), pp. 36–75.
3. L. Raphael (1996), 'Die Verwissenschaftlichung des Sozialen als methodische und konzeptionelle Herausforderung für eine Sozialgeschichte des 20. Jahrhunderts', *Geschichte und Gesellschaft* 22, pp. 165–93, quote on 166; see also his chapter in this volume.
4. 'Scientism', in W. A. Darity (ed.) (2008), *International Encyclopedia of the Social Sciences*, vol. 7, second edition (Detroit: Macmillan Reference), pp. 364f.
5. R. C. Bannister (1987), *Sociology and Scientism. The American Quest for Objectivity, 1880–1940* (Chapel Hill: University of North Carolina Press); D. Ross (2003), 'Changing Contours of the Social Science Disciplines', in T. M. Porter and D. Ross (eds.), *The Cambridge History of Science*, vol. 7: *The Modern Social Sciences* (Cambridge: Cambridge University Press), pp. 205–37.
6. J. Habermas (1970 [1964]), 'The Scientization of Politics and Public Opinion', in idem, *Toward a Rational Society. Student Protest, Science and Politics* (London: Heinemann), pp. 62–80.
7. Ibid., p. 65.
8. See the chapter by Elizabeth Lunbeck in this volume.
9. See the chapter by Stefan Schwarzkopf in this volume.
10. See S. Fitzpatrick (1993), 'Ascribing Class. The Construction of Social Identity in Soviet Russia', *Journal of Modern History* 65, pp. 745–70.
11. For this distinction, see N. Luhmann (1998), *Observations on Modernity* (Stanford: Stanford University Press); B. Ziemann (2012), 'Die Metaphorik der Gesellschaft. Soziologische Selbstbeschreibungen westeuropäischer Gesellschaften im 20. Jahrhundert', in L. Raphael (ed.), *Theorien und Experimente der Moderne. Europas Gesellschaften im 20. Jahrhundert* (Cologne: Böhlau), pp. 193–227.
12. On suicide, cf. U. Baumann (2001), *Vom Recht auf den eigenen Tod. Die Geschichte des Suizids vom 18. bis zum 20. Jahrhundert* (Weimar: Böhlau), pp. 202–26.
13. See the chapter by Julia Moses in this volume.
14. See P. Becker and R. Wetzell (eds) (2004), *Criminals and their Scientists. The History of Criminology in International Perspective* (Cambridge: Cambridge University Press), especially chapters in Part Two; R. Wetzell (2000), *Inventing the Criminal. A History of German Criminology 1880–1945* (Chapel Hill: University of North Carolina Press); S. Galassi (2004), *Kriminologie im Deutschen Kaiserreich. Geschichte einer gebrochenen Verwissenschaftlichung* (Stuttgart: Steiner).
15. M. Lengwiler (2000), *Zwischen Klinik und Kaserne. Die Geschichte der Militärpsychiatrie in Deutschland und der Schweiz 1870–1914* (Zurich: Chronos).
16. Cited in A. Nassehi (2000), 'Die Geburt der Soziologie aus dem Geist der Individualität', in T. Kron (ed.), *Individualisierung und soziologische Theorie* (Opladen: Leske + Budrich), pp. 45–67, here 48.
17. See Raphael (1996), 'Verwissenschaftlichung', pp. 178f.
18. P. Wagner (2003), 'Social Science and Social Planning during the Twentieth Century', in Porter and Ross (eds), *The Modern Social Sciences*, pp. 591–607.
19. U. Beck and W. Bonß (eds) (1989), *Weder Sozialtechnologie noch Aufklärung? Analysen zur Verwendung sozialwissenschaftlichen Wissens* (Frankfurt: Suhrkamp).

20. As an often-cited statement, cf. C. S. Maier (2000), 'Consigning the Twentieth Century to History. Alternative Narratives for the Modern Era', *American Historical Review* 105, pp. 807–31.
21. T. Schanetzky (2010), 'Aporien der Verwissenschaftlichung. Sachverständigenrat und wirtschaftlicher Strukturwandel in der Bundesrepublik 1974–1988', *Archiv für Sozialgeschichte* 50, pp. 153–67; C. E. Lindblom (1959), 'The Science of Muddling Through', *Public Administration Review* 19, pp. 79–88.
22. T. M. Porter and D. Ross, 'Introduction: Writing the History of Social Science', in Porter and Ross (eds), *The Modern Social Sciences*, pp. 1–10, here 9.
23. J. Vogel (2004), 'Von der Wissenschafts- zur Wissensgeschichte', *Geschichte und Gesellschaft* 30, pp. 639–60, here 644–6.
24. See, for instance, T. Judt (2005), *Postwar. A History of Europe since 1945* (London: Penguin); M. Mazower (1998), *Dark Continent. Europe's Twentieth Century* (London: Allen Lane); E. Hobsbawm (1994), *Age of Extremes. The Short Twentieth Century, 1914–1991* (London: Michael Joseph); A. Marwick (1998), *The Sixties. Cultural Revolution in Britain, France, Italy, and the United States, c. 1958–c. 1974* (Oxford: Oxford University Press); W. Chafe (1995), *The Unfinished Journey. America since World War II* (Oxford: Oxford University Press); J. T. Patterson (1996), *Grand Expectations. The United States, 1945–74* (Oxford: Oxford University Press).
25. H-U. Wehler (2008), *Deutsche Gesellschaftsgeschichte*, vol. 5: *Bundesrepublik und DDR 1949–1990* (Munich: C. H. Beck), pp. 110–18. But see the brilliant chapter on sociological self-descriptions by J. Harris (1993), *Private Lives, Public Spirit. Britain 1870–1914* (London: Penguin), pp. 220–50.
26. See the chapter by Martin Lengwiler in this volume.
27. See the chapter by Harry Oosterhuis in this volume; for penal policy, see V. Janssen (2007), 'From the Inside Out. Therapuetic Penology and Political Liberalism in Postwar California', *Osiris* 22, pp. 116–34, at 129.
28. See the chapters by Sarah Igo and by Anja Kruke and Benjamin Ziemann in this volume.
29. See M. Fahlbusch and I. Haar (eds) (2010), *Völkische Wissenschaften und Politikberatung im 20. Jahrhundert. Expertise und 'Neuordnung' Europas* (Paderborn: Schöningh); L. Raphael (2001), 'Radikales Ordnungsdenken und die Organisation totalitäter Herrschaft: Weltanschauungseliten und Humanwissenschaftler im NS-Regime', *Geschichte und Gesellschaft* 27, pp. 5–40.
30. M. Foucault (2002 [1969]), *The Archaeology of Knowledge* (Abingdon: Routledge), pp. 44ff., 55ff.
31. See, for instance, F. Keller (2001), *Archäologie der Meinungsforschung. Mathematik und die Erzählbarkeit des Politischen* (Constance: UVK); P. Sarasin and J. Tanner (eds) (1998), *Physiologie und industrielle Gesellschaft. Studien zur Verwissenschaftlichung des Körpers im 19. und 20. Jahrhundert* (Frankfurt: Suhrkamp).
32. M. Foucault (1991), 'Governmentality', in G. Burchell, C. Gordon, and P. Miller (eds), *The Foucault Effect. Studies in Governmentality* (Chicago: University of Chicago Press), pp. 87–104; T. Lemke (1997), *Eine Kritik der politischen Vernunft. Foucaults Analyse der modernen Gouvernementalität* (Hamburg: Argument).
33. N. Rose (1990), *Governing the Soul. The Shaping of the Private Self* (London and New York: Routledge); idem (1998), *Inventing Our Selves. Psychology, Power, and Personhood* (Cambridge: Cambridge University Press).
34. See the figures in G. Eghigian, A. Killen, and C. Leuenberger (2007), 'Introduction: The Self as Project. Politics and the Human Sciences in the Twentieth Century', *Osiris* 22, pp. 1–25, here 21.
35. See the chapter by Mathew Thomson in this volume; cf. also Ziemann (2007), *Katholische Kirche*, pp. 313f.
36. For an outline of his theory, N. Luhmann (1982), *The Differentiation of Society* (New York: Columbia University Press); with regard to science, see idem (1990), *Die Wissenschaft der Gesellschaft* (Frankfurt: Suhrkamp).

37. See H. Tyrell (1998), 'Zur Diversität der Differenzierungstheorie. Soziologie historische Anmerkungen', *Soziale Systeme* 4, pp. 125–7.
38. See Luhmann (1990), *Wissenschaft*, pp. 167–270; idem (1983), 'Insistence on Systems Theory: Perspectives from Germany', *Social Forces* 61, 987–98.
39. See the chapter by Julia Moses in this volume.
40. See S. Maasen and P. Weingart (1995/96), '"Metaphors" – Messengers of Meaning. A Contribution to an Evolutionary Sociology of Science', *Science Communication* 17, pp. 9–31; S. Maasen (2000), 'Metaphors in the Social Sciences. Making Use and Making Sense of them', in F. Hallyn (ed.), *Metaphor and Analogy in the History and Philosophy of Science* (Dordrecht: Kluwer), pp. 199–244.
41. S. Maasen, E. Mendelsohn, and P. Weingart (1995), 'Metaphors. Is there a Bridge over Troubled Waters?', in idem (eds), *Biology as Society, Society as Biology. Metaphors* (Dordrecht: Kluwer), pp. 1–8, at 2; on the 'bucket', see the chapter by Benjamin Ziemann and Chris Dols in this volume.
42. P. Wagner (2001), 'The Mythical Promise of Societal Renewal. Social Science and Reform Coalitions', in idem, *A History and Theory of the Social Sciences* (London: Sage), pp. 54–72; idem (1990), *Sozialwissenschaften und Staat. Frankreich, Italien, Deutschland 1870–1980* (Frankfurt: Campus), pp. 55f.
43. As a case study, see J. Platz (2010), '"Die White Collars in den Griff bekommen". Industrieangestellte im Spannungsfeld sozialwissenschaftlicher Expertise und gewerkschaftlicher Politik', *Archiv für Sozialgeschichte* 50, pp. 271–88.
44. M. G. Ash (2010), 'Wissenschaft und Politik. Eine Beziehungsgeschichte im 20. Jahrhundert', *Archiv für Sozialgeschichte* 50, pp. 11–46, here 13–15, quote on 13.
45. Luhmann (1990), *Wissenschaft*, p. 638.
46. E. Walter-Busch (2006), *Faktor Mensch. Formen angewandter Sozialforschung der Wirtschaft in Europa und den USA 1890–1960* (Constance: UVK), pp. 439–46.
47. T. M. Porter (1995), *Trust in Numbers. The Pursuit of Objectivity in Science and Public Life* (Princeton: Princeton University Press), p. 210.
48. P. Weingart (1983), 'Verwissenschaftlichung der Gesellschaft – Politisierung der Wissenschaft', *Zeitschrift für Soziologie* 12, pp. 225–41.
49. M. Lengwiler (2006), 'Zwischen Verwissenschaftlichung, Politisierung und Bürokratisierung: Expertenwissen im schweizerischen Sozialstaat', in S. Bundesarchiv (ed.), *Geschichte der Sozialversicherungen* (Zurich: Chronos), pp. 167–90, at 183ff.
50. Raphael, 'Verwissenschaftlichung', p. 178. For a sociological statement, see H. Nowotny (2005), 'Experten, Expertisen und imaginierte Laien', in A. Bogner and H. Torgersen (eds), *Wozu Experten? Ambivalenzen der Beziehung von Wissenschaft und Politik* (Wiesbaden: VS Verlag), pp. 33–44.
51. C. Albrecht (2004), 'Expertive versus Demonstrative Politikberatung. Adorno bei der Bundeswehr', in S. Fisch and W. Rudloff (eds), *Experten und Politik. Wissenschaftliche Politikberatung in Geschichtlicher Perspektive* (Berlin: Duncker & Humblot), pp. 297–308, at 306f.
52. See the chapter by Martin Lengwiler in this volume.
53. See the chapter by Mathew Thomson in this volume. Cf. also M. G. Ash (2003), 'Psychology', in Porter and Ross (eds), *The Modern Social Sciences*, pp. 251–74, at 273; B. Ziemann (2006), 'The Gospel of Psychology. Therapeutic Concepts and the Scientification of Pastoral Care in the West German Catholic Church, 1950–1980', *Central European History* 39, pp. 79–106.
54. See the chapters by Sarah Igo and Harry Oosterhuis in this volume.
55. See P. Wagner (2003), 'The Uses of the Social Sciences', in Porter and Ross (eds), *The Modern Social Sciences*, pp. 537–52, at 550.
56. See the chapter by Peter Becker in this volume.
57. P. Fuchs (1992), 'Gefährliche Modernität. Das zweite vatikanische Konzil und die Veränderung des Messeritus', *Kölner Zeitschrift für Soziologie und Sozialpsychologie* 44, pp. 1–11.

58. On risk and the 'risk society', see A. Giddens (1999), 'Risk and Responsibility', *The Modern Law Review* 62, pp. 1–10; U. Beck (1986), *Risikogesellschaft* (Frankfurt: Suhrkamp), translated as U. Beck (1992), *Risk Society* (London: Sage).
59. T. M. Porter (1986), *The Rise of Statistical Thinking 1820–1900* (Princeton: Princeton University Press).
60. M. Foucault (1978), 'About the Concept of the Dangerous Individual in 19th-Century Legal Psychiatry', *International Journal of Law and Psychiatry* 1, pp. 1–18; the definitive French version is 'L'évolution de la notion d' "individu dangereux" dans la psychiatrie du xixème siècle', in Foucault (1994), *Dits et écrits*, vol. 2 (Paris: Gallimard), pp. 443–64. Foucault seems to have developed this thesis from reading Raymond Saleilles, a turn-of-the-century French jurist who worked on accident liability in civil law and on penal reform. See R. Saleilles (1897), *Les accidents de travails et la responsabilité civile* (Paris: A. Rousseau); idem (1898), *L'individualisation de la peine* (Paris: Felix Alcan).
61. On *Causalhaftung*, see S. Andriopoulos (1996), *Unfall und Verbrechen. Konfigurationen zwischen juristischem und literarischem Diskurs um 1900* (Pfaffenweiler: Centaurus), pp. 65–68; G. Rümelin (1898), 'Culpahaftung und Causalhaftung', *Archiv für die civilistische Praxis* N.F. 38, pp. 285–316.
62. On *risque professionnel* and the 1898 law, see F. Ewald (1986), *L'état providence* (Paris: Grasset), pp. 282–317.
63. See A. Prins (1910), *La Défense sociale et les transformations du droit pénal* (Brussels and Leipzig: Misch et Throns), pp. 56–7.
64. Ewald (1986), *L'état providence*, pp. 409–16. The connection between *risque professionnel* concerning industrial accidents and 'dangerousness' (*état dangereux*) in criminal law is also explored in R. Harris (1989), *Murders and Madness. Medicine, Law, and Society in the* fin de siècle (Oxford: Clarendon Press).
65. Andriopoulos (1996), *Unfall und Verbrechen*, esp. pp. 1–2, 65–9, 85–6.
66. On the history of criminology, see Becker and Wetzell (2006), *Criminals and Their Scientists*; C. Emsley (2007), *Crime, Police, and Penal Policy. European Experiences 1750–1940* (Oxford: Oxford University Press), pp. 181–99. On Italy, M. Gibson (2002), *Born to Crime. Cesare Lombroso and the Origins of Biological Criminology* (Westport, CT: Praeger); on France, L. Mucchielli (1994), *Histoire de la criminology française* (Paris: Editions L'Harmattan). On Britain, N. Davie (2005), *Tracing the Criminal. The Rise of Scientific Criminology in Britain, 1860–1918* (Oxford: Bardwell Press). On Germany, P. Becker (2002), *Verderbnis und Entartung. Eine Geschichte der Kriminologie des 19. Jahrhunderts als Diskurs und Praxis* (Göttingen: Vandenhoeck); Wetzell (2000), *Inventing the Criminal*; S. Galassi (2004), *Kriminologie im Deutschen Kaiserreich*; T. Kailer (2011), *Vermessung des Verbrechers. Die Kriminalbiologische Untersuchung in Bayern, 1923–1945* (Bielefeld: Transcript). On the United States, N. H. Rafter (1997), *Creating Born Criminals* (Urbana: University of Illinois Press).
67. On Quetelet's work in criminal statistics, see P. Beirne (1993), 'The Rise of Positivist Criminology. Adolphe Quetelet's Social Mechanics of Crime', in idem, *Inventing Criminology* (Albany: State University of New York Press), pp. 65–110 (Quetelet quote on 88); on criminal statistics in general, see Emsley (2007), *Crime, Police, and Penal Policy*, pp. 117–34; Becker (2002), *Verderbnis*, passim; and Wetzell (2000), *Inventing the Criminal*, pp. 21–5.
68. D. Pick (1989), *Faces of Degeneration. A European Disorder, c. 1848–c. 1918* (Cambridge: Cambridge University Press); J. E. Chamberlin and S. L. Gilman (eds) (1985), *Degeneration. The Dark Side of Progress* (New York: Columbia University Press).
69. E. J. Engstrom (2003), *Clinical Psychiatry in Imperial Germany* (Ithaca, NY: Cornell University Press), pp. 174–98.
70. Engstrom (2003), *Clinical Psychiatry*; H-P. Schmiedebach (1997), 'The Mentally Ill Patient Caught between the State's Demands and the Professional Interests of Psychiatrists', in M. Berg and G. Cocks (eds), *Medicine and Modernity. Public Health and Medical Care in Nineteenth- and Twentieth-Century Germany* (Cambridge: Cambridge University Press), pp. 99–119.

71. Gibson (2002), *Born to Crime*; on Lombroso's international reception, see Becker and Wetzell (2006), *The Criminal and His Scientists*.
72. Wetzell (2000), *Inventing*, pp. 100–5, 233–94; L. Mucchielli (2006), 'Criminology, Hygienism, and Eugenics in France, 1870–1914. The Medical Debates on the Elimination of "Incorrigible" Criminals', in Becker and Wetzell (eds), *The Criminal and His Scientists*, pp. 207–29; Davie (2005), *Tracing the Criminal*, pp. 229–59.
73. On penal policy, see Emsley (2007), *Crime, Police, and Penal Policy*, pp. 228–45; R. F. Wetzell (2004), 'From Retributive Justice to Social Defense. Penal Reform in Fin-de-Siècle Germany', in S. Marchand and D. Lindenfeld (eds), *Germany at the Fin de Siècle* (Baton Rouge: Louisiana State University Press), pp. 59–77; M. Wiener (1990), *Reconstructing the Criminal. Culture, Law, and Policy in England, 1830–1914* (New York: Cambridge University Press); R. A. Nye (1984), *Crime, Madness, and Politics in Modern France* (Princeton: Princeton University Press); Gibson (2002), *Born to Crime*.
74. R. Wetzell (2009), 'Psychiatry and Criminal Justice in Modern Germany, 1880–1933', *Journal of European Studies* 39, pp. 270–89; C. Müller (2004), *Verbrechensbekämpfung im Anstaltsstaat. Psychiatrie, Kriminologie und Strafrechtsreform in Deutschland 1871–1933* (Göttingen: Vandenhoeck).
75. On criminal-biological examinations, see T. Kailer (2011), *Vermessung des Verbrechers*; O. Liang (2006), 'The Biology of Morality. Criminal Biology in Bavaria, 1924–1933', in Becker and Wetzell (eds), *Criminals and their Scientists*, pp. 425–46.
76. On the nexus between criminal justice and welfare, see D. Garland (1985), *Punishment and Welfare. A History of Penal Strategies* (Aldershot: Gower); W. Rosenblum (2008), *Beyond the Prison Gates. Punishment and Welfare in Germany, 1850 to 1933* (Chapel Hill: University of North Carolina Press).
77. For this debate, see R. Wetzell (2008), 'Die Rolle medizinischer Experten in Strafjustiz und Strafrechtsreformbewegung. Eine Medikalisierung des Strafrechts?', in A. Kästner and S. Kesper-Biermann (eds), *Experten und Expertenwissen in der Strafjustiz von der Frühen Neuzeit bis zur Moderne* (Leipzig: Meine-Verlag), pp. 57–71; Wetzell (2009), 'Psychiatry and Criminal Justice'.
78. On biological theories of crime, see N. Rafter (2008), *The Criminal Brain. Understanding Biological Theories of Crime* (New York: New York University Press).
79. M. H. Geyer and J. Paulmann (2001), 'Introduction: The Mechanics of Internationalism', in M. H. Geyer and J. Paulmann (eds), *The Mechanics of Internationalism. Culture, Society, and Politics from the 1840s to the First World War* (Oxford: Oxford University Press), pp. 1–25.
80. On the history of nineteenth-century congresses on social reform, see N. Randeraad (2011), 'The International Statistical Congress (1853–1876). Knowledge Transfers and their Limits', *European History Quarterly* 41, pp. 50–65; C. Leonards and N. Randeraad (2010), 'Transnational Experts in Social Reform, 1840–1880', *International Review of Social History* 55, pp. 215–39.
81. See S. Kesper-Biermann and P. Overrath (2007), *Die Internationalisierung von Strafrechtswissenschaft und Kriminalpolitik (1870–1930). Deutschland im Vergleich* (Berlin: Berliner Wissenschaftsverlag); D. Schauz and S. Freitag (eds) (2007), *Verbrecher im Visier der Experten. Kriminalpolitik zwischen Wissenschaft und Praxis im 19. und frühen 20. Jahrhundert* (Stuttgart: Steiner); T. P. Marques (2007), 'Mussolini's Nose. A Transnational History of the Penal Code of Fascism' (Diss., EUI Florence), pp. 76–91, 324–39; M. Kaluszynski (2006), 'The International Congresses of Criminal Anthropology', in Becker and Wetzell (eds), *Criminals and their Scientists*, pp. 301–16.
82. See R. O. Keohane and J. S. Nye (eds) (1981), *Transnational Relations and World Politics* (Cambridge, MA: Harvard University Press), p. xi.
83. P. Clavin (2005), 'Defining Transnationalism', *Contemporary European History* 14, pp. 421–39, especially 438.
84. M. Herren (2006), 'Sozialpolitik und die Historisierung des Transnationalen', *Geschichte und Gesellschaft* 32, pp. 542–59, esp. 544; M. Herren (2009), *Internationale Organisationen seit 1865. Eine Globalgeschichte der internationalen Ordnung* (Darmstadt: Wissenschaftliche Buchgesellschaft), esp. pp. 38–9.

85. P-Y. Saunier (2008), 'Les régimes circulatoires du domaine social 1800–1940. Trajectoires, projets et ingéniérie de la convergence et de la différence', *Genèses* 71, pp. 4–25.
86. On entangled history, see M. Werner and B. Zimmermann (2006), 'Beyond Comparison. *Histoire Croisée* and the Challenge of Reflexivity', *History and Theory* 45, pp. 30–50.
87. See also T. Etzemüller (2010), *Die Romantik der Rationalität. Alva & Gunnar Myrdal – Social Engineering in Schweden* (Bielefeld: Transkript).
88. On normalization, see M. Foucault (2007), *Security, Territory, Population. Lectures at the Collège de France 1977–1978* (New York: Palgrave Macmillan), pp. 55–79.
89. D. Peukert (1993), 'The Genesis of the Final Solution from the Spirit of Science', in T. Childers and J. Caplan (eds), *Reevaluating the Third Reich* (New York: Holmes and Meier), pp. 234–52; Z. Bauman (1989), *Modernity and the Holocaust* (Ithaca, NY: Cornell University Press).
90. See E. R. Dickinson (2004), 'Biopolitics, Fascism, Democracy. Some Reflections on Our Discourse about Modernity', *Central European History* 37, pp. 1–48.
91. S. Ward (2002), *Modernizing the Mind. Psychological Knowledge and the Remaking of Society* (Westport, CT: Praeger): E. Illouz (2008), *Saving the Modern Soul. Therapy, Emotions, and the Culture of Self-help* (Berkeley: University of California Press).
92. On the stages of 'citizenship', see T. H. Marshall (1950), *Citizenship and Social Class, and Other Essays* (Cambridge: Cambridge University Press); B. Turner (ed.) (1993), *Citizenship and Social Theory* (London: Sage); A. Fahrmeir (2007), *Citizenship. The Rise and Fall of a Modern Concept* (New Haven: Yale University Press).
93. E. Lunbeck (2003), 'Psychiatry', in Porter and Ross (eds), *The Modern Social Sciences*, pp. 363–77, quote at 367; E. Engstrom (2003), *Clinical Psychiatry in Imperial Germany. A History of Psychiatric Practice* (Ithaca, NY: Cornell University Press).
94. M. Ash (2003), 'Psychology', in Porter and Ross (eds), *The Modern Social Sciences*, pp. 251–74, here 252–5.
95. Ibid., pp. 255–9. On the scientization of childhood around 1900, see S. Mintz (2004), *Huck's Raft* (Cambridge, MA: Belknap Press), pp. 186f.; P. Stearns (2003), *Anxious Parents. A History of Modern Childrearing in America* (New York: New York University Press); A. Hulbert (2003), *Raising America. Experts, Parents, and a Century of Advice about Children* (New York: Alfred Knopf).
96. R. H. Wiebe (1967), *The Search for Order, 1877–1920* (New York: Hill and Wang).
97. J. Sealander (2003), *The Failed Century of the Child. Governing America's Young in the Twentieth Century* (New York: Cambridge University Press), pp. 19–31, quote at 30.
98. G. Eghigian, A. Killen, and C. Leuenberger (eds) (2007), *The Self as Project. Politics and the Human Sciences* (Osiris, vol. 22) (Chicago: University of Chicago Press); cf. P. Lerner (2003), *Hysterical Men. War, Psychiatry, and the Politics of Trauma in Germany, 1890–1930* (Ithaca, NY: Cornell University Press).
99. Sealander (2003), *Failed Century*, pp. 198–204; D. Kevles (1968), 'Testing the Army's Intelligence. Psychologists and the Military in World War I', *Journal of American History* 56, pp. 565–81; M. Sokal (ed.) (1987), *Psychological Testing and American Society, 1890–1930* (New Brunswick, NJ: Rutgers University Press). Most recruits were defined as unfit because of their insufficient weight, a fact not made known to the public, however.
100. S. Cohen (1999), *Challenging Orthodoxies. Toward a New Cultural History of Education* (New York: Peter Lang), pp. 249–71. On the psychologization of American childhood in the wake of the war, cf. E. Herman (2003), 'Psychologism and the Child', in Porter and Ross (eds), *The Modern Social Sciences*, pp. 649–62.
101. Ash (2003), 'Psychology', p. 267.
102. H. W. Schmuhl (1992), *Rassenhygiene, Nationalsozialismus, Euthanasie. Von der Verhütung zur Vernichtung 'lebensunwerten Lebens', 1890–1945*, second edition (Göttingen: Vandenhoeck and Ruprecht).
103. G. Cocks (1997), *Psychotherapy in the Third Reich. The Göring Institute* (New York: Oxford University Press); U. Geuter (1992), *The Professionalization of Psychology in Nazi Germany*, translated by R. J. Holmes (Cambridge: Cambridge University Press).

104. E. Herman (1995), *The Romance of American Psychology. Political Culture in the Age of Experts* (Berkeley: University of California Press), figure on p. 20; J. Capshew (1999), *Psychologists on the March. Science, Practice, and Professional Identity in America, 1929–1969* (New York: Cambridge University Press).
105. Ash (2003), 'Psychology', pp. 271–73.
106. A. Woolfolk (2004), 'The Dubious Triumph of the Therapeutic. The Denial of Character', in J. Imber (ed.), *Therapeutic Culture. Triumph and Defeat* (New Brunswick, NJ: Transaction), pp. 69–88; P. Rieff (1987 [1966]), *The Triumph of the Therapeutic. Uses of Faith after Freud*, second edition (Chicago: University of Chicago Press); E. Moskowitz (2001), *In Therapy We Trust. America's Obsession with Self-Fulfillment* (Baltimore, MD: Johns Hopkins University Press).
107. P. Stearns (2003), *Anxious Parents*; cf. K. Jones (1999), *Taming the Troublesome Child. American Families, Child Guidance, and the Limits of Psychiatric Authority* (Cambridge, MA: Harvard University Press); R. J. Plant (2010), *Mom. The Transformation of Motherhood in Modern America* (Chicago: University of Chicago Press).
108. C. McDonald and S. King (1996), *Sampling the Universe. The Growth, Development and Influence of Market Research in Britain since 1945* (London: NTC Publication), p. 1.
109. Ross (2003), 'Changing Contours of the Social Science Disciplines', in Porter and Ross (eds), *The Modern Social Sciences*, pp. 205–37, here 205.
110. Wagner (2003), 'The Uses of the Social Sciences', p. 537.
111. Ross (2003), 'Changing Contours', p. 205.
112. Ibid., p. 208.
113. Wagner (2003), 'The Uses of the Social Sciences', p. 545.
114. S. Herbst (2003), 'Polling in Politics and Industry', in Porter and Ross (eds), *The Modern Social Sciences*, p. 577.
115. On the importance and influence of techniques or technologies, see Raphael in this volume.
116. Herbst (2003), 'Polling in Politics and Industry', pp. 578–80; F. Keller (2001), *Archäologie der Meinungsforschung. Mathematik und die Erzählbarkeit des Politischen* (Constance: UVK), pp. 31–6.
117. Keller (2001), *Archäologie der Meinungsforschung*, p. 34.
118. See, for example, M. Bulmer, K. Bales, and K. Sklar (eds) (1991), *The Social Survey in Historical Perspective 1880–1940* (Cambridge: Cambridge University Press); J. Converse (2009 [1987]), *Survey Research in the United States. Roots and Emergence, 1890–1960* (New Brunswick, NJ and London: Transaction). Experiences of scientists and civil servants in the various colonies of the British Empire, for example, were brought back to the British Isles and discussed before established institutions (e.g. The Royal Statistical Society). B. Cohn (1987), 'The Census, Social Structure and Objectification in South Asia', in idem (ed.), *An Anthropologist among the Historians and Other Essays* (Delhi: Oxford University Press), pp. 224–54.
119. T. M. Porter (2003), 'Statistics and Statistical Methods', in Porter and Ross (eds), *The Modern Social Sciences*, pp. 246–7. Herbst (2003), 'Polling in Politics and Industry', pp. 580–84.
120. See the chapter in this volume by Sarah Igo, and Keller (2001), *Archäologie der Meinungsforschung*, pp. 23–4.
121. Raphael (1996), 'Verwissenschaftlichung des Sozialen', and his chapter in this volume.
122. See, for example, G. O'Hara (2006), *From Dreams to Disillusionment. Economic and Social Planning in 1960s Britain* (Basingstoke: Palgrave Macmillan); Wagner (2003), 'Social Science and Social Planning', pp. 591–607.
123. On different management styles, see K. Uhl (2010), 'Der Faktor Mensch und das Management. Führungsstile und Machbeziehungen im industriellen Betrieb des 20. Jahrhunderts', *Neue Politische Literatur* 55, pp. 233–54.
124. G. Bonazzi (2008), *Geschichte des organisatorischen Denkens* (Wiesbaden: Verlag für Sozialwissenschaften).

125. F. Taylor (1911), *Principles of Scientific Management* (New York; London: Harper and Brothers).
126. E. Walter-Busch (2006), *Faktor Mensch. Formen angewandter Sozialforschung der Wirtschaft in Europa und den USA, 1890–1950* (Constance: UVK Verlagsgesellschaft), pp. 312–46.
127. Wagner (2003), 'The Uses of the Social Sciences', p. 545.
128. Cf. Bonazzi (2008), *Organisatorisches Denken*, pp. 259–95.
129. See J. Platt (1996), *A History of Sociological Research Methods in America 1920–1960* (Cambridge: Cambridge University Press), pp. 58–60, 114–16, 258f., and U. Bröckling (2003), 'Schlachtfeldforschung. Die Soziologie im Krieg', in S. Martus, W. Röcke, and M. Münkler (eds), *Schlachtfelder. Codierung von Gewalt im medialen Wandel* (Berlin: Akademie), pp. 189–206.
130. C. Albrecht, G. C. Behrmann, M. Bock, H. Hofmann, and F. H. Tenbruck (1999), *Die intellektuelle Gründung der Bundesrepublik. Eine Wirkungsgeschichte der Frankfurter Schule* (Frankfurt am Main and New York: Campus), pp. 145–53.
131. McDonald and King (1996), *Sampling the Universe*, p. 3.
132. Wagner (2003), 'The Uses of the Social Sciences', p. 546.
133. R. Rosenberger (2004), 'Demokratisierung durch Verwissenschaftlichung? Betriebliche Humanexperten als Akteure des Wandels der betrieblichen Sozialordnung in westdeutschen Unternehmen', *Archiv für Sozialgeschichte* 44, pp. 327–55.
134. A. Doering-Manteuffel and L. Raphael (2008), *Nach dem Boom. Perspektiven auf die Zeitgeschichte seit 1970* (Göttingen: Vandenhoeck and Ruprecht).
135. S. Schwarzkopf (2011), 'A Radical Past? The Politics of Market Research in Britain, 1900–1950', in K. Brückweh (ed.), *The Voice of the Citizen Consumer. A History of Market Research, Consumer Movements, and the Political Public Sphere* (Oxford: Oxford University Press), pp. 29–50.
136. Mass-Observation in the UK is the most obvious example. See N. Hubble (2005), *Mass Observation and Everyday Life. Culture, History, Theory* (Basingstoke: Palgrave Macmillan).
137. R. Gries and S. Schwarzkopf (eds) (2010), *Ernest Dichter and Motivation Research. New Perspectives on the Making of Post-war Consumer Culture* (Basingstoke: Palgrave Macmillan).
138. Ross (2003), 'Changing Contours of the Social Science Disciplines', p. 206.
139. P. Bourdieu (1992), 'Meinungsforschung. Eine "Wissenschaft" ohne Wissenschaftler', in idem, *Rede und Antwort* (Frankfurt: Suhrkamp), pp. 208–16.
140. Raphael (1996), 'Verwissenschaftlichung', pp. 178f.
141. Cf. Igo's essay in this volume or Herbst's reference to the former finance minister of France, Jacques Necker (1732–1804), who proposed that public opinion was equivalent to the spirit of society, which was represented in the salons of the period. Herbst (2003), 'Polling in Politics and Industry', p. 577.
142. J. Moran (2008), 'Mass-Observation, Market Research, and the Birth of the Focus Group, 1937–1997', *Journal of British Studies* 47, pp. 827–51.
143. Herbst (2003), 'Polling in Politics and Industry', p. 588.
144. S. Herbst (1993), *Numbered Voices. How Opinion Polling has Shaped American Politics* (Chicago and London: University of Chicago Press).
145. Herbst (2003), 'Polling in Politics and Industry', p. 589.
146. D. J. Robinson (1999), *The Measure of Democracy. Polling, Market Research, and Public Life 1930–1945* (Toronto: University of Toronto Press), p. 8.
147. A. Kruke and M. Woyke (eds) (2010), *Verwissenschaftlichung von Politik nach 1945*, vol. 50 of *Archiv für Sozialgeschichte*.
148. Wagner (2003), 'The Uses of the Social Sciences', p. 546.
149. Cf. Emil Walter-Busch's essay in this volume.
150. See Walter-Busch's essay in this volume.
151. U. Brunnbauer, C. Kraft, and M. Schulze Wessel (eds) (2011) *Sociology and Ethnography in East-Central and South-East Europe. Scientific Self-Description in State Socialist Countries* (Munich: R. Oldenbourg).

2

Embedding the Human and Social Sciences in Western Societies, 1880–1980: Reflections on Trends and Methods of Current Research

Lutz Raphael

During the last 20 or so years, historical research on the social sciences and their use in society has developed from an emerging field of interdisciplinary research to an established branch of historical studies. This rise is closely linked to conceptual changes in dealing with these phenomena. Discourse analysis and social history approaches have replaced the classical internalist view of disciplinary histories, well established in the subfield of the history of sciences. Historians have been responsible for a large part of the new studies. This reflects the fact that the history of the multiple uses of 'scientific expertise' has become an integral part of 'general history'. The ways in which scientific knowledge has informed social or political practice and the ways in which it has been translated into local knowledge and has coloured the personal beliefs of the common man are now seen as vital for the understanding of Western societies in the last 150 years.

The new relevance attributed to the history of the applied social sciences has led to the creation of the neologism 'scientization of the social'. In English translation, the artificial character and constructivist background of this term are directly palpable, but in its German version – '*Verwissenschaftlichung des Sozialen*' – the term does not irritate the reader because it is simply a new application of a well-established linguistic procedure for designating an ongoing process. In this case, it serves to reconstruct one of the many basic processes typical in Western societies since the middle of the nineteenth century – on an equal footing with other dynamic changes like urbanization, industrialization or democratization. As such, 'scientization' is not just a casual element among the different stories concerning society, culture or politics but a larger process that has transformed an esoteric, academic knowledge about man in society into public categories, professional routines, and behavioural patterns. Embedding social sciences in Western societies, in this sense, means the creation of a new infrastructure of useful knowledge informing an ever-growing part of our lives. This infrastructure is multifaceted, being at the same time organizational, conceptual, and political.

Such an all-encompassing view needs to be split into specifics. It only makes sense if it takes into account the whole range of different disciplines involved in the process, from sociology and psychology to medicine. The very term 'human sciences' is a rather vague one, and it changes its meaning from one language to another. Its use 'is recent, unconsolidated, and contested' but very useful for the 'big picture'.[1] The human sciences have never been a well-ordered cluster of sharply defined disciplines, and the blurring of borders has been frequent in the case of the applied human sciences. There is no single accepted international designation – the French term '*sciences humaines*' is perhaps the best one – and the term 'behavioural and social sciences' encompasses a more or less comparable number of scientific disciplines but suggests an epistemic unity that has proved to be ephemeral. But all the disciplines have in common a desire to approach human actors in society from an empirical point of view, while creating their own methodology of empirical research. In the following pages the focus will be on sociology and psychology in particular, but smaller and more specialized disciplines like demography or demoscopy will be included. There are no clear-cut borders for the large field of the medical sciences. Many disciplines touch on social phenomena like occupational health medicine, social hygiene or psychiatry. The human sciences have incorporated into themselves a broad cluster of socio-biological ideas that have informed large segments of social and penal policies, and currents of social thought on race, inequality, and community that have had a lot of influence. Economics, however, is left somewhat in the margins in the arguments that follow, because mainstream modern economic theory has shaped its own field of research and its own concepts, which ignore the social as a reality in its own right. Nevertheless it is evident that the interplay between economic knowledge and the political has given birth to a secular process of embedding economic knowledge in society in a way parallel, and in some respects alternative to, that described in this book.[2]

All these disciplines started their careers, in both academe and society, at a time when the general assumptions of liberalism about the fabric of society had lost credence. Since the middle of the nineteenth century – and in an accelerated way since the 1880s – the liberal vision of reason, law, education, and individual freedom as universal principles automatically leading mankind towards a better future was tested by a generation of sceptical empiricists. These empiricists came from very different ideological and scientific backgrounds but generated the set of new social and behavioural sciences that still prevail today. The last decades of the 1880s mark a watershed, and are a starting point for the implementation of the social sciences.

The following essay will try to give a comprehensive view of this secular shift, and will start with a discussion of methods and concepts in order to analyse why, where, and when the human sciences made their increasing impact on Western societies in the century between 1880 and 1980. The next section will deal with problems of time and periodization and, last but not least, the text will present a kind of typology of human science-based interventions in Western societies since the end of the nineteenth century.

My last preliminary remark concerns the geographical range of the developments covered. 'Western society' is as a rather fuzzy notion. In this book it encompasses many but not all parts of Europe and also North America, especially the USA; but international history teaches us that these Western societies dominated much larger parts of the world, and the history of colonialism, imperialism, and decolonization is closely linked to our story. Even if the 'scientization of the social' ought to be seen as a chapter of twentieth-century global history, for pragmatic reasons this aspect will not be in the focus of our argument. However, we should be aware that one part of the story is an international one: the human sciences and their experts have spread all around the world, and international networks of social knowledge have been vital for the circulation and diffusion of ideas or models of social interventions and practices. Congresses, international journals, international associations, missions of experts abroad, and UN organizations are testimony to this international dimension.[3]

These transnational networks have developed in concurrence as well as in competition with a very strong and long-lasting pattern in which the new social sciences became embedded into the nation state. The nation state was regarded as the 'natural' container for all kinds of social interventions and expertise, and most of the knowledge practices discussed internationally among experts were strongly biased by national views and traditions of governance and society in their public debates.[4] Accordingly, the social sciences became an important element in the nationalization of Western societies because they represented and imposed standards of normality and deviation at the national level. As a result, in most Western societies, the human sciences have developed typically national styles. Such national particularities have come to the fore when social expertise and social science knowledge has been used in public institutions or discussed in public debates.[5] Even now, these national experiences are better known than their connections with international trends and transnational networks of ideas. Behind discussion of the social sciences in Western societies there is an implicit need for further studies on these transfers and common trends.

Elements of diffusion or roads of transfer

Embedding the human sciences in Western societies means many different things and concerns a range of very different aspects of social life, such as welfare institutions, statistical bureaux, psychological testing, and the use of opinion polls. In such a situation, methodological pluralism is the first remedy against reductionist fallacies. In different variants, the dominant approach in the field of research until now has been discourse analysis. There is no doubt about the fruitfulness and potential of discourse analysis, but it has its limits. These limits become clear when we look at the main dimensions we can separate out in an analytical survey of the roads social science knowledge took on its way into the fabric of Western societies. We may distinguish at least five different roads leading from academic scholarship into society.

'Ideas', 'discourses', and 'metaphors'[6]

Intellectual history and the history of scientific or political ideas have been – and still are – the backbone of historical studies dealing with the use of human sciences in society. Themes and methods are numerous, and the following chapters will highlight some prominent examples. In some way or another, they all deal with translations from the esoteric languages that more or less all of the newly formed human sciences created for their internal use into other languages, political, religious or colloquial. Metaphors and ideas play an important role in these translations. They are those single elements of discourses that are volatile and flexible enough to be transported to other discourses and that may generate new 'interdiscourses' linking two or more disciplines among the human sciences with the languages of political or moral debate. When we discuss 'ideas', the impact of these elements in these processes of diffusion and of translation become clear. Some examples may suffice: 'degeneration' (in its French and English meanings), the German notion of '*Bevölkerungstod*' ('the demographic death of a nation'),[7] 'mental health', 'integration' (in German, French, and English), and 'adaptation'/'assimilation' (in American, English, and French).[8]

It seems to me that this is the best-known way in which the social sciences became embedded in Western societies from the 1880s. The field of research surrounding it is near enough to the established history of the sciences, and there is sufficient written and publicized material available for scholars to get started. Along the way, many metaphors or ideas filtered into many languages, passed by and disappeared. These changes in languages or discourses do not simply follow the sequences of topical political debates or great ideological trends – they connect with them but do not simply follow or precede them. Thus it seems a fruitful approach to sharpen our periodization by a closer look at these cycles of intellectual modes or trends as they are linked to the applied human sciences either in general or particular. Such an enterprise requires a more systematic look at the different 'languages' or 'discourses' circulating among the different social and political actors at any given time.

Peter Wagner's concept of 'coalitions' around discourses is well suited for approaching these cycles. The coalitions are the results of a kind of double network – of 'discourses', on the one hand, and of actors coming together from different backgrounds in administration, political parties, academe or social classes, on the other.[9] 'Ideas' or 'metaphors' often serve as the attractive and mobilizing forces in these coalitions. Some examples may suffice to illustrate this point. In the period between 1880 and 1910, many observers have identified such a discourse cycle linked to a coalition of different actors coming from state administration, the new social sciences, and the different reform groups of conservatism, liberalism, and socialism. At its core, the discourse was that of the 'social problem', the German '*Arbeiterfrage*', and 'social reform'. From the first decade of the new century onwards, this very inclusive discourse tended to split, giving birth to a series of different and, later on, antagonistic discourses and discourse coalitions that continued to coexist throughout the interwar period in many Western countries. Many specialists in the complex intellectual history of these decades focus on the rise of another discourse coalition centred on demography and the political biology of the nation: eugenics, Social Darwinism, racism, various mental health movements, and the discovery of the family as the main object of social engineering represent this cycle, which ended as late as the 1960s.[10]

Another kind of umbrella discourse, drawing in many experts from different human sciences, senior administrative officials, and established parties in the Western World, continued from the end of the Second World War to the early 1970s. This cycle saw the rise and fall of the modernization discourse, which shaped the leading arguments on social planning and the adaptation of social structures and behavioural patterns to a new era of Western democracy and industrial society.[11]

Experts

All these cycles of ideas can be meticulously analysed and described, but their impact on society, its institutions or habits may be profound or non-existent. If 'scientization' should signify more than just a change in the semantics of political language in Western societies, it needs to be anchored in society itself. Experts – frequently, but not necessarily, academically trained professionals – make up one of those anchors that link the arguments of human sciences to the ground: the world of experts is the world of the 'men or women on the spot', the professionals doing their job. We need to look closely at experts and their work if we want to understand how social theories or research methods changed human realities. Experts have often been the hidden protagonists in the implementation of science-based programmes, and have paved the way for changes in how societies deal with social or human realities. Research has to cope with the fact that for quite a long period of the twentieth century their forms of applied knowledge had lesser intellectual status – especially in European societies. Disinterest in the particularities of their ways of thinking and the much greater audibility of academic voices in public debates are two obstacles resulting from this cultural tradition.

The study of expertise is at its most interesting in two different situations. The first situation is when the expertise is put into practice in a more or less 'taken for granted' mode, as a generally accepted way of intervening. The way social workers did their job after 1945 in most Western societies is a good example. We know much more about the academic disputes and the political controversies surrounding the curricula of the university social work degrees and such things than about the experts themselves and how they did their jobs. In the last decade, research has intensified, and we do now know about some expert groups and their practices in some detail,[12] but in many cases elementary prosopographic evidence – who did the jobs and with what kind of professional knowledge and rules of practice – is still far from being available. Perhaps we may find different 'generations' present in a professional field – explaining the coexistence of contrasting concepts and technologies? The times when the experts are occupied with a particular theme are often quite different from those of their contemporaries in other fields.

But even more exciting, and perhaps more revealing, is the second situation that invites closer study. This is the actual genesis of an expert group and, after formation, its frequent, but not guaranteed, evolution into a well-defined profession. I prefer to speak – following Bourdieu's hints and suggestions quite closely – of a field of expert knowledge that is created and defined as a result of conflicts over the best expertise and knowledge to be used in the field.[13] Conflict is vital here because the expertise has often come from different disciplinary backgrounds, and experts have entered into different coalitions with client groups and their interests. The recent study by Ruth Rosenberger on the experts on human resources in West German industry is an exercise in kind, reconstructing the emergence of this contested field of expertise as an interplay of different groups of experts, management, and trade unions, acting both inside and outside the enterprises and their individual plants.[14]

Clients and users

'Client' and 'user' are catch-all words that do not really fit the different groups that have come into contact with social engineering or scientific-based interventions. We should distinguish between those who choose the offers and promises of the social sciences to promote changes in their own fields of action or institutions and those who do not. Senior administrative officials, party leaders, the clergy, managers (and so on) have the capacity to choose. Generally, these client groups have had a strong bargaining position when faced with social scientists or experts: the battle for control of new techniques or new data and dispute over ultimate decisions concerning the implementation of these innovations is part of this story. Another story altogether is that of those clients who become the mere objects of social engineering – they do not enter into any discourse coalition as the first group often do. The critical history of eugenics, of criminology and of psychiatry has meticulously documented the dynamics of transgression of legal boundaries and of violence against groups at the margins of society. The new scientific approaches both furthered and legitimized these new aggressive, expert-based forms of dehumanization and social exclusion. The third story is that of social control and resistance by clients against the new technologies, of adaptations and compromises by both the experts and their clientele. We need to look very closely at the languages used in this kind of conflict and bargaining: did the new messages of psychology or sociology pass into the vernacular and, if so, how were they transformed? The history of social work belongs largely to this third type of story – just as does that of different groups of workers and employees in economic enterprises subjected to the testing and implementation of other forms of diagnosis and control.

Is there a chronology of changes in the relationship between clients and experts and their technologies? At a macro level, one could emphasize how the rise of secular education and educational level in Western societies and the diffusion of the basic concepts of human sciences affected the two groups of experts and clients, and show how, in a nutshell, the second group got more bargaining power. From the point of view of political history, one could emphasize the particular role of politicians and of the state administration in general. From the 1880s, the new human sciences were eager to enter into closer relations with the state, hoping for better funding and for political support for their various reform programmes. Moments of political crisis and mobilization, like the two world wars, instigated strong ties between the world of human sciences and political administrations. Politicians had, and still have, the power of decision on all kinds of social engineering programmes. In liberal and democratic systems, this power is more or less legally and constitutionally restrained, but a glance at authoritarian and totalitarian regimes shows how extensively the political class can use the human sciences for what they deem to be the better governance of their populations.

Techniques or technologies

Human scientists have invented a lot of new words, subsequently transformed into official categories of the current consensus or common notions of public debates; but they have constructed far more in the way of very specific devices for empirical research and direct intervention. These 'technologies' were targeted at clients, at social groups or at entire populations. The history of these techniques – testing, sampling, social surveying, interviewing, counting, classifying – is quite well documented as regards their 'invention' and their further refinement at the academic level. The introduction of statistical reasoning,[15] the development of probabilistic tools,[16] of survey techniques[17] and testing[18] have all been studied in detail. But knowledge of their practical use, their diffusion and their frequent adaptation to specific needs and local situations has begun to reach us, thanks to recent studies.[19] It is evident that, in practice, many of these procedures were not used in a 'scientific' way – they often produced evidence that was unreliable or which served as a mere legitimization for moral or political judgements on people. So long as these techniques stayed in the hands of the experts and were part of their discourses, there is ample evidence, so research opportunities are ideal for the historian; but what do we do when we would like to know more about the way these techniques were used by clients who heard or read about them via the media – the press, radio, TV or, nowadays, the Internet? Psychological testing and self-help books based on personality tests are striking examples.[20] In this case, the methods of historical investigation are still in the making and largely dependent on the source material available. I do not see any 'best' way to explore the history of the social uses of these techniques. What is the average speed and what are the speed limits on this road of diffusion of the human sciences? Can we answer these questions at least on the macro level?

One can identify, firstly, some periods during which innovations clustered – for example, about 1900 and during the late 1930s and 1940s – and secondly, periods when the diffusion of technologies made important leaps – during the two world wars – but that is not enough for constructing a chronology. Routinization in the general use of social science categories and of social statistics, in the calculation of risks and social data, the diffusion of testing and of therapeutic offers on the different levels of social life, may be regarded as an indicator of the passage of these technologies into the quasi-static time of collective habitus or basic habits of our civilization. At the end of the period examined in this book Western societies seem to have reached this point with many of these technologies, the therapeutic ones being the last to be generally accepted and to become everyday routines.

Institutions

The last dimension of the process of embedding the human sciences in society is the role of institutions. Institutions may be compared to motorways in the history of motor cars. They are the infrastructure that enables the new discourses of human sciences and their technologies to spread through society. Institutions are closely linked to experts and professions, often offering the economic framework for jobs and career opportunities. Through institutions, the whole process of bureaucratization in Western societies starts off. The creation of a continuous service designing and producing official statistics was one of the first preconditions for further steps in the practical use of the social sciences. Specific administrative branches for the regulation and control of social risk, of labour conditions or health were important next steps in the continuous embedding of the human sciences into the fabric of the modern state. The installation of specialized departments devoted to the application of social science knowledge inside the many administrative branches of modern welfare states was another important step in the embedding of the human sciences in Western societies. The psychological service of the German Labour Administration, active since the Weimar Republic, is one example; the medical and statistical departments of the social insurance organizations are another.[21]

But the study of institutions does not simply reveal a success story achieved through the implementation of applied social sciences. A very fruitful field of investigation is to check the resisting forces and the limits on the power of experts inside these organizations. We have a sufficient number of case studies about the use of human science knowledge and experts in the military, the Catholic Church, political parties and social insurance to show us that the success of the new experts was only relative and to reveal the limitations of their impact inside these larger and older organizations.[22] We meet other actors defending their own positions in the name of the specific profiles and traditions of their institutions. Institutions tend to produce a strong inertial tendency, stabilizing their external goals and their internal processes, with people defending their established working routines and interpretations. If this is true, we have to look at them as anchor places for continuity that may survive the different tides of theoretical explanation or discourse surrounding the practical use of social science knowledge.

Methodologically, the historical study of these institutions needs approaches other than discourse analysis, namely the study of conflicts and decisions, just to mention one classic approach in historical studies.

Here is a brief summing up of the first argument. An analytical look at some of the basic elements present in the complex plot of how the applied human sciences spread leads to a clear conclusion: we need a pluralism of methodological approaches and a combination of different perspectives in order to understand the whole story better.

Problems of periodization

My second argument concerns the dynamics of these phenomena and their relation to larger trends and periods in Western societies. I would like to start with some general assumptions needed for a comparative synthesis on the macro level.

First, the different fields and areas of the implementation and diffusion of social science knowledge follow their own temporalities. There are amazing differences between, for example, the start of diagnostic testing in the work sphere and the establishment of psychological and other social scientific knowledge and technologies in the management of personnel in enterprises.

Secondly, we must accept that there are considerable differences between national chronologies and the degrees to which the human sciences became embedded in the fabric of society during the first 50 years of our story – even if the boom years after 1945 tended to reduce these differences. However, since one of the key sectors for the implementation of social scientific knowledge – the institutions of the welfare states – followed their own national paths throughout the period under consideration, we cannot write an international overview simply from studying the experience of one country. Any 'big picture' must reflect the effects of nationalization in Western societies during the twentieth century.

My third remark on periodization concerns external pushes and opportunities. The two world wars created great opportunities for implementing new technologies of social engineering on a huge scale. They were the starting points for the founding of new institutions and services. The Cold War, linked with an unprecedented economic boom, assured the survival of the new social experts in their fields of activities once the Second World War was over and states had called a halt to their emergency interventions. These continuities are particularly strong in the case of the USA, the new hegemonic power of the Atlantic bloc.[23]

The fourth point is about the inner dynamics of these processes. There is no linear evolution, no constant expansion. Again and again we observe that, after a starting period of big programmes and huge promises, the impact of the new technologies and forms of knowledge turned out to be pretty limited, and that investments in expertise and professionalism remained low or just came to an end. In this sense, many stories of implementation are stories of half-successes only: the social engineering of consumers or workers is one striking example. Another aspect seems equally important: internal or external criticism is not a new phenomenon – as some sociologists liked to suggest when drawing a line between a first and a second reflexive wave in the use of scientific knowledge in society. The embedding of social science knowledge and its routines in public institutions – in public health or welfare services – guaranteed continuity; only where the implementation was in business and the workings of market economies – as in the case of consumer studies or political polling – do we perhaps detect an ongoing dynamic of growth.

I could end here and give in to the particularities of each field, each discipline, each situation, pleading in favour of micro-stories and a kind of kaleidoscopic view of trends between 1880 and 1980. If that approach is taken, two more main master narratives come to the forefront when things are summarized. The first is Foucault's modernity and its forms of governmentality – all the stories of 'scientization' fit more or less into his general scheme.[24] This gives us a kind of big picture and huge historical perspectives – since the eighteenth century the discourses flow and a great variety of 'disposifs' are created, but there is no substantial change.

The second master narrative is that proposed by some sociologists of scientific knowledge and its use in modern society, scholars like Beck or Weingart, who make a great distinction between the first and second time science was introduced into society, the Second World War being a kind of demarcation line.[25] Both approaches keep history low: the basic structure of the power–knowledge connection is part of what we call modernity and historians can simply add the contingent case studies.

Four configurations

No practitioner of contemporary history will be convinced that this kind of generic periodization is the only one possible and the most fruitful: it disconnects the micro level of case studies and the macro level of 'modernity' or 'modern' society. Historians need a synthesis closer to their facts and their source material, more embedded in the turbulence of modern times. This being so, a typological approach may be useful – one that tries to sum up and generalize common features in the different areas where human science knowledge was used during a certain time period. The different types can be regarded as specific configurations of the five different elements analytically distinguished previously – discourses or ideas, experts, clients, institutions, and technologies. I see at least four different configurations, and they are explained further.

'Social reform'

The first configuration is that of 'social reform'. We have mentioned its discursive elements and its social and political protagonists. This configuration was largely dominated by what Peter Wagner has called a discourse coalition. Its heyday was between 1880 and 1910, and it started from a general critique of liberal laissez-faire policies in social affairs, based on fresh scientific knowledge created by the new social and human sciences. The main targets of intervention were the working poor and the casualties of industrialization in general. The configuration is international on the level of discourses and experts, but we find great differences when it comes to the implementation of the new programmes of social intervention: here, in this period, national differences – especially those concerning the welfare state – deepened considerably. Nonetheless the similarities are striking when we look at the level of the experts and the new techniques: social inquiries, insurance, and quantitative social data formed a new basis for state intervention. Even the coalitions are often the same: the new social scientists and physicians at the universities, the new experts in the civil services, and social reformers in parliament and government coming from all major parties, conservative, liberal, and social democratic. Gradual reform and inclusion of the 'deserving' labouring poor was the common aim. Typically, expectations and ambitious programmes overshadowed the first steps towards implementation – which were still local or sectoral and often disrupted by practical, financial or political difficulties. Experts were still small in number and, socially, came from the rising class of new professionals, often well embedded in ruling-class networks, both in the political field and in the economy.

Generally, observers agree that from 1910 on, this configuration lost momentum. Critical voices were raised, both in the political field and equally among the new social experts of the human sciences; new discourses gained ground. This reminds us that this configuration was only the dominant one among several, and by no means did it represent the whole story of use of human science knowledge in the period between 1880 and the First World War.

Social engineering

It is common among historians of the welfare state and the social sciences to see the First World War as a catalyst for the spread of human sciences capable of implementation. Psychoanalytical and other psychiatric therapies were used to treat soldiers, the institutional grid of the welfare state grew larger, and its laws and services covered more people than before. However, the realities of the war and the unsettled peace prevented a simple return to the old social reform agenda. On the contrary, this period gave birth to a quite different configuration. Its core can be defined as social engineering understood as a drama of stabilization.

Thanks to a current research project by Thomas Etzemüller, we now have a better view of the international dimension of this type of embedding of social sciences, especially in Europe where the experiences and consequences of the First World War had the deepest impact on all actor groups relevant to our story.[26] In this configuration, social science-based interventions were guided and legitimized by arguments concerning national recovery in the broadest sense: the goal of achieving recovery was behind the demographic, eugenic, cultural, and social discourses put forward by experts coming from all human sciences. Earlier fears of degeneration, of the extinction of one's nation, or of the white race in general, gained new plausibility among the political classes.

A second metaphor had a brilliant career in these interwar years among the think tanks of applied human sciences: 'community'. Tönnies' concept of '*Gemeinschaft*' served as a kind of generic therapy against the instabilities of capitalism, industrial class conflict, and all the mental and social ills linked to individualism and 'deracination'. Community was not only an obsession of the German-speaking experts but was equally present in the plans and proposals of Swedish and British social experts and scientists.[27]

The last element that differentiates the intellectual style of experts and their political coalition partners from that of others is their perception of time. The many crises they discovered (in urbanism, in demography, in individual behaviour, in industrial relations, in families) were seen as phenomena that were on the edge of becoming irreparable; therefore social engineering was an emergency measure and a kind of heroic intervention. Professionals took on the role of prophets when they spoke in public, and their leaning towards radical solutions and authoritarian politics often exceeded the well-known histrionics of Germany. Technologies of intervention did not undergo a revolutionary change, but we can observe everywhere refinements in testing and social inquiries (sociography, for example) that permitted a closer look at old or new clients; and all of these tended to strengthen objectivist views among the experts. The use of cartography, icons or photography is typical of this new style of social engineering. Especially when it came to defending the 'national body' against hidden risks from within or from outside and to 'heal its sick parts', proposed therapies tended to become aggressive and, to put it more neutrally, insisted on efficiency (reaching the hard-to-reach). Science-based strategies of exclusion from the social body were easily backed by political ideologies like racism or, on the political left, by Bolshevik proletarianism.[28] We should, however, bear in mind that the other side – technologies for including majorities – was at the centre of this type of social engineering, and so its political allies ranged from social Catholicism and social democracy to conservatism, fascism, and National Socialism.

Again the Second World War dramatically changed the internal dynamics of this configuration. In Nazi Germany the radical wing profited from the politics of extermination and imperial expansion to turn their ideas into plans and to realize them. On the other hand, the realities of war, once again, gave new credit to the moderate variants of this kind of social engineering, with the result that, at least in Germany and France, experts and technologies survived till the early 1960s, adapting themselves to a new dominant configuration that, in the other Western democracies, had begun during the 1930s. Sweden developed a particular style of public social engineering, which allowed democratic participation but was strongly based on the authority of human science expertise.

Planned modernization

For the origins of the next configuration of embedding social science knowledge into Western societies we have to turn to the United States and again to Britain. In these nations, it was essentially the outcome of the Second World War that advanced social engineering as a technology of planned and peaceful modernization. Their interventionist devices became (as it were) intellectual 'bestsellers' for the Western hemisphere in general. The ideological foundations are well known. Modernization theory as a kind of general discourse of the human sciences from psychology to economics fitted the Western political ideologies of pluralism and pragmatism in their opposition to communist ideology during the Cold War.[29] Nearer to the experts and their interventions was the general spread of behaviourism and instrumental positivism as two other general discourses which, starting from the social science scene in the USA, spread throughout Western Europe. One of the leading metaphors was 'assimilation/adaptation'. The amalgamation of this new vocabulary with older versions of biological or medical metaphors such as 'pathologies' was ready-made.

The rise of this configuration was not solely for political reasons, even if it is true that the US exported this usable social knowledge as a tool for propagating or defending Western democracy in Europe. It profited from the invention of new technologies like sampling and polling and from new clinical therapies. The quantitative measurement of social realities was once again refined. International organizations of social experts profited greatly from the creation of the United Nations and UNESCO.

A second push came from the extension of the welfare state in Western countries. This part of the story is equally well known – it led to an increase in the number of professionals in a growing number of official agencies and services, and meant that human science expertise could become embedded in the routines of both parliaments and administrations. It may be useful to distinguish between the type of configuration we can find in Western Europe and North America and the variant we find in the so-called underdeveloped countries or in the Third World, whether these be European colonies, new independent states in Africa or Asia or the Latin American countries. Here again ambitions and radicalism among the modernizing missionaries increased, and social planning lost its link to liberal democracy as a political aim and as the controlling framework of social experts. In these countries, economic and demographic therapies – from heavy investment in core sectors of regional economies to birth control – overshadowed other technologies and expertise. One of the emerging fields of research is precisely to examine the relationship between these two variants – and first results from studies of French social engineering in the 1950s and 1960s show us that there were strong personal and intellectual ties between the two areas.[30]

The age of therapy?

The golden age of modernization lasted from the 1940s to the mid-1960s, but in some countries like the then West Germany its peak was in the early 1970s. Even here, internal and external criticism had started to undermine the certainties that social scientists, their supporters in politics, economy, and society in general had shared, and to mobilize intellectuals and clients (especially those in the welfare state services) against the dominant therapies and technologies. But this political criticism did not imply that social services or planning agencies were simply closed. On the contrary, if one looks at the number of new job opportunities, the 1970s must be regarded as the boom years for human sciences-based professionalism.

After a period of increased political and ideological dispute in the social sciences, new configurations took shape from the early 1970s. The end of planning and a growing political dispute about the welfare state and its costs opened a new era that may best be characterized by uncertainty about the ultimate goals of social intervention and a pluralism of approaches in the field of social science knowledge. On the one hand, social work and applied social sciences subscribed to the idea of protecting minorities and generally 'empowering' socially weaker and economically poorer clients to take up the new opportunities society and the economy offered them. On the other hand, neoliberal think tanks pleaded in favour of a retreat from welfare and a return to nineteenth-century models of self-regulation in society. Sociobiological arguments became popular again, and these have partly informed welfare politics in Western countries. The most spectacular turn could be observed in the United States, and since the 1980s the gap between the US experiences and Canadian and European ones has certainly widened.[31]

Profiting from both of the contrasting sociopolitical currents, a new emphasis on therapy has come to the fore. It is seen as a way of asserting the individual self and of exploring its potential for the benefit of all kinds of enterprises, economic, social, and scientific.[32] Here American society, where psychology and psychiatry have been strongly embedded since the Second World War, seems to have been the forerunner of a general trend towards an all-embracing 'therapeutic culture'. Whether there are some hidden patterns that may allow us to speak of one or two configurations is open to debate.

Conclusion

My last remark is about the moral or political commitment of the scholar doing research in this field. Nowadays, most of us are convinced constructivists and are mainly inspired by the path-breaking works and suggestive formula of Foucault. As a consequence, there is no need to deconstruct or revise a Whig history of the advancements of the human sciences in Western society. A kind of sceptical approach inspired by Luhmann's system theory or postmodernist criticism of rationalistic models of useful social knowledge dominates the debate, underlining the many unexpected effects and collateral damage the transfer of scientific knowledge produces in churches, politics or the management of our selves. This fruitful consensus on the ambivalence of the process needs to be tested by a more engaged debate about how social sciences are used in Western societies today, the standards maintained, and the risks in their implementation. Perhaps a stimulating way of looking at this field of research is still from the viewpoint of those who seek help or who are treated by the different technologies human sciences offer. In this way not only the collateral damage but also the positive effects may come to the fore – pushing into the background the 'official' history of the use of the social sciences, which no critical historian would want to trust.

Notes

1. R. Smith (1998), 'The Big Picture. Writing Psychology into the History of the Human Sciences', *Journal of the History of the Behavioral Sciences* 34(1), pp. 11–13, quote p. 5.
2. M. O. Furner and B. Supple (1990), *The State and Economic Knowledge. The American and British Experiences* (Cambridge: Cambridge University Press); A. Tooze (2001), *Statistics and the German State, 1900–1945* (Cambridge: Cambridge University Press); P. Miller (2003), 'Management and Accounting', in T. Porter and D. Ross (eds), *The Cambridge History of Science*, vol. 7: *The Modern Social Sciences* (Cambridge: Cambridge University Press), pp. 565–76.
3. See the issue 'Actions Sociales Transnationales', *Genèses* 71 (2008), esp. P.-Y. Saunier (2008), 'Les régimes circulatoires du domaine social 1800–1940. Projets et ingénierie de la convergence et de la différence', pp. 4–25; and 'Sozialpolitik transnational', *Geschichte und Gesellschaft* 32 (2006), pp. 437–516; D. T. Rogers (1998), *Atlantic Crossings. Social Politics in a Progressive Age* (Cambridge, MA: Belknap Press).
4. P. Wagner, C .H. Weiss, B. Wittrock, and H. Wollmann (eds) (1991), *Social Sciences and Modern States. National Experiences and Theoretical Crossroads* (Cambridge: Cambridge University Press).
5. S. E. Igo (2007), *The Averaged American. Surveys, Citizens, and the Making of a Mass Public* (Cambridge, MA: Harvard University Press); R. Lenoir (1992), 'L'Etat et la construction de la famille', *Actes de la recherche en sciences sociales* 91/2, pp. 20–32.
6. S. Maasen and P. Weingart (2000), *Metaphors and the Dynamics of Knowledge* (London: Routledge); D. E. Leary (ed.) (1995), *Metaphors in the History of Psychology* (Cambridge: Cambridge University Press).

7. T. Etzemüller (2007), *Ein immerwährender Untergang. Der apokalyptische Bevölkerungsdiskurs im 20. Jahrhundert* (Bielefeld: Transcript).
8. D. Pick (1989), *Faces of Degeneration. A European Disorder, 1848–1918* (Cambridge: Cambridge University Press); H. L. Kaye (1986), *The Social Meaning of Modern Biology. From Social Darwinism to Sociobiology* (New Haven: Yale University Press).
9. P. Wagner (1990), *Sozialwissenschaften und Staat. Frankeich, Italien und Deutschland 1870–1980* (Frankfurt am Main: Campus).
10. M. Gijswijt-Hofstra, H. Oosterhuis, J. Vijselaar, and H. Freeman (eds) (2005), *Psychiatric Cultures Compared. Psychiatry and Mental Health Care in the Twentieth Century: Comparisons and Approaches* (Amsterdam: Amsterdam University Press).
11. N. Gilman (2003), *Mandarins of the Future. Modernization Theory in Cold War America* (Baltimore, MD: Johns Hopkins University Press).
12. J. Verdès-Leroux (1978), *Le travail social* (Paris: Editions de Minuit); K. M. Bolte and F. Neidhardt (eds) (1989), *Soziologie als Beruf* (Baden-Baden: Nomos); L. Margolin (1997), *Under the Cover of Kindness. The Invention of Social Work* (Charlottesville and London: University Press of Virginia); P. Becker and R. Wetzell (eds) (2006), *The Criminals and their Scientists. The History of Criminology in International Perspective* (New York and Cambridge: Cambridge University Press).
13. P. Bourdieu (1993), *The Field of Cultural Production. Essays on Art and Literature* (New York: Columbia University Press); idem (1993), *Questions of Sociology* (Cambridge: Polity Press).
14. R. Rosenberger (2008), *Experten für Humankapital. Die Entdeckung des Personalmanagements in der Bundesrepublik Deutschland* (Munich: Oldenbourg).
15. A. Desrosières (1998), *The Politics of Large Numbers. A History of Statistical Reasoning* (Cambridge, MA: Harvard University Press).
16. L. Krüger, L. J. Daston, and M. Heidelberger (eds) (1987), *The Probabilistic Revolution*, 2 vols (Cambridge, MA: MIT Press); I. Hacking (1990), *The Taming of Chance* (Cambridge: Cambridge University Press).
17. M. Bulmer, K. Bales, and K. K. Sklar (eds) (1991), *The Social Survey in Historical Perspective, 1880–1940* (Cambridge: Cambridge University Press); J. M. Converse (1987), *Survey Research in the United States. Roots and Emergence, 1890–1960* (Berkeley: University of California Press); I. Gorges (1986), *Sozialforschung in Deutschland 1872–1914* (Frankfurt am Main: Suhrkamp).
18. G. Strube (ed.) (1977), *Die Psychologie des 20. Jahrhunderts*, vol. 5: *Binet und die Folgen* (Zürich and München: Kindler); K. Ingenkamp (1990), *Geschichte der Pädagogischen Diagnostik in Deutschland*, 2 vols (Weinheim: Beltz).
19. Igo (2007), *The Averaged American*; F. Keller (2001), *Archäologie der Meinungsforschung. Mathematik und die Erzählbarkeit des Politischen* (Konstanz: Universitätsverlag Konstanz); A. Kruke (2007), *Demoskopie in der Bundesrepublik Deutschland. Meinungsforschung, Parteien und Medien, 1949–1990* (Düsseldorf: Droste); M. M. Sokal (ed.) (1987), *Psychological Testing and the American Society* (New Brunswick, NJ: Rutgers University Press); G. Sutherland (1984), *Ability, Merit and Measurement* (Oxford: Clarendon Press); M. Lengwiler (2006), *Risikopolitik im Sozialstaat. Die schweizerische Unfallversicherung 1870–1970* (Köln, Weimar and Wien: Böhlau); E. Walter-Busch (2006), *Faktor Mensch. Formen angewandter Sozialforschung der Wirtschaft in Europa und den USA, 1890–1950* (Konstanz: Universitätsverlag Konstanz); S. J. Petri (2004), 'Eignungsprüfung, Charakteranalyse, Soldatentum. Veränderung der Wissenschafts- und Methodenauffassung in der Militärpsychologie des Deutschen Reiches, Großbritannien und der USA 1914–1945' (Groningen, Rijksuniversitet diss.).
20. A. P. Murphy (2004), *The Cult of Personality. How Personality Tests are Leading Us to Miseducate Our Children, Mismanage Our Companies, and Misunderstand Ourselves* (New York: Free Press).
21. D. Meskill (2003), 'Human Economies. Labor Administration, Vocational Training and Psychological Testing in Germany 1914–1964' (Ph.D. diss., Harvard University).

22. B. Ziemann (2007), *Katholische Kirche und Sozialwissenschaften 1945–1975* (Göttingen: Vandenhoeck and Ruprecht); Kruke (2007), *Demoskopie*; Lengwiler (2006), *Risikopolitik*; Petri (2004), 'Eignungsprüfung'.
23. C. Simpson (ed.) (1998), *Universities and Empire. Money and Politics in the Social Sciences during the Cold War* (New York: Free Press); E. Herman (1995), *The Romance of American Psychology. Political Culture in the Age of Experts* (Berkeley: University of California Press).
24. M. Foucault (2010), *The Government of Self and Others* (Basingstoke: Palgrave Macmillan); idem (2008), *The Birth of Biopolitics* (Basingstoke: Palgrave Macmillan).
25. U. Beck, A. Giddens, and S. Lash (1994), *Reflexive Modernization. Politics, Tradition and Aesthetics in the Modern Social Order* (Cambridge: Polity Press); P. Weingart (2001), *Die Stunde der Wahrheit. Zum Verhältnis der Wissenschaft zu Politik, Wirtschaft und Medien in der Wissensgesellschaft* (Göttingen: Velbrück Wissenschaft).
26. T. Etzemüller (2009), 'Social engineering als Verhaltenslehre des kühlen Kopfes', in idem (ed.), *Die Ordnung der Moderne. Social Engineering im 20. Jahrhundert* (Bielefeld: Transcript), pp. 1–39.
27. Y. Hirdmann (1997), 'Social Planning under Rational Control. Social Engineering in Sweden in the 1930s and 1940s', in P. Kettunen and H. Eskola (eds), *Models, Modernity and the Myrdals* (Helsinki: Renvall Institute for Area and Cultural Studies), pp. 55–80; G. Alcon (1985), *The Invisible Hand of Planning. Capitalism, Social Science, and the State in the 1920s* (Princeton: Princeton University Press); J. M. Jordan (1994), *Machine-Age Ideology. Social Engineering and American Liberalism, 1911–1939* (Chapel Hill: University of North Carolina Press).
28. L. Raphael (2001), 'Radikales Ordnungsdenken und die Organisation totalitärer Herrschaft. Weltanschauungseliten und Humanwissenschaftler im NS-Regime', *Geschichte und Gesellschaft* 32, pp. 445–66; T. Rüting (2002), *Pavlov und der Neue Mensch. Diskurse über Disziplinierung in Sowjetrussland* (München: Oldenbourg).
29. D. C. Engerman (ed.) (2003), *Staging Growth. Modernization, Development, and the Global Cold War* (Amherst, MA: University of Massachusetts Press); E. Latham (2000), *Modernization as Ideology. American Social Science and 'Nation-building' in the Kennedy Era* (Chapel Hill: University of North Carolina Press); A. Escobar (1995), *Encountering Development. The Making and Unmaking of the Third World* (Princeton: Princeton University Press); F. Cooper and R. Packard (eds) (1997), *International Development and the Social Sciences. Essays in the History and Politics of Knowledge* (Berkeley: University of California Press); Herman (1995), *Romance of American Psychology*.
30. A. H. Lyons (2006), 'The Civilizing Mission in the Metropole. Algerian Immigrants in France and Politics of Adaptation during Decoloniziation', *Geschichte und Gesellschaft* 32, pp. 489–516; S. Losego (2009), *Fern von Afrika. Die Geschichte der nordafrikanischen 'Gastarbeiter' im französischen Industrierevier von Longwy 1945–1990* (Cologne: Böhlau), pp. 237–72.
31. L. Wacquant (2004), *Deadly Symbiosis. Race and the Rise of Neoliberal Penality* (Cambridge: Polity Press).
32. J. B. Imber (ed.) (2004), *Therapeutic Culture. Triumph and Defeat* (New Brunswick, NJ: Transaction); N. Rose (1996), *Inventing Our Selves. Psychology, Power, and Personhood* (Cambridge: Cambridge University Press); F. Castel, R. Castel, and A. Lovell (1982), *The Psychiatric Society* (New York: Columbia University Press); Herman (1995), *Romance of American Psychology*, pp. 264–315; E. S. Moskowitz (2001), *In Therapy We Trust. America's Obsession with Self-Fulfillment* (Baltimore, MD: Johns Hopkins University Press).

Part I
Social and Penal Policy

3

Contesting Risk: Specialist Knowledge and Workplace Accidents in Britain, Germany, and Italy, 1870–1920

Julia Moses

In 1835, a horse-drawn cart full of mutton toppled over, breaking the thigh and dislocating the shoulder of the driver, and sparking off one of the most significant court cases in British legal history. Thomas Fowler, a Lincolnshire butcher, had ordered one of his employees to pack the cart before sending its ill-fated driver to market. Buckling under the weight of the load, the cart's front axle cracked along the way, causing the vehicle to overturn. Charles Priestley, the injured driver, was then accompanied to the nearest inn, where he was treated by two surgeons. It took him almost five months to recuperate. Priestley's expenses while unable to work proved costly, and his father supported him throughout this period.

Since Priestley was a minor at the time of the accident, his father brought a lawsuit against Thomas Fowler for full compensation. At first awarded compensation, Priestley later lost on appeal. Significantly, the court decided that an employer was not obliged to ensure that an employee travelled safely while working. The ruling echoed the view of courts across Europe at the time: as free agents, workers accepted all risks associated with their jobs because they had agreed to their contracts of employment.[1]

By contrast, just 60 years later, following intense legal debates and statistical analysis, the unfortunate butcher's assistant would have been legally guaranteed compensation from his employer. A decade later, someone else from his village would be guaranteed compensation for a lung disease related to mining, as a variety of physicians and occupational hygienists would have demonstrated the correlation between mining and particular illnesses. Another ten years on, around the time of the First World War, the village doctor might refer someone with Priestley's injuries to a clinic in order to be tested for muscle weakness – just in case he was exaggerating the damage to his leg.

From 1835, when Priestley was so severely injured, to the early 1920s, various forms of knowledge transformed how workplace accidents were perceived and addressed, ultimately turning the treatment of work-related injury and illness into one of the cornerstones of European welfare states. The role of specialist knowledge in shaping perceptions of workplace accidents was not, therefore, unique to Britain during this era. In fact, disciplines such as law, psychology, medicine, engineering, mathematics, and physics all played a role in determining how various states dealt with the issue. It was the very nature of workplace accidents that invited diverse areas of expertise to shape perceptions of these mishaps as both a social and a political problem. In the first part of the nineteenth century, social observers from across Europe noted that industrialization seemed to be accelerating the pace of everyday life, and especially the speed of work. As a consequence, accidents at work seemed more numerous – and, given the power of new steam engines and other new sources of energy, more deadly.[2] While measuring the correlation between work and accidents required specialist knowledge about statistics, physics, and engineering, determining how to address the social, physical, and psychological consequences of accidents necessitated the insights of professionals in other fields.[3]

By the end of the nineteenth century, governments across Europe sought to address workplace accidents with new legislation predicated on the findings of some of these specialists. What these laws had in common was that they required employers to pay compensation for accidents at work – regardless of what had caused the accident. Employers were exempted from compensating workers only in cases in which the injured person had wilfully contributed to his or her injury. At the root of these policies was the view that industrial accidents were the result of *occupational risk*, which was foreseeable and calculable, and *not* individual actions for which workers or employers could be held personally responsible.

The case of workplace accidents points towards a common response to industrialization across Europe, but these policies also emerged in a context of international connectedness. From the 1870s, governments across Western Europe began observing attempts at legislation abroad. From the 1880s, numerous international conferences on industrial accidents and hygiene convened in which governmental administrators and a variety of specialists associated with universities, insurance companies, and private institutes shared information about their research and developments in their home countries. By the 1900s, compensation for workplace accidents had become so widespread in Europe that various governments were able to sign reciprocity treaties that granted foreign labourers pensions even after they retired to their home countries. It was against this backdrop that Britain adopted a Workmen's Compensation Law in 1897, and Italy, which had only just begun to industrialize, adopted an Accident Insurance Law in 1898. By contrast, Germany had adopted an Accident Insurance Law as early as 1884, at a time when there were few relevant international conferences and exchanges related to workplace accidents. The policy was part of a broader social insurance package, which included sickness insurance and insurance for old age and disability, and it served as a crucial point of reference for Britain and Italy.

Although novel, each country's legislation on accidents built on shared earlier practices of regulating the workplace. Each state had enacted laws on workplace safety, for example, on fencing in machinery and banning women or children from particularly 'dangerous trades', between the 1830s and 1860s.[4] Moreover, all three governments established factory inspectorates that investigated workplaces falling under particular safety regulations.[5] A key role of factory inspectors and local officials, like that of later specialists administering social insurance and workmen's compensation systems, was to gather information on accidents, their causes, and consequences.

Far from being the product of the 'post-industrial era' of the late twentieth century and beyond, societies that mobilized around knowledge were, therefore, part and parcel of industrialization itself.[6] Moreover, the case of workplace accidents in Britain, Germany, and Italy at the dawn of the twentieth century reveals that the link between specialist knowledge and 'risk societies' has its roots at least in the late nineteenth century, if not earlier, rather than in the 1970s, as several sociologists have claimed.[7] Yet, the example of workplace accidents in the late nineteenth and early twentieth centuries does not demonstrate an early stage in what some scholars have seen as the growing preponderance of scientific knowledge for the organization and management of social and political life. That is, it does not serve to support arguments based on modernization theory about the ever-increasing influence of scientific expertise. Nor does it reveal what some historians have seen as a growing hegemony of governmental regulation that has been predicated on ostensibly objective arguments about risk.[8]

Instead, the case of workplace accidents in Britain, Germany, and Italy at the turn of the twentieth century reveals the extent to which specialist knowledge can remain contested and dynamic when addressing social issues. As we shall see further, and as suggested in the chapter by Lutz Raphael earlier in this volume, 'expert' knowledge did not simply 'penetrate' public and administrative discussions about the issue of accidents at work; it was also called upon and moulded by those involved in the accident problem.[9]

Compensation v. prevention: The legal problem of workplace accidents

The British, German, and Italian governments investigated the issue of workplace accidents against a backdrop of widespread political and legal debate about compensation. Already from the 1830s in Britain and Germany and the 1870s in Italy, jurists and legal experts had been debating how to approach cases involving compensation for workplace accidents.[10] The fundamental question underlying these cases was whether workers had consented to the risks inherent in their employment. If not, did employers owe compensation to workers who were injured while working?

The legal traditions and new social and political developments in each country proved crucial for answering this question. Over the course of the nineteenth century, liability law crystallized as an increasingly important field of jurisprudence, which had significant implications for perceiving the relationship between workers and employers in cases of industrial accidents.[11] In each country, legal discussions about liability for industrial accidents upheld a notion of industrial risk predicated on the freedom to close contracts; it assumed that every worker understood the risks associated with a particular job when agreeing to that job. Although largely unstated, what was implicit in this logic was a free-market understanding of labour: workers were free to choose any job in the labour market and, given the unlimited range of choice and their freedom to select, they could take whichever job they preferred. According to this logic, workers might even prefer to accept more dangerous jobs, such as those in mining or railways, as they would be remunerated accordingly. The free market of labour would, therefore, dictate the values of wages.

By the 1860s in Germany, and by the 1870s in Britain and Italy, consensus around these traditions began to falter rapidly, after jurists, following the demands of individual workers and trade associations, tested compensation claims for occupational accidents at court, thereby undermining free-market ideas about labour. This transformation in thinking about society, the economy, and the nature of the law in governing social relations had repercussions in the political sphere. Debates about whether to adopt an employers' liability law as a means of addressing industrial accidents emerged in all three countries over the course of the 1860s and 1870s. With the adoption of liability laws in Britain and Germany in 1871 and 1880 respectively, and with the increasingly loose interpretation of the liability clauses in the Civil Code in Italy, however, the issue of industrial accidents again came to the forefront. Crucially, these liability laws only required employers to compensate workers if employers themselves or workplace supervisors could be implicated in having caused a workplace accident; given this proviso, it proved impossible to claim compensation for the majority of them.

Legal debates surrounding the issue of employers' liability surfaced in a context in which workers were increasingly making claims for compensation under the new legislation, often with the financial assistance or legal advice of trade unions or other forms of organized labour. This was especially the case in Britain during the 1880s and 1890s, and Germany from the mid-1870s until the early 1880s.[12] In Italy, by contrast, workers claimed compensation less frequently under the liability clauses of the Civil Code. The strict burden of proof associated with these clauses was one of several deterrents to liability claims; and trials in Italy were known to be lengthy and expensive. The result was that workers rarely invoked these articles, and when they did bring claims to trial, they rarely succeeded in gaining compensation. In time, however, judges became increasingly flexible in their interpretation of the clauses.[13]

The court cases related to the new liability laws in Britain and Germany and the liability clauses in the Italian Civil Code brought great consternation in all three countries. In Germany, factory inspectors noted concerns about the mistreatment of workers by unscrupulous insurance companies, while in Britain employers cited the fees associated with these cases, which could prove costly in terms of both legal expenses and compensation.[14] In Germany, in particular, critics of the liability law pointed to these court cases as fomenting a kind of 'social war'.[15] In Italy, it was the central government that criticized the increasing number of claims under the liability clauses of the Civil Code. The Italian government echoed German concerns about the 'social war' associated with court cases by noting that, 'in this period of class agitation, in which the relations between capital and labour are anything but amicable, it is necessary to look for every way to promote peace'.[16]

These comments came at a time of growing public concern about working-class unrest and the rise of socialism. Around a third of the strikes in the 1880s had called for an improvement in working conditions, and the Italian Socialist Party had been founded just six years before the accident insurance law was enacted.[17] Similarly, in Germany, the law on accident insurance was adopted just six years after the government stamped out all organized socialist activity with its Anti-Socialist Law of 1878.[18]

In all three countries, jurists, legal experts, and employers began lobbying for compulsory compensation for workplace accidents instead of liability laws, which required workers to prove that their employers were responsible for workplace accidents. Representatives from these groups now argued for compulsory insurance as a means to remedy the expenses and aggravation associated with court cases involving employers' liability. The positive experience of several employers with mutual funds, especially those in the mining industries in Britain and Germany, led them to call for some form of compulsory insurance for accident compensation.[19] In Italy, it was the government that advocated compulsory compensation in the 1890s after several failed attempts, between 1879 and 1886, to adopt a liability law in parliament. These earlier efforts had been spearheaded by workers' organizations, socially orientated parliamentarians, jurists, and the government itself.[20]

Yet, workers' groups, especially in Britain, approached the possibility of compulsory compensation with scepticism. They instead emphasized the need for legislation that would help to *prevent* accidents rather than *compensate* for them. Often they argued that laws on compulsory compensation for workplace accidents would only encourage employers to be more careless in preventing accidents.[21] In Germany, members of the Social Democratic Party and the conservative and Christian Centre Party made similar arguments.[22] However, government factory inspectors, along with some employers, contended the opposite: laws requiring employers to compensate workers would indirectly enforce safety at work because employers would rather fence in machinery than pay compensation for an unknown number of potentially costly claims.[23] In Italy, in particular, the government joined with socially orientated parliamentarians in making these claims. Nonetheless, the Italian government also recognized the need to adopt protective legislation to work in conjunction with a law on compulsory compensation. The fact that Italy, unlike Germany and Britain, lacked comprehensive safety legislation and an established factory inspectorate at the time may have contributed to this thinking.[24] In fact, the Italian Accident Insurance Law of 1898 included provisions for both compensation *and* accident prevention.

Both arguments in favour of compensation and in favour of prevention highlighted a new role for specialist knowledge in determining the causes and consequences of workplace accidents. It was in this context, therefore, that claims that industrial accidents were statistically probable gained particular relevance. Moreover, it was in this setting that other countries' models for addressing the accident problem seemed to provide another form of expert proof, demonstrating the best way to deal with the issue.

Identifying workplace risk: The movement for compensation

Concerns about workplace accidents were not new by the time each country adopted a compensation policy. Factory inspection and the regulation of dangerous trades and dangerous equipment, especially steam boilers, had been in place in each state for decades. Each government – as well as concerned employers – relied on engineers, physicians and others with specialist knowledge to oversee workplace safety. Yet, the political coding about safety at work focused on *preventing* accidents and the maintenance of general hygiene rather than compensation. This emphasis on prevention and regulation supported existing legal arrangements in each state, allowing for a great deal of privacy in the agreement of labour contracts and in the running of individual workplaces. A substantial shift in thinking thus had to take place before compensation could be seen as the necessary solution to the accident problem.

The new government technology of statistics, which had begun to take hold in each state in the period between the 1830s and 1860s, proved fundamental for this conceptual shift.[25] Statistics helped to link particular industries to worker accident and fatality rates, thus engendering an idea of work-related risk that could be used to predict the probability of accidents in those industries. Representing the 'accident problem' statistically allowed governments to create homogenized notions of the 'risks' encountered by members of the workforce, and to imply that solutions were available to address such 'risks'. As Silvana Patriarca points out, 'statistics, like any representation, do not merely "reflect", but "supplement" reality and contribute to its making'.[26] The representation of accidents as statistically related to work thus took individual agency away from workers and attributed responsibility to industry. Since statistics could prove particular industries to be especially dangerous, they could be called on to indicate which industries might require special regulation.

It was the government in each state, in particular, that relied on accident statistics when calling for compensation policies. Focusing on the risk of workplace accidents, governments could evade issues of class and poverty as the source of social problems; the concept of workplace risk instead provided a seemingly universal and apolitical framework for understanding industrial accidents. Through statistics, therefore, the issue of workplace accidents no longer seemed a matter of personal fault or responsibility. Nor did it seem a problem for 'class' or 'social war' – even if the rhetoric of the 1860s and 1870s seemed to point in that direction; instead, statistics of workplace accidents demonstrated that the issue was an unavoidable aspect of everyday life. As several commentators at the time observed, accidents were part and parcel of 'modern industry': they were unavoidable, constantly increasing in number, and, most importantly, given the vast and complex nature of modern workplaces, it was almost impossible to attribute blame for them.[27]

By the late 1860s, the British, German, and Italian governments began to collect statistics that included accidents at work. The first comprehensive British statistics, collected for the 1869–70 period, focused on accidents in industries that were already perceived to be dangerous or that made use of dangerous equipment: shipping, mining, railways, factories, and roadways or workplaces involving steam boilers or explosives.[28] Comparable Italian statistics, first collected for the 1866–82 period from regional capitals throughout the kingdom, were more specific, focusing on 'violent deaths' of individuals in particular trades as diverse as agriculture, mining, and tailoring, and calculating death tolls according to causes as varied as suffocation and 'violence by animals'.[29]

However, the German government's statistical programme for industrial accidents proved to be much better suited to formulating the link between work, injury, and compensation. The main reasons for the success of the German programme in this regard were its longevity, its consistency, and the kinds of questions it asked. In 1873, the Prussian Ministry of Trade, which played a key role in devising the later national accident insurance legislation, drew up accident forms and asked for these forms to be filled out detailing accidents that happened within local government districts, if they caused fatalities or injury to workers. The Ministry of Trade was quite specific in its instruction, asking for the inclusion of only those injuries causing an 'inability to work' for at least eight days as a 'consequence' of the 'accident'. The form gave very specific categories of injuries, which could be listed, such as bone fractures, crushing, fire damage, hernia, concussion, and wounds. These forms were filled out annually by the local governments within Prussia.[30] Like both the British and Italian government initiatives in the gathering of statistics, the German data focused on industries that were already perceived to be dangerous. The collection of statistics in all three countries thus confirmed suspicions about the dangerousness of particular industries. Moreover, it helped to demonstrate how accidents could be linked to the risks of particular trades – not to worker negligence or inadequate preventive measures. The German forms went further, however, as they were much more specific about the consequences of accidents, and the consequences were, indeed, what compensation sought to address. It was this information that formed the backbone of the accident insurance law that Germany enacted in 1884.

The 1884 law spurred further interest abroad in the accident issue. While the German government looked to foreign information when reflecting on accidents as a problem – starting as far back as the 1830s – it did not prove influential for governmental thinking in the years immediately before the accident insurance law was adopted.[31] By contrast, foreign information guided both the British and Italian governments when making seemingly scientific and objective arguments about how to act on the accident issue. When the German government introduced statutory accident insurance in 1884, both the Italian and British governments monitored the wording of the legislation, news on its implementation, and extensions of and revisions to the law. German decisions about which industries should be included in the legislation and how particular consequences of accidents should be compensated proved especially important for shaping foreign discussions. This was particularly the case for Italy. For the next few years, the Italian government obtained copies of German accident statistics and compensation rates.[32] For administrators in both Britain and Italy, the German model of accident insurance provided an important point of reference when conceptualising how to proceed. In 1884, an Italian actuary working for the government wrote a detailed guide to the German system of social insurance, while in 1893, the government in London requested information from its consuls abroad about other countries' systems of employers' liability.[33]

While direct observation and legislative exchanges proved crucial for influencing governmental thinking about workplace risk, it was the international and domestic congresses on workplace accidents and industrial hygiene, which met every two to three years from 1889, that became the most important forums here. There were several of these, such as the International Congress on Accidents at Work, which was founded in 1889 and met regularly until the outbreak of the First World War; the International Congresses on Workers' Protection, which began in the same year; the International Association for the Legal Protection of Workers, which was founded in 1897 and met regularly at conferences; the International Labour Office, which was established in Basel in 1900 and formed a major international clearing house for information on occupational health and related issues; and the International Congresses on Professional Diseases, which met first in Milan in 1906. These points of contact proved important for shaping how governments perceived the problem of industrial accidents.[34] They also provided a significant opportunity for governments and others to mould perceptions of occupational risk and propose solutions to that problem. For example, the French government, whose Corps of Mines had long been interested in the issue of workplace accidents, helped to establish the International Congress on Accidents at Work and played a major role in shaping discussions at its first few meetings. Private associations involved in accident prevention also played an important part in organizing these international points of contact, as was the case with the Association of Italian Industrialists for the Prevention of Industrial Accidents, which was founded in 1894 and was modelled on a similar organization based in Mulhouse, in Alsace.[35]

It was through these connections, and with the assistance of the universal language of statistics, that an international 'public sphere' focusing on the issue of workplace risk was established.[36] The burgeoning 'science of work', which uncovered the problem of work-related fatigue, and – as demonstrated in Martin Lengwiler's chapter in this volume – the budding area of actuarial science both played particularly important roles in the arena. Official government representatives, academics, and actuaries met regularly at these congresses in order to discuss topics ranging from mortality rates in particular industries to therapeutic treatment for the injured.[37] These meetings of specialists, in particular, helped to consolidate a common scientific idiom for discussing workplace risk. This idiom would continually inform the treatment of and discussions about the accident problem in Britain, Germany, and Italy after each state had adopted its policy on compensation.

Specialist knowledge and compensation in practice

After each state had adopted a compensation law based on the common concept of workplace risk – Germany in 1884, Britain in 1897, and Italy in 1898 – new problems emerged. Each country's policy required employers to provide the entire amount of compensation for workers, though only Germany and Italy required employers to buy insurance and pay its premiums. Britain, which had enacted a workmen's compensation policy, allowed employers to pay out of their own pockets or through insurance. In Germany, which established a corporatist social insurance system that was state-supervised, central government played a major role in administering compensation, especially through the Imperial Insurance Office, brought into being in 1885 and serving too as the final court of appeal in claims cases. By contrast, in Italy, employers could choose private insurance or buy a policy through the semi-public National Accident Insurance Fund that had already been founded in 1883. Moreover, in Italy, as in Britain, central government oversaw the basic workings of the compensation systems but was not involved in adjudicating. By introducing new legislation (e.g. on occupational illnesses) or by revising existing compensation laws, however, both governments, like that in Germany, could substantially shape how occupational risk was interpreted and, as a consequence, how compensation was administered. It was at this stage, therefore, that expertise from law, medicine, and the natural sciences became especially important for answering questions about compensation. The seemingly apolitical understanding of workplace risk that underpinned the laws appeared to depoliticize decisions about interpreting them. Nonetheless, the differing opinions of specialists invoked in compensation cases and discussions about revising compensation policies demonstrates the degree to which workplace risk remained contested and – even if only latently – politically charged after the compensation system had been introduced to tackle the accident problem.

The task of interpreting the compensation policies was dealt with primarily at a local level in Italy and Britain, through courts of cassation and tribunals respectively. Both jurists and medical experts played important roles here as did, on occasion, insurance experts and university scientists. Owing to the structure of the German system of accident insurance, in which the Imperial Insurance Office acted as the final court of appeal, the central government interpreted the law directly. When answering questions related to compensation, the German administration relied on its own and affiliated staff, including lawyers in the Imperial Insurance Office and physicians from various clinics. It was in these circumstances that, for example, the government commissioned Hermann von Helmholtz, the renowned physicist and current president of the Physical-Technical Imperial Institute in Berlin, to determine whether a farmer working in a potato field was more likely to be struck by lightning than a worker employed indoors. His answer – that lightning was an occupational risk for workers employed in open fields – was instrumental in broadening the scope of the law to include what insurance experts and government officials alike had previously viewed as an uninsurable 'act of God', rather than a foreseeable – and insurable – form of 'risk'.[38]

'Acts of God' did not, however, prove the most perplexing issue for the compensation systems. Instead, the most significant test for the laws was the debate about whether an accident was simply 'something that happens', as its Latin root suggests, indicating a momentary event, or whether it could be a longer process. The Italian accident insurance law explicitly identified the kind of accidents that could be compensated as those deriving from a 'violent cause', which seemed to indicate the latter. By contrast, the British and the German policies both assumed that the meaning of an accident was transparent, so they did not further define the term.

It was in this context that occupational illnesses, which usually develop over a longer period of time, proved problematic when there was a strict legal definition of accidents as brief events. Nonetheless, Italian courts, like those in Britain and Germany, recognized that instantaneous poisoning or infection could be considered an 'accident' in terms of the compensation policies, and medical knowledge proved particularly important for these decisions. Thus, blood poisoning or the inhalation of toxic fumes could be compensated.[39] Similarly, the acquisition of a disease that struck suddenly was also compensated without a need to alter the format of the accident insurance or workmen's compensation laws. Malaria in Italy was the most common case of this kind of 'accident', but occasional bubonic plague outbreaks among dock workers also helped to expand legal interpretations of accidents.[40] According to the same reasoning, German sailors who contracted tropical diseases while working abroad were also awarded compensation.[41] So, too, were British textile workers who contracted the anthrax bacillus from the wool they were handling.[42]

These cases would prove crucial for later government debates about whether and how to compensate further occupational illnesses. Influenced by factory inspectors' reports of respiratory conditions associated with textile industries, accounts by miners debilitated by respiratory and ocular illnesses and other accounts presented to a special commission on occupational disease, the British government modified its workmen's compensation law as early as 1906 – within ten years of its enactment – by including a provision for adding certain diseases and conditions within the scope of the law as necessary.[43] The German government added a similar clause in its 1911 codification of the social insurance laws. The issue of compensating occupational illnesses was indeed so internationally pressing that the German government monitored the British legislation closely and considered including the same list of diseases that had been adopted in Britain.[44]

The question of how to treat occupational diseases – at least as a legal and actuarial issue – was, of course, not merely a matter for intergovernmental observation; it was also one of the major problems debated in the numerous domestic and international conferences on industrial hygiene and social insurance during this period. It seems surprising, therefore, that after much discussion in parliament and the administration, including widespread concern that they might be lagging behind developments abroad, the Italian government ultimately declined to include occupational illnesses within its accident insurance law. Instead, Italy only recognized hernias, deafness and similar ailments, which were seen by specialists in the burgeoning Italian field of medico-legal studies as possible consequences of identifiable accidents. Italian courts therefore maintained a relatively strict understanding of the wording of the accident insurance law, finding 'violent cause' particularly difficult to ascribe to the development of occupational illness. The lack of legislation on sickness insurance in Italy, along with widespread concerns among parliamentarians and employers alike about the costs of including occupational illness within the Italian compensation policy, may have influenced these decisions.[45]

Indeed, it was not only scientific circles that influenced governmental thinking about industrial diseases at this time. Organized labour also noted the connections between ailments and certain trades, and began calling for some degree of compensation.[46] The Social Democrats in Germany, Labour in Britain and the Socialists in Italy began invoking revisions to the accident insurance and workmen's compensation laws to take industrial diseases into account.[47] In response, groups of industrialists called for narrower readings of the law.[48] Moreover, international bodies became increasingly involved in these debates. At its 1897 meeting in Brussels, the International Congress for Workers' Protection Legislation sought to highlight 'trades dangerous to health' with a view towards legislative action.[49] At its meeting in Rome, the eighth International Congress on Workers' Insurance inquired whether industrial diseases should be granted compensation, but it came to mixed conclusions on the matter.[50] By 1912, the International Labour Office, which was founded in 1900 in Basel and served as a precursor to the International Labour Organization, began pressuring the German government on the issue of industrial diseases. As part of its lobbying endeavours, the organization published a booklet in which it listed toxins related to the workplace and demanded protective legislation.[51]

As the case of occupational illnesses indicates, it was during these discussions about interpreting the law that the concept of occupational risk seemed to become commonly accepted, not only by 'scientists of work', jurists, physicians and other specialists, but also by government administrators as well as the potential beneficiaries of each country's compensation system. This agreement about the meaning of workplace risk appeared to stunt explicitly political debate about whether particular risks should be addressed with compensation. Indeed, participants in this discourse found that they could rely on specialist knowledge in order to make seemingly neutral, objective claims about compensation.

The veneer of neutrality, however, broke down in particular instances in each state. In the early 1900s, as courts became more lenient over compensating accidents, and as legislators began to include new classes of workers within the remit of the compensation laws, concerns about fraud became more and more prevalent. In Germany and Italy 'accident neurotics', 'pension addicts', and 'simulators' were targeted as symbols for a possible failure of the compensation laws. In part, the acknowledgement within medical science that traumatic accidents could cause both physiological and psychological effects – accident neurosis – and the recognition of insurance companies and courts that both should be taken into account when granting compensation, helped to fuel concern about whether the consequences of some accidents were real or imagined and how best to compensate individuals who might be 'addicted' to compensation itself.[52] In both cases, magistrates, insurance companies, and medical experts saw small pensions for minor accidents as the root of much fraud committed either by opportunistic individuals or 'pension addicts', as certain claimants came to be known in Germany. Either eliminating smaller pensions altogether or distributing lump-sum payments instead of weekly pensions were each seen as possible solutions, and a variety of informed specialists was available to provide opinions that supported the differing views on the issue.[53]

By contrast, in Britain, 'accident neurotics' or others suffering from psychological ailments related to accidents were never discussed within government. A likely reason for this development was that the science of work that had begun to flourish on the continent during the last decades of the nineteenth century did not find similar support in Britain until after the outbreak of the First World War.[54] In Britain, concerns about fraud, and especially malingering, emerged in the discussions leading up to the enactment of the first Workmen's Compensation Act in 1897. These surfaced again in discussions about revising the law, when the government sent a circular to county court judges asking about the effects of the 1897 Act. Certain judges were quite critical of workers for their 'unblushing perjury' and 'frivolous claims', while others instead pointed to unscrupulous insurance companies inducing workers to accept small lump-sum payments instead of the weekly payments to which they were entitled.[55] Nonetheless, discussions of fraud subsided not long after the enactment of the laws. Having enlisted medical examiners throughout Britain in order to adjudicate in cases of medical disputes, the government outsourced possible problems of fraud to experts, refusing to respond to questions regarding specific cases whenever they were addressed to the relevant ministry in London.[56] By refusing to engage directly in debates about fraud – even if government-commissioned specialists dealt with the question on the ground – the British government achieved some success in keeping the issue apolitical.

During and immediately following the First World War, however, all three governments found that questions related to workplace risk had become politically charged owing to the special circumstances of the war and the related economic crisis in each state. The governments saw that they needed to address special cases of risk and workplace accidents because the typical accidents of working life had diminished in proportional significance. Some of these special cases became loaded with a political salience that transcended immediate debates on interpreting the compensation laws. Thus, the issue of 'pension addiction' in Germany and 'pension vampires' in Italy became more vehement at this time because malingerers had come to be seen as deviant elements in the national communities forged during war and its aftermath.[57]

Similarly, as workers at home became involved in the war effort, the previously tenuous border between work and war, each with its separate risks as well as its separate forms of organization, began to disappear. Workers and insurance bodies in all three countries increasingly called for compensation for injuries and death directly related to the war, such as those resulting from the bombing of munitions factories or exposure to noxious chemicals used by the military. Associating the consequences of the war with the government itself, they looked to government with their concerns about compensation.[58] In these situations, calling on specialist knowledge in order to justify compensation no longer seemed necessary. The corporatist merger between industry and the state that was forged during this period proved crucial for expectations within civil society as well as governmental responses to those demands. In each country, therefore, special wartime decrees ensured that workers injured in 'dangers of war', such as aerial strikes or exposure to noxious chemicals, would receive compensation for these 'workplace accidents'.[59]

Severe inflation was the issue, during the war and for several years afterwards, that caused all three governments to intervene most strongly in the existing compensation schemes, ignoring the actuarial logic behind their policies. Having left many accident pensioners without enough money to live on, inflation brought individuals and interest groups to protest on the street or write pleading letters directed at government, demanding action.[60] The administrations in Britain and Germany accordingly responded to these public demonstrations and pleading letters with fixed pension rises that continued even after the Armistice.[61] By contrast, the Italian government used the economic crisis – accompanied by widespread domestic disapproval of the war – as an opportunity to extend compensation to its millions of uninsured agricultural workers. While extending accident insurance to these workers had been heatedly discussed since the late 1890s, and although several expert commissions had drawn up reports on the question, it was only through a special military decree that the Italian government was able to tackle the issue.

Although specialist knowledge seems to have been consigned to the margins of compensation policy at this time, for the most part, these developments were temporary. Moreover, the war proved a watershed for the involvement of specialist knowledge in two spheres: the treatment of the disabled and industrial management. The war experience was enormously influential on governmental responses to disability resulting from injury. Anxieties about workforce size as well as disaffected veterans after the war led all three countries to pursue retraining programmes for injured soldiers. These concerns overflowed into discussions about the 'soldiers of industry' who had fallen victim to accidents at work.[62] Interestingly, however, one of the greatest disabilities suffered by war veterans, psychological disturbances in the forms of 'hysteria', shell shock or war neurosis, did not affect administrators' decisions about compensating workplace accidents. Indeed, retraining efforts as well as practices observed in the social insurance sphere, such as lump-sum payments for 'accident neurotics', were recommended for these cases of veterans.[63]

Secondly, concerns during the war about maintaining workplace efficiency meant that accident prevention came back to the forefront of governmental agendas, as it had been in the mid-nineteenth century – before discussions about compensation seemed to take precedence over the issue of prevention. After adopting a compensation policy, each state continued to build its factory inspectorate and enact new safety legislation – despite the worries of many workers that compulsory insurance policies, in particular, would encourage employers to become lax about safety.[64] However, accident insurance and workmen's compensation were administered as separate matters from safety legislation. Even in Italy, where the accident insurance law included a list of basic safety regulations, the government focused its efforts on the compensation aspect of the policy in subsequent years. The experience of the war, therefore, marked a turning point, as the administrators involved in compensation focused increasingly on issues related to safety. In each country during the war, but especially in Germany and Britain, the budding field of industrial management began to flourish. As a result, each state issued a number of safety decrees that focused on issues ranging from tying one's hair back – in the case of the many new women workers in factories – to wearing protective goggles.[65] After the war, the area of industrial management, including the related field of industrial psychology, continued to advance. As a result, the focus on prevention persisted for the next few years. For example, a spate of new posters, based on American cartoons from the Safety First movement, provided all three governments with a new model for addressing the accident problem into the 1920s.[66] While the paradigm of occupational risk continued to inform compensation policies – and claims in court cases – the evolution of specialist knowledge meant that the issue of prevention seemed to emerge from the ashes.

Conclusions

The case of workplace accidents points to five more general conclusions about the role of specialist knowledge in the treatment of social problems at the turn of the twentieth century. First and foremost, specialist knowledge filtered into policy discussions through the specific optic of *legal* reasoning: at its heart, the issue of workplace accidents was a matter for employment law. Therefore all medical, technical, and other considerations about policy needed to be framed in terms of what had been the treatment of labour contracts under common law and the continental law codes and, later, according to what would become the public law of accident compensation. Psychiatric reports, thus, only mattered if they spoke the language of compensation law. The case of workplace accidents demonstrates, therefore, the need to include legal reasoning within the basic paradigm of 'scientization'.

Secondly, the most important factor for the *adoption* of these policies was the availability of seemingly objective information that each government could rely upon in order to claim a *Sachzwang* – that is, a 'constraint to act rationally' – about the issue of workplace risk.[67] Maintaining a *Sachzwang* about occupational risk proved especially necessary in light of the divergent view, which was popular particularly among workers, that accidents should be prevented in the first place rather than compensated after they had occurred. Ostensibly objective specialist information could, therefore, turn the issue of occupational risk into a matter that appeared to transcend politics; occupational risk seemed a neutral category that rose above class politics, gender distinctions, and regional disparities. Its apparently apolitical nature meant that the decision to adopt a policy on compensation seemed the only rational choice. The evidence of statistics, which proved the likelihood of workplace accidents in particular industries, and, for Britain and Italy, the example of foreign successes in addressing the accident problem only furthered arguments in favour of compensation.

Thirdly, scientific ways of reasoning about industrial accidents continued to be important after each state had adopted a compensation policy. The central governments and the administrative and judicial apparatus for the compensation laws consistently relied on legal and medical expertise as well as specialist knowledge from the natural sciences in order to address new issues inherent in the laws. It was during this stage that transnational sources of expertise became particularly useful for each government when addressing questions about the meaning of an accident and the form of industrial risk. It was also during this stage that the potential beneficiaries of each compensation system learned to employ specialist knowledge in order to defend their claims in court and, in the case of extending the laws, in parliament.

Fourthly, the role of specialist knowledge in the administration of accident compensation was neither unilinear, nor, once in place, stagnant. With their reliance on seemingly objective, 'expert' knowledge, the systems of accident compensation appeared apolitical. Nonetheless, in moments of perceived crisis, the compensation systems, along with the expert knowledge upon which they were based, could seem politicized, as in the cases of 'pension addiction' in Italy and Germany. Moreover, in times of crisis, such as during the First World War, specialist knowledge could be pushed to the margins in favour of pragmatic solutions – even if only temporarily. Not least, innovations in specialist knowledge – as in the case of industrial management during the war – meant that particular fields of expertise could come to the fore, changing perceptions of ongoing problems.

Finally, it is important not to overemphasize the importance of specialist knowledge for the treatment of social questions such as the problem of workplace accidents. In Britain, Germany, and Italy, as elsewhere at this time, the creation of policies on accident compensation was contingent upon a complex interplay of factors. Pragmatic decision-making based on governmental structures, including the relationship between central government and local and regional governments (which deserves further elaboration elsewhere), and relations with citizens also played roles in defining and treating the accident problem during this period. The significance of individual factors differed for each state and varied over time. In order to explain these differences, we therefore have to turn our attention to historically specific political and social developments within each country.

Notes

1. *Priestley* v. *Fowler* (1837) 3M & WI.
2. For example, J. Radkau (1993), 'Industrialiserung des Bewußtseins und moderne Nervosität: Zur Mythologie und Wirklichkeit der Neurasthenie im Deutschen Kaiserreich', in D. Milles (ed.), *Gesundheitsrisiken, Industriegesellschaft und soziale Sicherungen in der Geschichte* (Bremerhaven: Wirtschaftsverlag NW), pp. 363–85; W. Schivelbusch (2004), *Geschichte der Eisenbahnreise. Zur Industrialisierung von Raum und Zeit im 19. Jahrhundert*, third edition (Frankfurt: Fischer), p. 15.
3. A. Rabinbach (1996), 'Social Knowledge, Social Risk, and the Politics of Industrial Accidents in Germany and France', in D. Rueschemeyer and T. Skocpol (eds), *States, Social Knowledge, and the Origins of Modern Social Policies* (New York: Russell Sage), pp. 48–89; F. Ewald (1986), *L'État providence* (Paris: B. Grasset); G. Eghigian (2000),

Making Security Social. Disability, Insurance, and the Birth of the Social Entitlement State in Germany (Ann Arbor: University of Michigan Press); A. Killen (2006), *Berlin Electropolis. Shock, Nerves and German Modernity* (Berkeley: University of California Press).

4. A variety of German and Italian states enacted such legislation before national unification. See, for example, B. L. Hutchins and A. Harrison (1911), *A History of Factory Legislation*, second revised edition (London: P. S. King and Son); Angelika Pensky (1987), *Schutz der Arbeiter vor Gefahren für Leben und Gesundheit. Ein Beitrag zur Geschichte des Gesundheitsschutzes für Arbeiter in Deutschland* (Bremerhaven: Wirtschaftsverlag); Ministero di agricoltura, industria e commercio (MAIC), Ufficio del lavoro (1904), *L'ispezione del lavoro. Studi sull'organizazione di vigilanza per l'applicazione delle leggi operaie* (Rome: Tip. Naz. G. Bertero).
5. M. Karl (1993), *Fabrikinspektoren in Preussen. Das Personal der Gewerbeaufsicht 1854–1945* (Opladen: Westdeutscher Verlag). On the function of the police in this capacity, see L. von Stein (1888), *Handbuch der Verwaltungslehre*, vol. 1: *Der Begriff der Verwaltung und das System der positiven Staatswissenschaften*, third edition (Stuttgart: J. G. Cotta'schen Buchhandlung), pp. 184–5; MAIC (1904), *L'ispezione del lavoro*, pp. 9–10; J. Pellew (1982), *The Home Office, 1848–1914. From Clerks to Bureaucrats* (East Brunswick, NJ: Associated University Presses), pp. 122–49.
6. N. Stehr (1994), *Arbeit, Eigentum und Wissen. Zur Theorie von Wissensgesellschaften* (Frankfurt: Suhrkamp); J. Vogel (2004), 'Von der Wissenschafts- zur Wissensgeschichte. Für eine Historisierung der "Wissensgesellschaft"', *Geschichte und Gesellschaft* 30, pp. 639–60, here at 640.
7. A. Giddens (1999), 'Risk and Responsibility', *The Modern Law Review* 62, pp. 1–10; U. Beck (1992), *Risk Society. Towards a New Modernity* (London: Sage); N. Luhmann (1993), *Risk. A Sociological Theory* (New York: A. de Gruyter), pp. 23–7.
8. Ewald (1986), *L'État providence.*
9. See also Vogel (2004), 'Von der Wissenschafts- zur Wissensgeschichte', p. 654.
10. M. Eckardt (2004), 'The Evolution of the German Tort Law in the Nineteenth Century. An Economic Analysis of the Evolution of Law', in O. Volckart (ed.), *The Institutional Analysis of History* (Munich: *Homo Oeconomicus*, Special Issue 21), pp. 83–116; D. J. Ibbetson (2001), 'The Tort of Negligence in the Common Law in the Nineteenth and Twentieth Centuries', in E. J. H. Schrage (ed.), *Negligence. The Comparative Legal History of the Law of Torts* (Berlin: Duncker and Humblot), pp. 229–71; I. Piva (1980), 'Problemi giuridici e politici della "responsabilità"', *Rivista degli infortuni e delle malattie professionali* 5–6, pp. 649–56.
11. Ibbetson (2001), 'The Tort of Negligence'; W. Ernst (2001), 'Negligence in 19th-century Germany', in Schrage (ed.), *Negligence*, pp. 341–59; M. Taruffo (1980), *La giustizia civile in Italia dal '700 ad oggi* (Bologna: Mulino), pp. 107–50.
12. For example, Parliamentary Papers (PP) 1884 (151) lxiii.139; Geheimes Staats-Archiv Preußischer Kulturbesitz (GStAPK): I.HA.120BB. Abt. VIII. Fach. 4. No. 1. Vol. 1: 'Promemoria betreffend Versicherung der Arbeiter gegen Unfälle und Beschädigung während der Arbeit beim Bergbau, bei der Industrie und sonstiger Gewerbethätigkeit, sowie bei der Landwirtschaft', 30 April 1880. On legal aid, see W. Steinmetz (2002), *Begegnungen vor Gericht. Eine Sozial- und Kulturgeschichte des englischen Arbeitsrechts (1850–1925)* (Munich: R. Oldenbourg), pp. 371–461.
13. A. Lonni (1982), 'Fatalità o responsabilità? Le jatture degli infortuni sul lavoro. La legge del 1898', in M. L. Betri and A. Gigli Marchetti (eds), *Salute e classi lavoratrici in Italia dall' unità al fascismo* (Milan: Franco Angeli), pp. 737–62, here 749; Piva (1980), 'Problemi giuridici', p. 655; C. Biondi (1903), 'La Legge sugli infortuni del lavoro nel diritto e nella medicina', *Rivista di Diritto e Giurisprudenza* 4, pp. 4–6.
14. Bundes-Archiv Berlin-Lichterfelde (BArch): R1501: 100402: 4: Annual Report on Factory Inspection, 1878; PP 1886 (192) viii.1, 'Report from the Select Committee on the Employers' Liability Act (1880) Amendment Bill', pp. 66–7, 110.
15. *Schwäbischer Merkur*, 1881, Nr. 20.

16. Atti Parlamentari della Camera dei Deputati (A. P. Cam.) XVI/4 (1889–90), Doc. Nr. 116, 'Disegno di legge presentato alla camera dei deputati dal ministro di agricoltura, industria e commercio (Miceli), nella seduta dell'8 febbraio 1890. Provvedimenti per gl'infortuni sul lavoro', 4.
17. V. Foa (1973), 'Sindacati e lotte sociali', in *Storia d'Italia* (Turin: Einaudi), vol. 5: *I documenti*, p. 1790.
18. See F. Tennstedt (2001), '"Bismarcks Arbeiterversicherung" zwischen Absicherung der Arbeiterexistenz und Abwehr der Arbeiterbewegung', in H. Matthöfer, W. Mühlhausen and F. Tennsteadt (eds), *Bismarck und die Soziale Frage im 19. Jahrhundert* (Friedrichsruh: Otto-von-Bismarck-Stiftung), pp. 51–87.
19. GStAPK: I. HA. 120 BB. Abt. VIII. Fach. 4. No. 1. Vol. 1: 'Promemoria', 30 April 1880; PP: 1877 (285) X.551: 'Report from the Select Committee on Employers' Liability for Injuries to their Servants; together with the Proceedings of the Committee, Minutes of Evidence, Appendix' (25 June 1877), pp. 35–7.
20. For example, Third Congress of the Confederation of Lombardian Workers, Varese, 1883; A. P. Cam., Sessione 1878–1879. Discussioni. 17 March 1879, pp. 4958–61; Documenti, Nr. 191, p. 2; A. P. Cam. Doc. Nr. 73A, 8 April 1884, p. 4.
21. For example, 'Editorial' (1897), *The Amalgamated Engineers' Journal*, n. s., 1, pp. 2–5, here 4; 'Labour's Death Roll: Employers' Liability versus Insurance: A Trades Union Plebiscite: What the Leaders of a Million Workmen Say' (8 February 1897), *Daily News*.
22. *Stenographische Berichte des Deutschen Reichstags* (SBDR), 14 January 1885, vol. 79, p. 607; 16 January 1885, vol. 79, pp. 650–1.
23. For example, PP 1882 XVIII, 'Report of Inspector Redgrave', pp. 3–4; BArch R1501: 100379: 58: Minutes of the Volkswirthschaftsrath, 29 January 1881.
24. For example, A. P. Cam. Doc. Nr. 73, 19 February 1883, pp. 6–8. See also A. P. Cam. Discussioni, XV/1, 18 May 1885, p. 13786.
25. See T. M. Porter (1986), *The Rise of Statistical Thinking, 1820–1900* (Princeton: Princeton University Press); I. Hacking (1990), *The Taming of Chance* (Cambridge: Cambridge University Press).
26. S. Patriarca (1996), *Numbers and Nationhood. Writing Statistics in Nineteenth-century Italy* (Cambridge: Cambridge University Press), p. 5, fn 15.
27. For example, *Atti per l'istituzione della cassa nazionale di assicurazione per gli operai contro gli infortuni sul lavoro* (1884), 1: *Relazione del Comm. Luigi Luzzatti* (Rome: stabilimento tipografico dell' opinione), p. 8; T. Bödiker (1884), *Die Unfallgesetzgebung der europäischen Staaten* (Leipzig: Duncker and Humblot), pp. 1–2; PP 1877 (285) X.551, p. v.
28. PP 1871 (488) LVI.591: 'Accidental Deaths...'. Less comprehensive statistics on factory accidents had been compiled since 1845.
29. 'Tavole delle morti accidentali...' (1884), *Atti per l'istituzione della Cassa Nazionale* 1, Appendix 1.
30. GStAPK: I. HA. 120BB. Abt. VII. Fach 1. nr. 4. vol. 4, pp. 111–13: Minister of Trade and Commerce to the Collective Royal Governments, Police Presidia, and Landdrostereie, 30 June 1873.
31. GStAPK: I. HA. Rep. 120BB. Abt. VII. Fach 1. nr. 16. vol. 1, p. 2: On behalf of the Chancellor of the North German Confederation to Itzenplitz, Prussian Minister of Trade and Commerce, 5 May 1868.
32. For example, 'Le associazioni professionali nell'anno 1886' (31 July 1888), *Bollettino di notizie sul credito e la previdenza* 6, pp. 829–40; PP 1886 (c. 4784) LXVII.571: 'Reports by Her Majesty's Representatives Abroad on the Laws Regulating the Liability of Employers in Foreign Countries'.
33. U. Mazzola (1885), 'L'assicurazione degli operai nella scienza e nella legislazione germanica', *Annali del credito e della previdenza* 14; Foreign Office (1894), *Report on the Question of Employers' Liability in Germany* (London: Eyre and Spottiswoode).
34. For example, É. Gruner (ed.) (1889–90), *Exposition universelle internationale de 1889. Congres international des accidents du travail*, 2 vols (Paris: Librairie polytechnique Baudry

et Cie); Gruner (1909), *Actes du VIII Congrès International des Assurances Sociales ... Rome 12–16 Octobre 1908*, 3 vols (Rome: Imprimerie Nationale de J. Bertero & C.).

35. J. Moses (2012), 'Policy Communities and Exchanges across Borders. The Case of Workplace Accidents at the Turn of the Twentieth Century', in D. Rodogno, B. Struck, and J. Vogel (eds), *Shaping the Transnational Sphere, 1830–1950* (New York: Berghahn); R. Gregarek (1995), 'Le mirage de l'Europe sociale. Associations internationales de politique sociale au tournant du 20e siècle', *Vingtième siècle* 48, pp. 103–18; M. Herren (1993), *Internationale Sozialpolitik vor dem Ersten Weltkrieg. Die Anfänge europäischer Kooperation aus der Sicht Frankreichs* (Berlin: Duncker and Humblot); R. Ricci (1896), 'Per gl'infortuni del lavoro', *Nuova Antologia*, fourth series, 63 (16 June), pp. 705–21, here at 712.
36. See G. Leclerc (1982), *L'Observation de l'homme. Une histoire des enquêtes sociales* (Paris: Seuil), p. 82.
37. On the 'science of work', see A. Rabinbach (1990), *The Human Motor. Energy, Fatigue, and the Origins of Modernity* (New York: Basic Books).
38. BArch R89: 15110: unnumbered report from Hermann von Helmholtz, President of the Physical-Technical Imperial Institute in Berlin, to Dr Bödiker, President of the Imperial Insurance Office, 16 October 1888; unnumbered: Ia 12166. See also 'Rekurs-Entscheidungen des Reichs-Versicherungsamts. Unfall durch Blitzschlag' (1898), *Arbeiterversorgung* 15, pp. 69–71.
39. For example, *Thompson* v. *Arlington Coal Company, Ltd* (84 LT 412); BArch R 1501: 100753: 266–7: IA 4780/12 – 15B: *Frieda Peukert* v. *Hamburg Baugewerks-BG*, 20 February 1913.
40. For example, *Rosa Esposito* v. *Mutua Infortuni*, Court of Appeal of Naples, 29 July 1904, cited in *Giurisprudenza giudiziaria*, vol. 1 (Rome, 1906), pp. 305–9.
41. BArch R89: 343: unnumbered: Ia 7071/01.18: *Dependants of Captain Heinrich Braue* v. *the See-Berufsgenossenschaft*, 13 December 1901.
42. *Brintons Ltd* v. *Turvey* (1905), AC 230.
43. PP 1907 (CD. 3495). XXXIV.1045: 'Report of the Departmental Committee on Compensation for Industrial Diseases. Report presented to both Houses of Parliament by Command of His Majesty'.
44. BArch: R89: 15127: unnumbered letter from the Secretary of the Interior to the President of the Imperial Insurance Office, 4 April 1913. The clause was only invoked in 1925, when Germany created a list of compensable diseases analogous to Britain's.
45. A special commission investigated the matter in 1902. See *Atti della Commissione incaricata di studiare le cause ed i provvedimenti preventivi delle malattie degli operai nelle industrie* (1902), special issue of *Annali del Credito e della previdenza.*
46. For example, BArch R1501: 100753: 163: Zentralstelle des Gewerkvereins christlicher Bergarbeiter Deutschlands to Secret governmental advisor Dr Würmeling, 7 August 1912; Camera dei deputati, Archivio Storico (CdAS), legislatura XXII, Sessione 1904–9, Bta. 860: Representatives of the National League of Cooperatives, of the General Confederation of Labour, of the Italian Federation of Mutual Aid Societies, and of the Legal-Medical Advisory Offices of the Chambers of Commerce of Milan and Genoa, n. d., c. 27 April 1908; The National Archives, Kew (NA) HO 157: 6: 280: E. Blackwell, Under Secretary of State at the Home Office, to the Secretary of the Amalgamated Society of Railway Servants, 23 May 1911.
47. For example, 'Sociales. Die Vergiftungen in Betrieben und das Unfallversicherungsgesetz' (30 January 1902), *Vorwärts*; A. P. Camera, 27 March 1901; 4 *Hansard* HC, vol. 154 (26 March 1906), cols 900–6.
48. For example, CdAS: legislatura XXII, Sessione 1904–9, Bta. 860: unnumbered: 'Relazione illustrativa dei voti del congresso degli industriali italiani per la riforma della legge sugli infortuni del lavoro, Roma 1908', p. 16.
49. *Congrès international des accidents du travail et des assurances sociales. Quatrième session tenue à Bruxelles ... 1897. Rapports, procès-verbaux des séances et communications présentées au Congrès, publiés par les soins du Comité belge d'organisation* (Brussels, 1897).

50. *Actes du VIII Congrès International des Assurances Sociales.*
51. BArch R89: 15127: unnumbered: Director of the Office International du Travail to the Imperial Insurance Office, 18 November 1912.
52. See Eghigian (2000), *Making Security*, pp. 238–44; Schivelbusch (2004), *Geschichte*, Chapter 9; E. Fischer-Homberger (1975), *Die traumatische Neurosen. Vom somatischen zum sozialen Leiden* (Bern: H. Huber).
53. For example, Prof. Dr Rumpf (1913), 'Über nervöse Erkrankungen nach Eisenbahnunfällen', *Zeitschrift für Bahn- und Bahnkassenärzte* 5, p. 5; E. Vecchietti (1912), *Rilievi ed appunti sulla esecuzione della legge infortuni* (Naples: Tip. editrice cav. aurelio tocco), pp. 10–11.
54. Rabinbach (1990), *Human Motor*, pp. 275–6.
55. NA: HO157. 3: 232–3: Sir K. E. Digby to all County Court Judges, 15 October 1900; NA: PIN12. 1: unnumbered: Report from Dr F. Stevenson, 30 October 1900, to Sir K. E. Digby.
56. For example, NA: HO 157. 3: 88: Under-secretary of State of Home Office to Mr S. Harris, 7 July 1899.
57. For example, 'I vampiri: alcuni documenti' (1918), *Rassegna di assicurazioni e previdenza sociale* 5, pp. 390–1; P. Kaufmann (1924), *Zur Umgestaltung der deutschen Sozialversicherung* (Berlin: Verlag der Reichsdruckerei), p. 19.
58. For example, BArch: R89: 11158: 302: Verband Südwestdeutscher Industrieller to the Imperial Insurance Office, 31 March 1916.
59. 'Unfallversicherung. Recht und Rechtsübung auf dem Gebiete der Unfallversicherung mit Bezug auf den Krieg' (1916), *Monatsblätter für Arbeiterversicherung* 10, pp. 101–32, here 105; 'Decreto luogotenenziale 31 ottobre 1915, n. 1577, riguardante i soprapremi pei rischi di guerra nelle assicurazioni degli operai contro gli infortuni sul lavoro' (10 November 1915), *Gazzetta ufficiale*, 275; NA: HO 157: 9: 483: Under Secretary of State in the Home Office to London Line Mills, Ltd, 16 July 1917.
60. For example, NA: PIN 11. 2: unnumbered: David Brown to W. Horne, Esq., 25 July 1917; BArch: R 3901. 4547: 125: Robert Nagel to the "Welfare Minister", 1 September 1921; Archivio centrale dello stato (ACS): PCM 1921. Bta. 629. Fasc. 6. No. 2: 10: Telegram from Alfredo Bianconi to Prime Minister Giovanni Giolitti, 5 January 1921.
61. NA: PIN 11. 4: unnumbered: Home Office memorandum, untitled, 7 October 1919.
62. NA: LAB 2. 225. ED 17097. 175. 1918: unnumbered: Memorandum by Sir Herbert Austin, K.B.E.: 'Scheme for Employment of Disabled Men', 19 November 1918; P. Kaufmann (1920), *Wiederaufbau und Sozialversicherung. Vorschläge zur Änderung der Reichsvericherungsordnung*, second edition (Berlin: Georg Stilke), pp. 16–17; 'Il lavoro agricolo e gli invalidi di guerra' (1916), *Rassegna di assicurazioni e previdenza sociale* 3, pp. 1356–8.
63. See Eghigian (2000), *Making Security*, pp. 244–6; P. Lerner (2003), *Hysterical Men. War, Psychiatry, and the Politics of Trauma in Germany, 1890–1930* (Ithaca, NY: Cornell University Press).
64. In Germany, there was a slight delay in adopting new safety measures after the enactment of the accident insurance law until the Chancellor Otto von Bismarck left office and a so-called new course in social policy was pursued. H-J. von Berlepsch (1987), *'Neuer Kurs' im Kaiserreich? Die Arbeiterpolitik des Freiherrn von Berlepsch 1890 bis 1896* (Bonn: Neue Gesellschaft).
65. Health of Munitions Workers Committee (1918), *Final Report* (London: Harrison and Sons), p. 16; BArch R89: 11308: 2: Circular from the Imperial Insurance Office to the Committees of the Berufsgenossenschaften, 20 December 1916; 'Il lavoro delle donne e dei fanciulli' (1916), *Rassegna di assicurazioni e previdenza sociale* 3, pp. 1454ff.
66. For example, 'Manifesti murali e prevenzione della cecità in dipendenza d'infortuni sul lavoro' (1922), *Rassegna della previdenza sociale* 9, pp. 115–18; G. Müller (1925), 'Unfallverhütung und Gewerbehygiene', *Reichs-Arbeitsblatt*, n. s., nichtamtl. Teil 5, pp. 505–14. On the British movement, see M. Esbester (2006), '"No Good Reason for the

Government to Interfere". Business, the State and Railway Employee Safety in Britain, circa 1900–1939', *Business and Economic History On-line* 4, pp. 1–19.

67. This is a connotative translation. W. Steinmetz (1995), 'Anbetung und Dämonisierung des "Sachzwangs". Zur Archäologie einer deutschen Redefigure', in M. Jeismann (ed.), *Obsessionen. Beherrschende Gedanken im wissenschaftlichen Zeitalter* (Frankfurt: Suhrkamp), pp. 293–333.

4
Politics Through the Back Door: Expert Knowledge in International Welfare Organizations

Martin Lengwiler

Social policy and the institutions of the welfare state belong to the core areas affected by the process of scientization since the late nineteenth century.[1] Among important agents of scientization that have so far escaped historiographical scrutiny are international organizations and transnational networks.[2] This neglect reflects the general state of welfare history. Until now, the history of the welfare state and its institutions has been mainly understood within the context of national states, populated by national actors.[3] Welfare history is still, in the words of the historian Christoph Conrad, a 'bastion of the national', characterized by national case studies and comparative analyses based on purely national examples.[4] Only in recent years have cross-border transfers and processes of transnationalization and internationalization attracted scholarly interest. These are now increasingly seen as productive or even driving elements in modern welfare history.[5]

Recent historiography has highlighted four fields in which inter- or transnational expert networks and organizations have had a crucial impact on the development of the welfare state. First, international expert networks shaped national welfare debates even before the foundation of the first social insurance schemes in the 1880s. Madeleine Herren has pointed out that not only *social* movements, such as the labour and women's rights movements, but also *scientific* enterprises and institutions like world exhibitions and international scientific congresses formed the emerging international social policy networks of the mid-nineteenth century.[6]

A second field of research has examined the border-crossing learning processes of Western actors in colonial contexts. Research in colonial history has shown that European colonial powers used their colonies as 'laboratories of modernity', not least in social policy fields. Many members of the French and British civil service began their careers in the colonies, receiving important qualifications there, particularly in public health. Colonial influences were also important for the German Empire, in the form of colonial missionary organizations that fed back their experiences into religious welfare institutions in Germany. Also in the United States, the development of public health crucially profited from administrative experiences from the colonial occupation of the Philippines between 1904 and 1945.[7]

Thirdly, recent research has focused on the bilateral sociopolitical learning and demarcation processes that went on between European welfare states, in particular between Germany, Britain, and France. Here, studies have concentrated primarily on the period in which modern welfare states were formed – the decades from the 1880s to the Second World War.[8]

A fourth field of research covers the processes of Europeanization and globalization in social policy, and the significance of international organizations in the development of modern welfare states. The International Labour Organization (ILO), one of the comparatively well-examined institutions, is now seen as a trendsetter in social policy, in particular for its contributions to international labour law and migration regulation in the interwar and early postwar period.[9] Other studies have examined the Europeanization of social policy, in particular by the European Union (and its predecessors) or by the Council of Europe. They conclude that convergence to the oft-cited 'European social model' has not (yet) gone far, and that the influence of the European Union on the social insurance systems of its member states is still rather modest.[10] Nevertheless, sociological studies concede that the 'post-national constellation' (Leibfried/Mau), emerging with continuing globalization from the 1970s, has indeed affected national welfare states – though not in the form of erosion or destabilization of their social security systems, as once feared. On the contrary, social insurance has often been expanded, in particular to protect against the increased risks of unemployment in a liberalized global economy. This development is paralleled by an expansion of supranational ways of coordinating welfare systems, especially among members of the European Union.[11]

Based on these emerging outlines of a transnational history of the welfare state, this chapter tries to assess the role of expert knowledge in the shaping of social policies by international organizations and transnational networks. This will be achieved in two steps. The first part of the essay takes an actor- and organization-centred approach, and gives an outline of the relevant organizations and networks engaged in the field of social policy. In the second part, I discuss the institutional mechanisms from which expert knowledge was able to profit for building up its influence on social policy-making – it is in this latter part that the actual processes of 'scientization' will be addressed. While the first part is based mainly on a review of the secondary literature, the second part focuses on two empirical case studies: on the International Congress of Actuaries (ICA) on the one hand; and on the International Labour Organization (ILO) and the closely related International Social Security Association (ISSA) on the other.

The ICA was the main congress for the international community of actuaries. Their discipline – actuarial science – was based on a combination of mathematical (mainly probabilistic theory), statistical and demographical knowledge, and was seen, in the context of social insurance as well as in private insurance, and at least until the 1970s, as representing the supreme discipline of all insurance-related sciences. Other disciplines, for example, from the social sciences, only played a marginal role in the construction and management of social insurances.[12] The ICA mirrored the heterogeneous institutions of modern welfare systems; representatives of both private and public insurance organizations belonged to the actuarial community. The ICA was therefore a platform where the commercial insurance industry and statutory social insurance departments were both present.[13] The ILO and the ISSA also belonged, for much of the twentieth century, and at least until the 1970s, to the most prominent and influential international organizations engaged in social policy issues – the ILO as a mainly political institution, the ISSA as an expert organization.

Geographically, I will mainly focus on Western Europe, which can easily count as an exemplary area of welfare systems. Although the period covered includes the whole twentieth century, my focus will be on the postwar decades between 1945 and about 1990. In this era, most Western European welfare states witnessed a unique boom and converged on a common type of comprehensive social security system.[14] How far this development was informed by scientific expertise is the central question of the following argument.

On a theoretical level, I will refer to two traditions: one reflects actor-centred approaches, pointing out the relevant agents for social political transfers; the other analyzes the systems of norms and values surrounding the development of welfare states. Both perspectives – actors; norms and values – are crucial for understanding the effects of expert knowledge on welfare history. In an actor-centred perspective, Christoph Conrad has underlined the multiple layers of actor constellations, pointing at *organizations* of internationalization like the ILO, international and transnational expert *communities* and *networks*, or *institutional actors* such as border-crossing languages, legal systems, and transfers in international finance.[15] Patricia Clavin has stressed the relevance of civil society networks and academic communities for border-crossing transfers and their affinities for universalistic value systems. These actors had a constitutive role in the processes of transnationalization of social policy in the nineteenth and twentieth centuries.[16] Other authors have pointed at the spatial dimensions in which international and transnational actors moved. Pierre-Yves Saunier, for example, has suggested making an analytical distinction between, on the one side, mechanisms of import and export at a national level and, on the other side, mechanisms of transfer situated in a transnational space, such as interactions within international or transnational organizations.[17] Similarly, Johannes Paulmann and Martin H. Geyer have used the concept of 'transnational border spaces' to analyse intercultural transfers. These border spaces are situated not only at geographical borders but also within international organizations and networks.[18]

Recent research has also highlighted the significance of universal systems of norms and values for processes of transfer and transnationalization in social policy. Among the particularly powerful norms and values are fundamental ethical and political values such as the ideas of human rights and global development, but also the notion of academic expertise (as a normative system for gaining objective knowledge).[19] Historian Akira Irye, for example, has coined the concept of 'cultural internationalism' to point out the contribution of common ideas of order, progress, and modernization, propagated by civil society actors, to the process of globalization since the nineteenth century.[20] On a more general level, sociologist John W. Meyer has developed his 'world polity' theory, a framework for conceptualizing the role of cultural values and norms for modern processes of globalization. Among the crucial norms, Meyer counts, for instance, Christian values, ideas of progress and justice, human rights, social security, scientific knowledge, the idea of education, and modern concepts of health and physical well-being. Remarkably, this approach understands globalization as a process *prior* to nationalization – and not the other way round, as most theories of globalization would have it – granting national actors a central role in the way globalization progressed. According to Meyer, national actors were the basic units of the globalization process and converged by referring, independently from each other, to common and universal norms and values.[21]

Who spoke? Changing constellations of organizations and networks during the twentieth century

As indicated previously, social policy debates were driven, as early as the mid-nineteenth century, by international and transnational actors.[22] Most of them were loosely organized in multilateral associations or networks, with the ambition of establishing an international community to collect and compare knowledge – often in the form of mathematical and statistical expertise – on national institutions and policies. Since the Great Exhibition in London in 1851, world exhibitions had regularly offered a platform for the launch of international scientific associations.[23] Their congresses then developed a life of their own. By the end of the nineteenth century, a whole series of congresses focused on the 'Social Question' and related issues in social policy. They included the Congress for Hygiene, the Congress for Demography, the Congress for Social Insurance (founded in 1886) and the Congress for Actuaries (1895). Although these expert networks did not wield any formal power and often served as a platform for nationalistic forms of self-promotion, they still triggered important knowledge transfers from the international to national levels and between national actors.[24]

The First World War was a watershed in the development of transnational networks of social policy experts. Many congresses, especially in the politicized field of social policies, did not survive the political, nationalistic antagonisms of the war and the following interwar period. The International Congress of Social Insurance, for example, perished after the war; its organizing body, the Permanent International Committee for Social Insurance, simply broke apart amid continuing resentments between the victorious and the defeated powers. Only the more technical networks, such as the International Association of Actuaries, survived the polarizations of the interwar period.[25]

The interwar period also saw the rise of more formal expert communities embedded in the newly founded international organizations, notably the League of Nations and the International Labour Organization (ILO), itself a sub-organization of the League. The ILO stood in the tradition of international social policy actors such as the socialist Second International and the expert-based International Association for Labour Legislation. With its self-proclaimed status as a leading institution in social policy and its tripartite structure bringing together trade unions, employers' associations, and government representatives, the ILO quickly became the epicentre for international debates on social policy, especially in the 1920s and early 1930s, under the directorship of the charismatic Albert Thomas. The ILO not only acted in the field of social policies – its core competence – but also in the related areas of migration and economic policies. It also served as a safe haven for the stranded representatives of international networks that did not survive the world war, such as the Permanent International Committee for Social Insurance, whose members often became technical advisors to the ILO.[26]

Historically the ILO continued two important traditions of nineteenth-century internationalism: political internationalism, as represented by the international labour and women's movements, and scientific internationalism, represented by the international scientific congresses. Against this background, the ILO's influence was based on two organizational specificities: its tripartite structure and its strong reliance on expert knowledge. The tripartite structure often helped to broker innovative compromises and shape debates on social policies in an effective way before they entered national realms. For its deliberations on social security, the International Labour Office, the administrative body of the ILO, established its own expert group, the Social Insurance Section, as early as 1921. In the first years, the group acted as a corresponding committee to answer specific questions. From the 1930s it constituted itself as a permanent body, pursuing general questions by actively initiating comparative research projects. In technical questions, such as legal, economic or actuarial issues, the Social Insurance Section was the primary advisor to the ILO.[27]

Another testimony to the significance of expert knowledge in the ILO was the foundation, in 1947, of the International Social Security Association (ISSA), an expert organization dedicated to the study of social security systems. It was sponsored by the ILO and quickly became its most important technical think tank. Originally, the ILO had wanted to join forces with other international social policy organizations and persuade them to support the expansion of its own expert group. But rivalries with competing organizations, in particular the International Association for Labour Legislation, had led in 1927 to the foundation of the International Conference of National Associations of Friendly Societies and Sickness Funds, an expert organization formally independent of the ILO that transformed itself, in 1947, into the ISSA.[28]

In the 1950s and early 1960s, the ISSA expanded its activities by bringing out an increasing number of research publications and by founding a series of specialized committees on issues such as actuarial theory and the statistics of social security, family allowances, unemployment insurance, work accident, and old age insurance. Further fields of engagement included the protection of migrant workers and labourers in the agricultural workforce.[29] The research activities of the ISSA were still very much descriptive and geared to practical issues: most projects consisted of collecting comparative data on the social security institutions of the ILO's member states.[30] As the association studied roughly the same issues as the Social Insurance Section of the International Labour Office, it was only a matter of time before the ISSA became an institutional member of the Section. Also, the ISSA and the ILO often shared the same staff. The long-time general secretary of the ISSA, the Austrian Leo Wildmann, combined his position at the ISSA (from 1949 to 1974) with a functionary position at the International Labour Office.[31] In their work, the ISSA and the ILO often joined forces too, for example, in 1956 when they organized the constitutive first meeting of the International Conference of Social Security Actuaries and Statisticians.[32]

From the mid-1960s, the ISSA strengthened its research activities by founding a five-member Study Group with eminent participants such as Ida Merriam, an old New Dealer and research director at the US Social Security Administration; Richard Titmuss, who held the first British chair in Social Policy (at the London School of Economics) and was a leading scholar in the field; and Henning Friis, another eminent scholar in Welfare Studies and director of the Danish National Institute for Social Research. The Study Group was sponsored by the Ford Foundation, at that time the biggest philanthropic organization in the world.[33]

Both the ILO and the ISSA remained central international actors in social policy until at least the 1960s.[34] Since then, however, the ILO has gradually lost its former status. In the process of decolonization, most newly founded African and Asian states became members of the ILO – a process that changed the political majorities of the organization, shifted the focus of activities towards the Third World, and increased the political polarizations within it. This development resulted in a severe institutional crisis at the end of the 1970s. Unhappy with the politicization of the ILO, the United States temporarily withdrew its membership (and its crucial funding).[35] In these unstable years, the role of the ILO was taken up by other international organizations, in particular by the Organization for Economic Cooperation and Development (OECD) and the European Economic Community (later the European Community and ultimately the European Union).

The OECD's foundation was an indirect offspring of the postwar Marshall Plan and its managing organization, the Organization for European Economic Cooperation (OEEC). By the end of the 1950s, the OEEC has gradually developed into a platform for coordinating economic policies among Western industrial nations and was kept in existence, even after the end of the Marshall Aid programme, but under the new name of the OECD.[36] Thus since its foundation in 1961, the OECD acted primarily as an international think tank for the formulation of macro-economic policies and the promotion of trade liberalization among Western industrial nations.[37] Especially in the 1960s and 1970s, it also served as the main body, joined in 1975 by G7, for coordinating national economic policies between its member states. Social policy and social insurance issues became relevant for the OECD from the mid-1970s. This development must be seen in the context of emerging monetarist convictions, following the collapse of the Bretton Woods system and the shift from Keynesian to monetarist paradigms. The new policies highlighted the significance of monetary and fiscal discipline, and the need for structural adaptations by governments to maintain low levels of inflation.[38] The structure of the labour market, labour legislation, and the amount of government social expenditure began to be seen as crucial conditions for economic growth and productivity.

The OECD's approach to policy-making is characterized by the paradox that, unlike other international organizations, its influential status is not based on any formal authority over its member states, such as economic or political power. Instead, the OECD acts as a 'talking shop', bringing together 'communities of influence' and providing consensus and cooperation in highly topical policy areas.[39] The organization operates on two levels: on the one hand, bringing together representatives of national civil services in the form of committees and working parties or working groups in order to coordinate national policy-making at an international level;[40] on the other hand, by publishing comparative policy reviews and providing normative advice based on statistical and other research. But it makes no binding regulations. Political scientist Klaus Armingeon argues that the efficacy of the OECD's highly consistent recommendations – in other words, its power to shape national policy orientations directly – is rather low. The authority of the organization is instead based on an indirect, 'soft' power.

Only as an expert-driven organization and an 'epistemic community' is the OECD successful in convincing policy actors through scientific expertise and thus establishing a common value system at the civil servant level of the member states. One way in which it achieves this is through the statistical 'outlooks' it regularly publishes. By playing this 'idea game' the OECD succeeds in spreading a common definition of economic problems and shared strategies for policy solution and thus influences long-term policy orientations among its member states.[41] In this way – to give two examples – the organization heavily influenced the idea of a 'technological gap' between Europe and the United States in the 1960s, spurring a wave of political measures to promote national innovation systems in Europe; and it fostered the perception of an international crisis in the developed welfare states in the 1970s and 1980s – a crisis based on a perceived conflict between the need for economic development and the detrimentally high level of public spending on social security.[42]

The heyday of the OECD's general influence was from the 1960s to the 1980s. In the 1990s, the European Union (EU) increasingly rivalled the OECD's status as the authority in economic and social policy, at least in the European context. However, the OECD kept its eminent role because it remained an influential think tank that advised the EU and G7, for example, in the fields of employment strategies and labour market reforms.[43] The EU also adopted some of the statistical instruments of the OECD.[44]

Although the priorities of the European Economic Community (EEC) were in economic policy – to promote the Common Market – and not in social policy, issues of social protection were nevertheless important for European integration, at least from the early 1960s.[45] The EEC knew that the lack of a coordinated form of social protection would generate an obstacle for migrant workers and thus harm the project of a single market. Therefore, the EEC treaty of 1957 aimed already at improving the 'living and working conditions' of the European workforce. As early as 1962, the EEC organized a Europe-wide conference on social security, jointly prepared with the International Labour Office and the ISSA.[46] However, the EEC had no formal authority to intervene in national social policy-making.

This changed only with the Maastricht Treaty of 1992, when the European Union received authority to legislate in the fields of labour law and labour protection (but only in a very limited way on social insurance).[47]

Under these circumstances, the activities of the EEC and the EU focused more on coordinating than on directly influencing national social policies. In this, expert knowledge was a crucial resource for European integration. Like the ILO and the OECD, the EEC engaged in comparative research on the social insurance systems of its member states in order to understand the diversity of European welfare states and to be able to single out institutions and practices of social protection it would recommend. To this end, in 1969, the EEC and the Council of Europe founded the European Institute for Social Security, with offices situated in Leuven, Belgium. The Institute first set out to stimulate scientific cooperation among national welfare experts by convening workshops and conferences, and by publishing the results of these debates. Since the late 1970s, the Institute has started to pursue its own research projects, often funded by the European Commission and the Council of Europe.[48]

The different forms of collaboration between the EEC, the ISSA, the ILO, and the OECD represent another important trend in the postwar period: increasing interactions between international and supranational organizations, often based on the exchange of expert knowledge. These emerging organizational networks formed another effective basis for social policy-making on an international level. As an agency of the United Nations, the ILO was anyway in close contact with other UN agencies, such as the World Health Organization (WHO), the Food and Agriculture Organization (FAO), and the United Nations Educational, Scientific and Cultural Organization (UNESCO).[49] The ISSA in particular operated as a platform for bringing together diverse international and supranational organizations engaged in social security. Its general assemblies usually hosted delegations from the UN, the WHO, and the EEC.[50] The ISSA was also a consultant on social security issues, for example, advising the EEC and the Council of Europe on issues such as the protection of migrant workers and the organization of pension and family allowance systems.[51] There was also a regular collaboration between the ISSA and the International Congress of Actuaries, operated through the ISSA's committee of social security actuaries and statisticians who cooperated regularly with the actuarial congress.[52] Other international organizations were also in regular contact, for example, the Council of Europe with the EEC, or both of them with the ILO.[53] There were even first attempts towards an institutionalization of these interorganizational networks. In 1971, the ISSA and the EEC founded a study group of all national institutions of social security in the EEC member states.[54]

Mechanisms of scientization in international organizations and networks

What were the main factors for the spread of expert knowledge and the process of scientization in these international organizations and networks? Based on research into the debates at the International Congress of Actuaries (ICA) and around both the ILO and the ISSA, I would like to highlight three crucial factors: the technical need for expert knowledge; its symbolic value; and the use of expert knowledge as a way of bypassing difficult political decisions.

First, expert knowledge was in high demand for merely technical reasons. The massive expansion of welfare systems significantly increased their dynamic and complexity. Insurance-related sciences, such as actuarial theory and the sociology of social security, were needed to cope with the rising complexity of social insurance schemes, guarantee their present stability, and secure sustainable development in the medium term.[55] Of course, complexity had been an issue in insurance-related research already in the interwar period. Actuaries tried, for example – often in vain – to develop probabilistic calculations to assess the development of risks of unemployment, workplace accidents, and disabilities. These topics were a current theme at the the International Congresses of Actuaries, for example, at the International Congress of Actuaries 1934 in Rome and again 1937 in Paris, where the application of mathematical and probabilistic calculus on the risk of unemployment was intensely discussed, however mainly in a critical way.[56] This work continued into the postwar years, often with only small steps in progress.[57] However, the complexity of social insurance systems increased in an accelerated form over two stages: the first between 1945 and the mid-1960s, the second in the 1970s and 1980s.

In several European countries, the years after the Second World War were marked by fundamental welfare reforms, which aimed at integrating disparate branches of social insurance into a coherent system of social security. These reforms were often accompanied, or followed, by substantial expansions of welfare benefits. Indirectly, this had the effect of considerably increasing their economic and political relevance, mainly because social security expenditure now grew to a significant proportion of government budgets and of gross domestic product. In other words, expenditures on social security now took on importance in the development of public spending and national economies.

This process, its causes and effects, was a current theme at the International Congress of Actuaries and in the International Social Security Association from the 1950s. One reason, discussed at the ICA, was that social security benefits, in particular the expensive old-age pensions, were increasingly calculated so as to replace people's established standards of income at retirement age. As wages rose more quickly than general inflation, the financing and benefit plans of social insurances had to be reformed. In the 1960s and early 1970s, most Western European countries introduced index-linked benefit plans that automatically adapted at least to the level of inflation, often even to the higher level of wages. At the ICA, such index-linked plans were being discussed already in the early 1960s.[58] Other factors for the rising social security costs, as discussed at the ICA, included direct expansions in welfare benefits – for example, the inclusion of new groups such as the self-employed and the agrarian workforce into social insurance schemes – and the increasing costs of public health institutions. The congress also debated the macro-economic effects of the rising budget for social security costs, in particular the inflationary tendencies caused by the growing social expenditures.[59]

Inflation posed a particular problem for the calculation and the investments of financial reserves in social insurances. Actuaries in the early 1950s were eager to avoid a scenario of fundamental crisis like the one in the inflationary years after the First World War when pension systems suffered or even collapsed.[60] In the 1960s, rising rates of inflation, particularly in Britain and Italy, but also in France and other European countries, affected not only pension systems but also, among other schemes, accident and disability insurances.[61]

The international actuarial community was also concerned about the difficulties posed by demographic trends. In particular, the demographic changes caused by the baby-boom generation had to be correctly assessed so that the long-term financial commitments of pension systems could be recalculated. In a pay-as-you-go system – adopted by most public insurance schemes for old age in Western Europe – the baby-boom brought a temporary relief for pension systems. But actuaries at the ICA in the 1950s and 1960s were well aware of the long-term burdens that would arise once the fertility rate decreased again.[62]

The increased complexity of insurance systems called for more sophisticated technical knowledge. Accordingly in the 1960s, research on social security systems considerably expanded, not only at the national level, by national institutions, but also at the international level, in particular at the ISSA.[63] Much expertise focused on developing instruments for planning social security systems. Planning offered guidelines for political decisions in the present, as well as prognostic outlooks for the future. It was also used, by the ISSA, as a synthesizing technique, in order to detect 'blind spots' in social protection and thus to develop further instruments of social security.[64] In addition, for international organizations planning offered a topic over which the political polarizations of the Cold War could partly be overcome. Especially in the field of social security, experts of both socialist and capitalist countries carried out a productive international discussion about the concepts and practices of welfare state planning. With their long-term experience in economic planning, the socialist countries usually took the lead in these international debates. Participants at an ISSA conference in 1972 agreed that the planning experience of societies with a market economy was still lagging behind that of countries with planned economies, although the consensus was that there was no disagreement on the basic questions concerning welfare state planning.[65]

Of course, the emergence of planning approaches goes back to the first part of the twentieth century. In public administration, they were first used on a large scale in the war economies of the First World War (especially in Germany) and, of course, in the planned economies of the totalitarian states of the interwar period, first in the Soviet Union under Lenin and Stalin, later in Fascist Italy and National Socialist Germany. The war economy of 1939–45 brought another boost for the development of management and planning techniques such as operations research and systems theory, in particular in Britain and the US. With the rise of Keynesian economic policy, popularized by the OECD, organizations of the United Nations and other bodies, planning techniques were widely adopted by public administrations of the postwar democracies in Western Europe, notably in Britain, France, and Germany, and of course in the US.[66]

With the support of international organizations, a series of special planning instruments was developed in the 1950s and 1960s for social policy-making: cost-benefit analyses, cost-effect analyses, budgeting systems (in particular the integrated Planning, Programming and Budgeting System, PPBS), and indicator-based planning (which formalized the use of planning for strategic policy decisions) – to name just a few.[67] Apart from actuarial theory and statistics, it was the discipline of sociology that profited most from the planning boom. In the early 1970s, quantitative sociological approaches formed the core of the interdisciplinary field of planning sciences. The ISSA, for example, held an international conference on planning in 1970 at which one hundred sociologists, mainly from Eastern and Western European countries, attended. The conference addressed a variety of topics such as the differences between planning practices in socialist countries (with planned economies) and capitalist countries (with market economies), personnel planning in national health services, or the role of opinion research in the planning of social security organizations.[68]

In the 1950s and 1960s, planning approaches usually helped support a controlled expansion of welfare systems. Since the 1970s, however, planning has been increasingly affected by emerging debates about a welfare state crisis, by the related logic of scarce resources and by a perceived need for social security reforms. In fact, the 1970s were a very ambivalent and contradictory period in international welfare policy debates, characterized by two opposing policy strategies: on the one hand, a move to further expand social insurances – not least as a reaction to the recession of 1974/5 – and, on the other hand, an impulse to start cutting back on the financial resources of the welfare state. In this sense, the 1970s brought social security debates to a new degree of complexity. Although in the emerging debate about a fundamental crisis of the welfare state there were voices criticizing the possibility of scientific prognosis and planning, expert knowledge was not fundamentally put into question. Ultimately it rather profited from the changed situation.

The second phase of expansion of welfare states in the 1970s affected those areas where in the 1960s and 1970s expert networks (such as the ISSA) and concerned policy actors still diagnosed a backlog. Already in the 1960s, many countries had heavily invested in their public health systems, although, by the end of the decade, the cost-effectiveness of the health systems was being questioned.[69] Moreover, new groups had gradually entered the clientele of social insurance – a process that was also closely analyzed and planned by social security experts. Such groups included ones previously marginalized in social policy such as migrant workers and the agricultural workforce. Several Western European countries, in particular countries in Southern Europe, accordingly expanded their social security systems.[70] Another group systematically disadvantaged was that of women as a whole. Women's insurance status often depended on their husbands, which resulted in widespread discrimination against single women and those who had been divorced.[71]

The recession of 1974/5 brought a double blow for most Western European welfare states. On the one side, the shrinking workforce meant that social security contributions, a significant part of the earnings of social insurances, declined. On the other side, expenditures rose, in particular for unemployment insurance but also for early retirement schemes introduced to alleviate the rise of unemployment.[72] This situation spurred a series of international expert discussions about the need for social security reforms and the directions they should take, so as to cope with the changing economic and political conditions of welfare policies. Already in 1976, the International Conference of Social Security Actuaries and Statisticians, an expert organization closely related to the ISSA, had agreed that research would be an adequate instrument to assess the efficiency and effectiveness of social security programmes.[73] Consequently, the ISSA organized a series of research projects and related conferences on the assessment of social security programmes (1975), new pension and retirement schemes (1976), and social security in times of economic recession and inflation (1977). For these endeavours the ISSA could rely on an effective network of its own research staff and of national 'research correspondents', often affiliated with national social security organizations. By 1977, around 40 correspondents participated at these international research collaborations, forming a powerful transfer tool between the international and national levels.[74]

In all these debates expert knowledge, as argued before, helped states cope with the rising complexity of social security systems in postwar Europe. Beyond this direct impact, expert knowledge also profited indirectly from increased difficulties experienced in managing these systems. Disciplines like actuarial theory and statistics also had the *symbolic effect* of enhancing the public's trust in the stability of social insurances, increasing the acceptability of social government interventions, and generally persuading the public that collective social risks were controllable and related social instabilities manageable. In this way, expert knowledge also operated, in the words of Theodore Porter, as a 'technology of trust', an epistemic resource particularly important in socially and politically controversial fields such as social policy.[75] The planning discourse, in particular, was a symbolic communication of this kind, convincing policy-makers and the public of the long-term controllability of their expanding welfare systems, especially in times of perceived crises, such as the mid-1970s and early 1980s.[76] Similarly during the Cold War, in the political confrontation between Western and Eastern Europe, science often proved a useful common language to promote or accomplish a rapprochement in the field of social security.[77] How effective the symbolic added value of the planning rhetoric was is illustrated by the unease that the extensive public confidence in welfare planning caused among some social security experts. Vladimir Rys, a long-term collaborator with the ISSA, and from 1974 its general secretary, warned in an article of 1974 that public trust in planning policies was disproportionate and would inflate planning into a 'panacea to cure all social ill'.[78]

A third mechanism that helped diffuse expert knowledge in international organizations and networks was the use of scientific knowledge as a way of bypassing difficult political decisions. International organizations like the ILO and the OECD did not have much formal power to intervene in fields like welfare policy, where most responsibility lay with national authorities. In this situation, the sheer persuasiveness of scientific expertise offered compensation for the lack of political authority these international organizations had. In other words, the back door of scientization was a way in for the 'soft power' of international experts.[79] At the ILO and the ISSA it was common procedure to delegate controversial topics to an expert committee for further research on the matter.[80] It is not a coincidence that the expansion of expert committees within the ISSA in the 1960s and 1970s coincided with a period in which the political integration of welfare systems was delayed or even blocked.[81] The foundation in 1969 of the European Institute for Social Security was, for example, a direct reaction to the difficulties of political integration within the EEC.[82] The international activism of American Welfare Studies experts like Ida Merriam and Paul Fisher, both directors at the US Social Security Administration, was not least motivated by the increasing political stagnation in their national institutions during the 1970s.[83] Also, from an academic perspective, the political obstacles to integrating the disparate national welfare systems proved a crucial motor for building up and funding the field of Comparative Welfare Studies in the social sciences – a development welcomed, of course, by researchers like Richard Titmuss.[84]

Conclusion

To conclude, I would like to highlight two points about the process of scientization provoked by international organizations. The first concerns the effects of international scientific activities and debates on *national* policy actors. Although this chapter has only focused on the international level, a tentative assessment of this transfer is still possible – though mainly from the perspective of international actors. For international organizations and networks, expert knowledge remained, throughout the twentieth century, a crucial means of getting their message across to national policy-makers – not least because other political instruments that would have been more powerful and effective were not at hand. In social security matters, where the relevant organizations were national and international actors that were rather powerless compared to their national colleagues, the influence of international organizations like the ILO, the ISSA, and the OECD relied on weak instruments, such as epistemic influences, and not on political forms of power. In this sense, international expert organizations and networks were more successful in shaping the conceptual framework in which welfare policy was defined – the political *Zeitgeist* – than in directly shaping national policy-making practices. Over the twentieth century, the organizational centres of these international expert networks shifted, from the International Congress of Actuaries, which dominated the field until the early interwar period, over the International Labour Organization (and its associated International Social Security Association), whose influence was pivotal until the 1960s, to the OECD and the European Union, which rose to prominence in the 1970s and 1980s.

The second conclusion refers to the *geographical significance* of the cases examined. The development of international expert organizations and networks in social policy, as outlined in this essay, is a mainly Western story. In the mid-1970s, the ISSA carried out a survey on the research activities of social security institutions in its member states. The result showed that practically the whole worldwide budget for research on social security was spent by the industrialized nations. Countries of the Third World had hardly any resources for research projects.[85] This is an important reminder that – seen from a global perspective – the process of scientization, although it appeals to universal norms and values, has ultimately been a very regional phenomenon.

Notes

1. D. Rueschemeyer and T. Skocpol (1996), 'Introduction', in D. Rueschemeyer and T. Skocpol (eds), *States, Social Knowledge, and the Origins of Modern Social Policies* (Princeton: Princeton University Press), p. 3; P. Wagner and B. Wittrock (1996), 'Social Science and the Building of the Early Welfare State. Toward a Comparison of Statist and Non-statist Western Societies', in Rueschemeyer and Skocpol (eds), *States, Social Knowledge, and the Origins of Modern Social Policies*, pp. 90–113; L. Raphael (1996), 'Die Verwissenschaftlichung des Sozialen als methodische und konzeptionelle Herausforderung für eine Sozialgeschichte des 20. Jahrhunderts', *Geschichte und Gesellschaft* 22, pp. 165–93.
2. I am following the classic distinction in political science between internationalism, based on intergovernmental interactions; transnationalism, consisting of non-governmental actors; and supranationalism, characterized by international authorities with governmental powers. For a historical perspective, see P. Clavin (2005), 'Defining Transnationalism', *Contemporary European History* 14(4), pp. 421–39.
3. E. Eichenhofer (2006), 'Europäisierung sozialer Sicherung', *Geschichte und Gesellschaft* 32, pp. 517–41, at 517.
4. C. Conrad (2006), 'Vorbemerkung', *Geschichte und Gesellschaft* 32, p. 437 (introduction to special issue on 'social policy transnational').
5. See the exemplary contributions to the special issue on 'social policy transnational' of *Geschichte und Gesellschaft* 32(4) (2006).
6. M. Herren (2006), 'Sozialpolitik und die Historisierung des Transnationalen', *Geschichte und Gesellschaft* 32, pp. 549–55; P. Weindling (1995), *International Health Organisations and Movements, 1918–1939* (Cambridge: Cambridge University Press).
7. W. Anderson (2006), *Colonial Pathologies. American Tropical Medicine, Race and Hygiene in the Philippines* (Durham, NC: Duke University Press); B. Stuchtey (2005), 'Introduction: Towards a Comparative History of Science and Tropical Medicine in Imperial Cultures since 1800', in B. Stuchtey (ed.), *Science Across the European Empires* (Oxford: Oxford University Press), pp. 15–26; D. van Laak (2004), 'Laboratorien der Moderne', in S. Conrad and J. Osterhammel (eds), *Das Kaiserreich transnational, Deutschland in der Welt 1871–1914* (Goettingen: Vandenhoeck and Ruprecht), pp. 257–79.
8. For example, E. P. Hennock (2007), *The Origin of the Welfare State in England and Germany. Social Policies Compared* (Cambridge: Cambridge University Press); A. Mitchell (1991), *The Divided Path. The German Influence on Social Reform in France after 1870* (Chapel Hill: University of North Carolina Press); P. Wagner, C. Didry, and B. Zimmermann (2000), *Arbeit und Nationalstaat. Frankreich und Deutschland in europäischer Perspektive* (Frankfurt am Main: Campus).
9. For a survey of the research on the ILO, see J. Van Daele (2010), 'Writing ILO Histories. A State of the Art', in J. Van Daele, M. Rodríguez García, G. van Goethem, and M. van der Linden (eds), *ILO Histories. Essays on the International Labour Organization and its Impact on the World During the Twentieth Century* (Bern: Peter Lang), pp. 13–40; for the ILO's role in social policy debates in the interwar period, see S. Kott (2010), 'Constructing a European Social Model. The Fight for Social Insurance in the Interwar Period', in Van Daele et al.

(eds), *ILO Histories*, pp. 173–96; T. Cayet, P-A. Rosental, and M. Thébaud-Sorger (2009), 'How International Organisations Compete. Occupational Safety and Health at the ILO, a Diplomacy of Expertise', *Journal of Modern European History* 2, pp. 173–94; see also C. Guinand (2003), *Die Internationale Arbeitsorganisation und die soziale Sicherheit in Europa (1942–1969)* (Bern: Peter Lang).

10. P. Baldwin (1996), 'Can we Define a European Welfare State Model?', in B. Greve (ed.), *Comparative Welfare Systems. The Scandinavian Model in a Period of Change* (Basingstoke: Palgrave Macmillan), pp. 29–44; E. Eichenhofer (2007), *Geschichte des Sozialstaats in Europa. Von der 'sozialen' Frage bis zur Globalisierung* (Munich: Beck), pp. 95f.; Eichenhofer (2006), 'Europäisierung sozialer Sicherung'; and, as more sceptical assessments, the contributions to H. Kaelble and G. Schmid (2004), *Das europäische Sozialmodell. Auf dem Weg zum transnationalen Sozialstaat* (Berlin: Edition Sigma).
11. S. Leibfried and S. Mau (2008), 'Welfare States. Construction, Deconstruction, Reconstruction', in S. Leibfried and S. Mau (eds), *Welfare States. Construction, Deconstruction, Reconstruction*, vol. I: *Analytical Approaches* (Cheltenham: Edward Elgar), pp. xxiv–xxviii; see also: Eichenhofer (2006), 'Europäisierung sozialer Sicherung', pp. 516f.
12. For example, I. Hacking (1990), *The Taming of Chance* (Cambridge: Cambridge University Press), pp. 3–5.
13. M. Lengwiler (2010), 'Signifikanzen der Sicherheit. Intendierte und nicht intendierte Erkenntnisse der sozialstaatlichen Prognostik (1918–1970)', in J. Vogel and H. Hartmann (ed.), *Zukunftswissen. Prognosen in Wirtschaft, Politik und Gesellschaft seit 1900* (Frankfurt am Main: Campus), pp. 33–54.
14. Baldwin (1996), 'Can we Define a European Welfare State Model?'
15. Conrad (2006), 'Vorbemerkung', pp. 440–3.
16. Clavin (2005), 'Defining Transnationalism', p. 425.
17. P.-Y. Saunier (2004), 'Circulations, connexions et espaces transnationaux', *Genèses* 57, pp. 110–26.
18. J. Paulmann (2004), 'Grenzüberschreitungen und Grenzräume. Überlegungen zur Geschichte transnationaler Beziehungen von der Mitte des 19. Jahrhunderts bis in die Zeitgeschichte', in E. Conze, U. Lappenküper, and G. Müller (eds), *Geschichte der internationalen Beziehungen. Erneuerung und Erweiterung einer historischen Disziplin* (Köln: Böhlau), pp. 183–96; M. Geyer and J. Paulmann (2001), 'Introduction: The Mechanics of Internationalism', in M. Geyer and J. Paulmann (eds), *The Mechanics of Internationalism* (Oxford: Oxford University Press), pp. 1–7.
19. M. Herren (2009), *Internationale Organisationen seit 1865. Eine Globalgeschichte der internationalen Ordnung* (Darmstadt: Wissenschaftliche Buchgesellschaft), pp. 12ff., 23ff.; J. W. Meyer (2005), *Weltkultur. Wie die westlichen Prinzipien die Welt durchdringen* (Frankfurt am Main: Suhrkamp).
20. A. Iriye (1997), *Cultural Internationalism and World Order* (Baltimore, MD: Johns Hopkins University Press), pp. 3–6.
21. Meyer (2005), *Weltkultur*, pp. 8f., 47–84, 109–18.
22. Herren (2006), 'Sozialpolitik und die Historisierung des Transnationalen', pp. 549–55.
23. H. Meller (1995), 'Philanthropy and Public Enterprise. International Exhibitions and the Modern Town Planning Movement, 1889–1913', *Planning Perspectives* 3, pp. 295–310.
24. Herren (2009), *Internationale Organisationen seit 1865*, pp. 38–40.
25. *Internationale Revue für Soziale Sicherheit*, 20(2) (1967), p. 8.
26. Kott (2010), 'Constructing a European Social Model', p. 181; Cayet, Rosental, and Thébaud-Sorger (2009), 'How International Organisations Compete'; C. Guinand (2008), 'Zur Entstehung von IVSS und und IAO', *Internationale Revue für Soziale Sicherheit* 61(1), pp. 93–111.
27. Kott (2010), 'Constructing a European Social Model', pp. 182f.; *Internationale Revue für Soziale Sicherheit* 22(4) (1969), pp. 567–70.
28. M. Stack (1967), *Vierzig Jahre im Dienste der Sozialen Sicherheit* (Geneva: ISSA), pp. 15–19. The predecessor of the ISSA, the Conférence internationale de la mutualité et des

assurances sociales (CIMAS), founded in 1927, was dominated by associations of sickness funds and friendly societies, and had little interest in scientific expertise; see Stack (1967), *Vierzig Jahre*, pp. 9f.

29. ISSA (1986), *Im Dienste der Sozialen Sicherheit. Die Geschichte der Internationalen Vereinigung für Soziale Sicherheit 1927–1987* (Geneva: ISSA), p. 41.
30. Stack (1967), *Vierzig Jahre*, pp. 33f.
31. *Internationale Revue für Soziale Sicherheit* 27(4) (1974), pp. 620f.; *Internationale Revue für Soziale Sicherheit* 30(2) (1977), pp. 124f.
32. *Internationale Revue für Soziale Sicherheit* 22(4) (1969), p. 573; Stack (1967), *Vierzig Jahre*, pp. 30–3.
33. ISSA (1986), *Im Dienste der Sozialen Sicherheit*, pp. 43f.; V. Berghahn (2004), *Transatlantische Kulturkriege. Shepard Stone, die Ford-Stiftung und der europäische Antiamerikanismus* (Stuttgart: Steiner), pp. 183–5.
34. Guinand (2003), *Die Internationale Arbeitsorganisation*.
35. D. Maul (2007), 'Der transnationale Blick. Die Internationale Arbeitsorganisation und die sozialpolitischen Krisen Europas im 20. Jahrhundert', *Archiv für Sozialgeschichte* 47, pp. 351–6. For the role of the ILO in social policy debates in the late 1970s and early 1980s: M. Leimgruber (2012, forthcoming), 'Facing the Emergence of the "Crisis of the Welfare State". The ILO and the OECD in Comparative Perspective (1975–1985)', in S. Kott and J. Droux (eds), *Globalizing Social Rights: The International Labour Organization and Beyond* (Basingstoke: Palgrave Macmillan).
36. R. Griffiths (ed.) (1997), *Explorations in OEEC History* (Paris: OECD).
37. R. Woodward (2009), *The Organisation for Economic Co-operation and Development (OECD)* (London: Routledge), pp. 18–20.
38. For a detailed account, see Leimgruber (2012, forthcoming), 'Facing the Emergence of the "Crisis of the Welfare State"'; see also Woodward (2009), *The OECD*, pp. 24–8.
39. M. Marcussen (2004), 'Multilateral Surveillance and the OECD Playing the Idea Game', in K. Armingeon and M. Beyeler (eds), *The OECD and European Welfare States* (Cheltenham: Edward Elgar), pp. 13–31, at 17; S. Sullivan (1997), *From War to Wealth. Fifty Years of Innovation* (Paris: OECD), p. 6.
40. Marcussen (2004), 'Multilateral Surveillance', pp. 21–9.
41. Ibid., pp. 17f.; Woodward (2009), *The OECD*, pp. 15–17.
42. Leimgruber (2012, forthcoming), 'Facing the Emergence of the "Crisis of the Welfare State"'; for the debate on the 'technology gap', see B. Godin (2005), *Measurement and Statistics on Science and Technology, 1920 to the present* (London: Routledge), pp. 32–46.
43. Woodward (2009), *The OECD*, pp. 29–32.
44. Godin (2005), *Measurement and Statistics on Science and Technology*, pp. 32–46.
45. B. Schulte (2004), 'Die Entwicklung der Sozialpolitik der Europäischen Union und ihr Beitrag zur Konstituierung des europäischen Sozialmodells', in H. Kaelble and G. Schmid (eds), *Das europäische Sozialmodell. Auf dem Weg zum transnationalen Sozialstaat* (Berlin: Edition Sigma), pp. 75, 78, 84.
46. H. F. Zacher (1994), '25 Jahre Europäisches Institut für Soziale Sicherheit', *Zeitschrift für ausländisches und internationales Arbeits- und Sozialrecht* 8, pp. 254–66, at 256.
47. Eichenhofer (2007), *Geschichte des Sozialstaats in Europa*, pp. 65–80.
48. Zacher (1994), '25 Jahre Europäisches Institut für Soziale Sicherheit', p. 257.
49. G. A. Johnston (1970), *The International Labour Organisation, its Work for Social and Economic Progress* (London: Europa Publications), pp. 78–82; for a conference jointly organized by the ILO, the UN, and the WHO, see *Bulletin der Internationalen Vereinigung für Soziale Sicherheit* 10–11 (1952), pp. 149ff.; Cayet, Rosental, and Thébaut-Sorger (2009), 'How International Organisations Compete', pp. 174–94.
50. For example, in 1967, see *Internationale Revue für Soziale Sicherheit* 20(2) (1967), p. 119.
51. *Internationale Revue für Soziale Sicherheit* 32(4) (1979), p. 550; ISSA (1986), *Im Dienste der Sozialen Sicherheit*, p. 41.
52. *Bulletin der Internationalen Vereinigung für Soziale Sicherheit* 8–12 (1965), p. 565.

53. All three organizations, for example, coordinated their efforts for the protection of migrant workers in Europe: see *Internationale Revue für Soziale Sicherheit* 31(3) (1978), pp. 315–23.
54. *Internationale Revue für Soziale Sicherheit* 26(3) (1973), p. 367.
55. For a historical survey of the boom years, see L. Raphael (2004), 'Versuch einer historischen Distanzierung einer "klassischen Phase" des europäischen Wohlfahrtsstaats', in Kaelble and Schmid (2004), *Das europäische Sozialmodell*, pp. 51–74.
56. *Transactions of the International Congress of Actuaries* (1934), pp. 170–5; *Transactions of the International Congress of Actuaries* (1937), pp. 47–54.
57. Lengwiler (2010), 'Signifikanzen der Sicherheit'; *Transactions of the International Congress of Actuaries* (1957), pp. 512f.
58. *Transactions of the International Congress of Actuaries* (1964), pp. 26f., 168f., 325f., 336f.
59. P. Fisher (1978), 'Die Krisis der Sozialen Sicherheit. Ein internationales Dilemma', *Internationale Revue für Soziale Sicherheit* 31(4), pp. 423–26, at 424.
60. *Transactions of the International Congress of Actuaries* (1951), pp. 264ff., 342ff., 436ff., 495f. The congress discussed the advantage of pay-as-you-go systems in contrast to funded systems, because funded systems accumulate high reserves that are particularly exposed to the risk of inflation. See also *Transactions of the International Congress of Actuaries* (1960), pp. 301f.
61. *Transactions of the International Congress of Actuaries* (1964), pp. 222f.; ibid., pp. 315f.; I. Merriam (1976), 'Die Ziele der Forschung und Bewertung im Bereich der Sozialen Sicherheit', *Internationale Revue für Soziale Sicherheit* 29(1), pp. 12f.
62. *Transactions of the International Congress of Actuaries* (1957), p. 467; see also *Transactions of the International Congress of Actuaries* (1960), pp. 229f.
63. For the national level, see *Internationale Revue für Soziale Sicherheit* 21(2) (1968), p. 171; see also *Internationale Revue für Soziale Sicherheit* 20(1) (1967), pp. 27–34; ISSA (1986), *Im Dienste der Sozialen Sicherheit*, p. 42.
64. *Internationale Revue für Soziale Sicherheit* 27(2–3) (1974), p. 335; *Internationale Revue für Soziale Sicherheit* 29(1) (1976), pp. 4f.
65. *Internationale Revue für Soziale Sicherheit* 25(4) (1972), pp. 531f.
66. *Internationale Revue für Soziale Sicherheit* 27(2–3) (1974), pp. 329f.; V. Rys, 'Soziale Indikatoren und ihre Anwendung auf die Soziale Sicherheit', *Internationale Revue für Soziale Sicherheit* 26(1–2) (1973), pp. 147–58.
67. *Internationale Revue für Soziale Sicherheit* 27(2–3) (1974), pp. 338–41; Rys (1973), 'Soziale Indikatoren', pp. 144–73.
68. *Internationale Revue für Soziale Sicherheit* 23(4) (1970), pp. 685–9.
69. With respect to the British National Health Service, see *Internationale Revue für Soziale Sicherheit* 25(1–2) (1972), pp. 81–92; also Zacher (1994), '25 Jahre Europäisches Institut für Soziale Sicherheit', pp. 261–3; *Internationale Revue für Soziale Sicherheit* 33(3–4) (1980), pp. 303–5.
70. *Internationale Revue für Soziale Sicherheit* 33(3–4) (1980), pp. 295–371; ISSA (1986), *Im Dienste der Sozialen Sicherheit*, p. 42.
71. *Internationale Revue für Soziale Sicherheit* 29(2) (1976), pp. 127–66; Fisher (1978), 'Die Krisis der Sozialen Sicherheit', p. 426; *Internationale Revue für Soziale Sicherheit* 30(3) (1977), pp. 283–301; for the EU-policies in this area, see *Internationale Revue für Soziale Sicherheit* 33(3–4) (1980), pp. 307–9.
72. Fisher (1978), 'Die Krisis der Sozialen Sicherheit', pp. 424–6.; *Internationale Revue für Soziale Sicherheit* 29(3) (1976), pp. 352–7; *Internationale Revue für Soziale Sicherheit* 30(3) (1977), pp. 295–301; on the early retirement schemes, see Fisher (1978), 'Die Krisis der Sozialen Sicherheit', pp. 427f.; *Internationale Revue für Soziale Sicherheit* 30(3) (1977), pp. 283–301.
73. *Internationale Revue für Soziale Sicherheit* 29(3) (1976), pp. 347–53.
74. ISSA (1986), *Im Dienste der Sozialen Sicherheit*, p. 42.
75. T. Porter (995), *Trust in Numbers. The Pursuit of Objectivity in Science and Public Life* (Princeton: Princeton University Press), pp. 145–7, 186–9, 217–31.

76. See, for example, *Internationale Revue für Soziale Sicherheit* 21(2) (1968), p. 171; with reference to the politicized debate about benefit abuses, claiming to depoliticize this debate through expert knowledge: *Internationale Revue für Soziale Sicherheit* 33 (1980), pp. 452–545.
77. *Internationale Revue für Soziale Sicherheit* 20(2) (1967), pp. 119f.; *Internationale Revue für Soziale Sicherheit* 20(3) (1967), pp. 271–99; *Internationale Revue für Soziale Sicherheit* 25(4) (1972), pp. 531f.
78. *Internationale Revue für Soziale Sicherheit* 30(2) (1977), pp. 124f.; *Internationale Revue für Soziale Sicherheit* 27(2–3) (1974), p. 329.
79. For the late nineteenth century, see M. Herren (2000), *Hintertüren zur Macht. Internationalismus und modernisierungsorientierte Aussenpolitik in Belgien, der Schweiz und den USA, 1865–1914* (Munich: Oldenbourg).
80. For example, at an ISSA conference in 1973 when debating the possibilities of converging national systems of social insurance within the European Economic Community; the differences between the systems of family allowances, pensions, and sickness insurance proved too big, so that the participants at the conference could only agree on initiating more comparative research. *Internationale Revue für Soziale Sicherheit* 26(3) (1973), pp. 366–77.
81. For the ISSA's expert committees, see *Internationale Revue für Soziale Sicherheit* 22(4) (1969), pp. 565–92.
82. Zacher (1994), '25 Jahre Europäisches Institut für Soziale Sicherheit', p. 257.
83. For Fisher, see *Internationale Revue für Soziale Sicherheit* 31(4) (1978), p. 423.
84. Titmuss argued that the diversity of welfare systems in Europe could only be overcome by a thorough scientific analysis of the differences and the opportunities for convergence; see R. Titmuss (1967), 'Die Beziehungen zwischen der Einkommenssicherung und den Leistungen der Sozialdienste – ein Überblick', *Internationale Revue für Soziale Sicherheit* 20(2), pp. 61–72, at 61.
85. *Internationale Revue für Soziale Sicherheit* 30(2) (1977), pp. 148f.

5

Rationalizing the Individual – Engineering Society: The Case of Sweden

Thomas Etzemüller

The 'scientization of the social' noted by Lutz Raphael can also be observed in twentieth-century Sweden.[1] Industrialization, modernization processes, and the global economic crisis created even in the North a need for experts and technologies to guide an increasingly complex society. As in Germany, the level of ideas was dominated by the utopia of a socially harmonious 'national community', which shaped both politics and the work of social experts. Since the purpose of social policy was always to keep the 'national body' 'healthy', it was thus at all times population policy. However, Sweden was characterized by a feature that set it apart from National Socialism, namely a tradition of self-evaluation that was more than a century old. While in Germany in the 1930s a rhetoric of the 'deed' (*die Tat*), of 'decisionism' as a radical rupture, was able to gain political influence, Sweden bet on permanent *calibration*. Although Swedish experts employed the rhetoric of radicalism in an almost inflationary manner, in actuality they rejected every form of revolutionary transformation of Swedish society. In essence, the only radical thing was their hypostatization of a rational approach to problems; from their perspective, all of the solutions they proposed were decidedly free of ideology.

Added to this attitude is the fact that the Swedish population has been, since the early modern period, the object of the state's efficient hold on its subjects; at the same time, though, the population, organized into 'national movements' (*folkrörelser*), became a negotiation partner for this very state. The citizens were continuously asked to participate in the measures through which the 'national body' was to be regulated. The upshot is that we are dealing with a radically utilitarian community for which the conceptual dualism of 'state' and 'society' makes little sense. Sweden, to use Michel Foucault's phrase, is the perfect 'normalizing society'.[2] It focuses less on standardization and disciplining than on the systematic production of knowledge in order to differentiate between 'normality' and problem areas, and to lay the groundwork for interventions in the social body. This kind of society – unlike the 'Third Reich', for example – is simultaneously totalizing in its technocratic grip on the population and anti-totalitarian in its political practice. Where the totalitarian society is coercive, the normalizing society generates and designs programmes in order to structure the social and material order in such a way that people will condition *themselves*.[3]

The first part of the present essay delineates the contours of this normalizing society as a backdrop to the second part, which will outline – using the example of a few Swedish experts (the chief focus being on Alva and Gunnar Myrdal) – how this totalizing/anti-totalitarian thinking is supposed to be put into practice.

Characteristics of Swedish society

Sweden's road to modernity was much like that of the continent. Industrialization did not begin until the late nineteenth century, but then proceeded rapidly. In its wake, social structures, labour conditions, and gender roles were transformed; added to this were urbanization, the beginnings of a consumer society, and the supposed demographic crisis. As was the case throughout Europe, for many Swedes the world seemed chaotic and menacing at the beginning of the twentieth century. Moreover, the political conditions were unstable. To be sure, the Social Democratic Workers' Party of Sweden (SAP) had succeeded in gaining a foothold in the political system well before the First World War, and in 1917 it was able to participate in the national government for the first time. However, until 1932 bourgeois and Social Democratic governments alternated in rapid succession. Then, the Social Democrats, as already in 1924 and 1930, won the national elections, this time with more than 40 per cent of the votes, and once again took the reins of the nation's fate in a coalition with the conservative Farmers' Party. This would prove the beginning of more than forty years of governance by the Social Democrats. They addressed the economic–social problems with decidedly Keynesian policies, and in a remarkably short time they laid the groundwork for what has ever since been considered one of the world's model welfare states.

Welfare state

However, the foundation of this welfare state programme had already been created in the late nineteenth century, when liberal politicians sought to reform the outdated poor relief through social legislation. Social aid was to service as an aid to self-help, to enable the recipient to return to a life of gainful employment. In the process, a distinction was always drawn between 'deserving' and 'undeserving' benefit recipients, and only a minimum benefit was granted so as to prevent a 'recipient mentality'. This kind of social aid was to deter those who did not need it, to provide merely a safety net of last resort for the truly needy, and increasingly also to function as coercive education for alcoholics, juvenile delinquents or 'asocials'. In spite of this paternalistic tilt, important sociopolitical institutions and laws go back to this period. Largely (but not entirely) new in subsequent social democratic policies was the generalization of social benefits. In principle, these benefits were to guarantee all citizens basic material security according to standardized, uniform assessment criteria, be tax-financed, and exert a prophylactic effect. The focus was initially on mothers, children, and 'respectable' workers, who were to be assisted in difficult phases of life that were precisely circumscribed and temporary – motherhood, childhood, unemployment. It was a long way still to the complete inclusion of all population groups, but a universal welfare state moved within reach in the second half of the 1930s.[4]

'People's home'

To this day, one metaphor has served as the guiding image for this idea of the welfare state: the 'people's home' (*folkhem*). The term is not a social democratic invention, but was used by conservative intellectuals beginning in the late nineteenth century. Yet it was Per Albin Hansson, the popular Minister President from 1932 to 1946, who seized the power of interpretation in a speech to the Riksdag (parliament) in 1928. In a few sentences Hansson adumbrated a social ideal that is still valid today:

> Communality and the feeling of solidarity form the foundation of the Home. The Good Home does not know any privileged or disadvantaged individuals, no spoiled children and no stepchildren. There one person does not look down on another, nobody tries to gain advantages at the expense of others, and the strong does not oppress the weak and exploit him. In the Good Home there is equality, circumspection, cooperation, and helpfulness. Translated to the large national and citizens' home, it would mean the tearing down of all social and economic barriers that now divide the citizens into the privileged and the disadvantaged, into rulers and dependants, into rich and poor, the well-off and the impoverished, exploiters and exploited.[5]

Not only did this metaphor embody the utopia of a socially just society, it also helped at the time to solve a weighty ideological problem affecting all social democratic parties, namely the question concerning revolution or reform. The Swedish Social Democrats had already bid farewell to their class warfare programme in the 1920s, but for that reason they were in need of a new, potent terminology. With the 'People's Home' they were able, from the 1930s, to contest the elections credibly and successfully as a 'People's Party'. Not only were workers exploited (by manufacturers), went the theory, but the entire nation should be seen as a victim of the capitalist system. In spite of formal and political equality, social and economic conditions in Sweden continued to be characterized by glaring injustice; democracy was therefore not yet complete. As the strategy for a solution, the SAP offered its voters the creation of the welfare state as a way of providing the people with material security over the short term. By contrast, the party had long since abandoned a nationalization of enterprises, regarding state economic planning as a more effective instrument. In this way society would harness the productive forces of capitalism, gradually displace it, and realize the goals of socialism over the long term through evolutionary means.[6]

Evaluation

Although the Social Democrats articulated their policy in a radical language, they pursued a moderate course in order to draw in their opponents. To be sure, there was always political resistance against far-reaching plans of nationalization, but even private entrepreneurs, thanks to the Swedish culture of consensus, were willing to compromise with the workers' movement. Beginning in the 1930s, the Social Democrats were able to integrate broad segments of the Swedish population because they promised a realization of the People's Home via consensus. In this way the SAP proved – in contrast to the situation in Continental Europe – that the parliamentary system and a social democratic party, too, were capable of an effective economic policy that overcame the effects of the global economic crisis. Therein was grounded the sustained attractiveness of the 'Swedish model',[7] which was to be glorified a few decades later as flexible and pragmatic 'functional socialism'.[8]

The political pragmatism of social democracy corresponds to that of the political system as a whole. The tradition of an efficient bureaucracy in Sweden goes back to the seventeenth century, when the desperately poor country had to finance an effective military machine. Then in the early nineteenth century the institution of the 'Statens offentliga utredningar' (SOU or Official Government Reports) was created, which became *the* symbol of an efficiently working state machinery visible to everyone. These reports are created in a standardized process: every imbalance in the machinery of the Swedish state that becomes sufficiently apparent attracts a commission of inquiry that examines the problem intensively and makes recommendations for adjustment. The commission in question is appointed by a minister. In its smallest form it is composed of a single expert, who concludes his examination after a brief period with a relatively short report. In the case of deeper problems, it is made up of one or several chairpersons, who define several areas of inquiry and bring in the relevant experts; such commissions can work for more than ten years and generate reports numbering into the double-digits. What is characteristic is that the experts come not only from the fields of science, the economy or politics, but that representatives of social groups are regularly included in the inquiry or are asked for their expert opinion on the completed reports. This is meant to ensure that no single political camp dominates the work, that all segments of the population affected by a report are included, and – simultaneously – that a problem will be grasped in its complexity by placing it in relationship to the most varied factors. The material is transformed into statistics and graphs, after which detailed, alternative solution paths are designed and weighed by assessing future trends. Extensive inquiries conclude with a special, summarizing final report. The goal of every inquiry is to *transform* society; the SOU embody a permanent dynamic of reform, which productively picks up even conservative thinking because it ensures against *exaggerated* changes. The countless volumes embody almost physically a thinking that aims at the permanent self-evaluation of the political system.

Population question

The Population Commissions of 1935 (*Befolkningskommissionen*) and 1941 (*1941 års befolkningsutredning*) show in exemplary fashion how state, social policy, and *folkhem* were interconnected.[9] Like many countries in Europe, Sweden, too, feared a dramatic population decline in the 1930s because birth rates were dropping continuously. The war heightened concerns about the 'demographic catastrophe'. The two commissions were charged with determining how the demographic collapse could be averted. The *Befolkningskommission* and the *Befolkningsutredning* were made up of physicians, sociologists, politicians, economists, household experts, statisticians, educators, and so on. By 1938 and 1945, respectively, they presented 17 reports intended to thoroughly examine all concerns of Swedish society at the macro and micro levels: the depopulation of the country, future labour needs in agriculture, the projected population trend of various regions and the cities, Sweden's economic geography, the distribution of smaller and medium-size industrial enterprises and of the various occupational groups in the country, the transportation network, crèches, pre-schools, school lunches, the living conditions of different social strata, the family structures in the country, the relationship between family income and expenses, standards of food, clothing, and housing, the degree to which housework was technologized, cooking habits, maternal mortality, and the incidence of illness among children in families or crèches.[10] Sweden was treated as an all-encompassing whole, in which the most varied elements were to be optimally arranged with respect to one another so they could function together. No area seemed unimportant: the false decisions by a director of a steel plant and a poorly trained domestic helper at the sink were studied in the same painstaking manner to make it possible to plot the possible state of the country's shortcomings.

The population question imparted a new quality to the 'scientization of the social' in Sweden. Countless actors and institutions involved in social policy in the 1930s found here a crystallization point for their efforts to reform the country. This created a climate in which the detailed mapping (*kartläggning*) of the entire nation seemed all but necessary in order to be able to rebuild the land in a common, positive endeavour. In this way the boundary between 'private' and 'public' was blurred. For if it came down to literally the smallest movements of people, the latter had to alter their behaviour should it prove to be socially counterproductive. From this perspective, Swedish social and population policy assumed a different kind of importance. For social policy arose not merely from ideas about caring for the citizens of the state. To be sure, the social democratic welfare state committed itself to helping people in need and to improve their *conditions* of life. At the same time, however, it saw it as its task to regulate how they *lived* their lives, whether through extensive housing inspections, the rationing of alcohol or abortions. A broad palette of measures and interventions was available to oversee and correct the social behaviour of individuals.[11] Social policy remained a paternalistic, moral regimen. And it was deployed as an instrument of modernizing policy in order to turn the population into a 'healthy' and 'new' populace and to equip it for the demands of industrial modernity. That is why it also had specific consequences in terms of population policy.

Eugenics

If one situates Swedish population policy – in terms of discourse analysis – within the European population debate since the late eighteenth century, one encounters similarities.[12] With respect to social policy, it was legitimate also in Sweden to exclude 'asocials' or individuals who were merely insufficiently adjusted. They supposedly were a potential burden on social funds that were still not very extensive and a threat to the biological stock of the 'national body', because they had an above-average number of children and passed on their psychic, genetic, and social defects; this was a demographic development that would intensify to the point of biological collapse. That is why eugenics was not simply the 'dark flipside' of Sweden's social policy but, in the eyes of contemporaries, part of an unsentimental health programme that was supposed to benefit everyone.[13] Like the other Scandinavian countries, Sweden therefore passed two sterilization laws (1935 and 1941), which regulated the compulsory sterilization of 'feebleminded' persons.[14] All told, between 1935 and 1975 more than 60,000 individuals were sterilized in Sweden, about 20,000 of them against their will.[15]

The crimes of National Socialism that became apparent early on barely influenced the discussion in Sweden. Nobody related the crimes of the National Socialists to the ever more extensive practice of sterilization in Sweden, for it seemed only too obvious that a supposedly helpful eugenics had been 'abused' by the German regime. No people were killed in Sweden, thus nobody had to deal with crimes of that nature. No one questioned the right to life – 'merely' the right to *give* life.[16] Some of its democratic neighbours – Denmark, Norway, and Finland – had passed their own sterilization laws earlier than Sweden. Swedish politicians had drafted only general framework laws, which were supposedly oriented simply towards practical needs, not ideologies. The sterilizations were carried out by sober practitioners in clinics, overseen by incorruptible medical authorities.[17] Here, too, nobody seemed to be guilty of wrongdoing. With that, the sterilization question had been removed from both political debate and ethical discussion. After all, eugenics clearly served purely 'humanitarian' goals, namely protecting society from biologically and/or socially 'threatening elements'. When the sterilization law was amended in 1976 and based thereafter exclusively on voluntary participation, it happened without any public debate. The media picked up the issue only at the end of the 1990s, and a state commission of inquiry charted in detail the Swedish sterilization practice and the suffering of victims.[18]

'Society'

What, then, does 'society' (*samhälle*) mean in Swedish? If we start from the influential distinction by Ferdinand Tönnies between an 'organic' *community* and an 'atomistic' *society*,[19] the term should be situated more on the side of 'community'. Both society in the narrower sense and the state merge into the *samhälle*. The term is, on the one hand, quite undifferentiated because it is no counter-term to the 'state',[20] and, on the other hand, comprehensive because it fuses people and institutions into a collective: individuals, families, neighbourhoods, and localities are cells of a large communal organism.[21] This *samhälle* is the result of a grass-roots democratic process on multiple levels that has been going on over a century, which has given rise to a corporatistic and collectivist society in which (still today) all societal groups are in principle closely woven into society's mechanism of self-direction.[22] One should mention the countless study groups (*studiecirklar*) in which the Swedes continue their education; the large popular movements – workers' movements, Free Churches, abolitionists, but also cooperatives, the gymnastics movement, and so on – in which hundreds of thousands of Swedes organized themselves and learned to exert political influence; the state inquiries (SOU), on which the specific social groups concerned were (and are) asked to provide well-founded comments; or the establishment of corporatist institutions.

This is the reason why it would never have occurred to Swedish politicians – even in the 1930s – to *mandate* behaviour, since Swedish society generated internally, in a continuous evolutionary process of self-evaluation, those principles of utility to which individuals were expected to submit because that was the reasonable thing to do. No system was further from a totalitarian dictatorship, and hardly any other system allowed the state so readily to carry out profound interventions in the integrity of individual people if they refused to embrace utilitarianism, to the detriment of society. In this way the linkage of welfare state, *folkhem*, and eugenics in the 1930s and 1940s could readily be seen as democratic. The *folkhem* metaphor transformed traditional everyday and political practices of the Swedes into a potent image. In conjunction with the SAP's welfare state project, there arose the utopia of a just and harmonious society, which was to be gradually realized with the tried-and-true, rational techniques of steering society.

By this means the negative social consequences of modernity were to be absorbed, while the people were to be simultaneously put in the physical and psychological condition to appropriate modern life. The Swedes *wanted* to be a modern nation – not even a pre-modern rhetoric was popular. Traditions were preserved (highly successfully in terms of museum pedagogy) so as not to lose sight of the nation's origins. But this modernity was perceived as an ambivalent phenomenon: the boundary between utopia and disintegration, between healthy and sick, and between normal and abnormal seemed precarious and had to be defended with all means. That is why Sweden was not a totalitarian but a radical normalizing society, and why exclusion was an inseparable part of sociopolitical inclusion in society. Thus in the *folkhem* there was 'no governing; rather, the common good was realized with paternalistic authority, with a loving hand – the Leviathan as paterfamilias'.[23]

The example of Alva and Gunnar Myrdal

Following this general sketch, I will take a close-up look at the role of experts within Swedish normalizing society. The example of Alva and Gunnar Myrdal can be used to illuminate more closely the interconnection of science, social policy, and a sociopolitical utopia and its implementation. I will concentrate on three areas: housing, child rearing, and consumption; the women's question factors into all three as a fourth area. First, however, I will provide some biographical information.

Alva and Gunnar Myrdal

At the beginning of the 1930s, Alva and Gunnar Myrdal had a longer period of searching behind them. Both had been born around the turn of the century into families of the lower middle class. Alva Myrdal was able to pass the A levels (*studentexamen*) in the provincial town of Eskilstuna only with great effort and thanks to her tenacious will. Over the years she pursued a patchwork of studies at university, which never led to the planned dissertation. Gunnar Myrdal studied law in Stockholm. The two met in 1919 and married in 1924. In their letters they quickly began to imagine themselves as the ideal couple, more rational and progressive than most other marriages, an intellectual symbiosis equal to none. Alva admired her husband's genius and subordinated herself to him without giving up her own independent intellectual life. Gunnar admired his wife as a first-rate intellectual 'sparring partner' who followed her own path. In the end, though, he insisted on the traditional gender roles. Already in the early years this attitude led to intense conflicts, which tested the marriage time and again.[24]

Moreover, Gunnar felt anything but self-confident. For a while the future Nobel Prize winner and potential General Secretary of the UN dreamed of a 'career' as a judge and mayor of the small town of Mariefred, until Alva – allegedly – pushed him to study economics. In 1927 he attracted attention in Sweden's small intellectual circles with his dissertation.[25] In 1929, both of them – each with a Rockefeller fellowship – went, first to Great Britain, and then to the United States. Great Britain and the Fabians did not leave any larger traces in their thinking. The United States initially struck them as uncivilized. They were shocked by the social effects of the global economic crisis on people, and Gunnar considered the customs and habits of his American colleagues ridiculous. But the better they got to know the country, the more they became staunch (if critical) prophets of the American way.

With their return from a stay in Geneva, their search came to an end. They became public intellectuals and represented like few of their contemporaries Sweden's departure into modernity. The media quickly packaged them as the exemplary embodiment of an 'American' couple, as the paradigm of an emancipated, modern marriage. Writing their texts jointly, they were seen as talented, knowledgeable, polarizing speakers who took a position on all sociopolitical questions of Swedish society. It was from this perception that the two derived most of their public and political impact. Conversely, they used the media with great skill and purpose for their concerns. They were controversial,[26] but by the beginning of the 1930s they had found their role as social engineers.[27]

Social engineering

The concept of *social engineering* was supposedly coined by William Tolman. He meant an ideal, reciprocal relationship of capital and labour, whereby the asserted reciprocality boils down in the end to a conditioning of workers in the interest of entrepreneurs. Christine Frederick saw something similar in *household engineering,* namely the scientization and rationalization of the household in order to relieve overworked housewives – and as a chance to prove how indispensable they were.[28] Of course, the genealogy of the concept goes back not only to Henry Ford and Charles Taylor; Ferdinand Tönnies also come into play here, since social engineers pursued the rationalization of life in order to replace the 'atomized' *society* through an 'organic', harmonious *community*. This probably concerned the entire Western world;[29] Sweden was the ideal laboratory of such ideas of order, and the Myrdals were among the most adept lab technicians. They stood for the departure to modernity, which they wished to guide deliberately.

'Crisis in the population question'

In 1934, Alva and Gunnar published a book that caused a stir, and made them known beyond Sweden including the neighbouring Scandinavian countries.[30] In *Kris i befolkningsfrågan* ('Crisis in the Population Question'), they – and especially Alva Myrdal – developed their large questions of social policy. The book outlined the dream of a normalizing society. On the surface the book was about the 'population crisis', which was being heatedly discussed at the time also in Sweden;[31] the real topic was the organization of society. According to their diagnosis, women were no longer having children because they did not wish to give up their professions or jobs, and families remained childless because they could not afford offspring. That is why society had to be radically reformed: first, production should be socialized in order to make the economy more efficient and boost the state's revenues; secondly, a systematic policy of redistribution in favour of potential families had to be initiated; thirdly, rationalized housing construction should relieve working women of housework and thereby make marriages more harmonious; fourthly, child rearing should be collectivized to provide relief for women, but also to teach children social behaviour from an early age and to systematically supervise them with an eye towards undesirable developments; fifthly, consumption should be socialized, for only sensible consumption minimized the waste of resources from the production of baubles and luxury goods. At the end there would be a 'new man', who would overcome the 'false individualism' of modernity, the throng of 'atomized egotists'; a healthy, collective fellow citizen who was able to fit – subjectively harmoniously and objectively in the optimal way – into the social life of society. In this way there would arise, in contrast to the National Socialist subjugation, a true community. Only it offered the guarantee of a growing number of qualitatively desirable children. Eugenically undesirable offspring was to be prevented through abortions and – if necessary, extensive – sterilization campaigns. Only those children who had a chance of good living conditions should be allowed to come into the world.[32]

The intensive media debate in Sweden, as well as Danish and Norwegian translations of the book, turned it into the trigger of the previously mentioned government inquiry into the population question and an equally important study of the housing question, in which the Myrdals played a leading role, and through which they were able to exert considerable influence on practical social policy.[33] What were these reforms supposed to look like in detail?

Collective housing

Sweden owed its ability to cope successfully with the global economic crisis of the 1930s to an extensive investment policy. An important element of this policy was government-supported housing construction. And in this area so-called functionalism played an important role. Functionalist construction was the symbol of the dynamism, clarity, and rationality of modern life, and it was to transform people in this sense.[34] It is therefore no surprise that the Myrdals saw in housing construction a crucial element of their sociopolitical blueprint.

Together with the architect Sven Markelius, Alva Myrdal designed a 'collective house' in order to realize what she had previously planned theoretically with her husband. The advertising brochure shown in Figure 5.1 illustrates the plan of the house and the (still fictitious) life forms of its residents.[35] 'The inferno of the individual household' is highlighted – in cell after cell, every woman is cooking for herself alone (left). Next to them (right) is a collective house, which on the outside has the same cell structure, but in the eyes of these architects this is a positive, because 'rational', form of serialization that will contribute to a sensible social order.

What took time and prevented women from gainful employment was collectivized on the ground floor: cooking, laundry, and child rearing. The individual units, by contrast, were tailored to the individual habits and desired furnishings of various residents: small families, singles, couples without children; workers, single-mother opera singers or intellectual life partnerships. The laundry went to the laundry room via chutes, children were sent to the 'collective nursery'

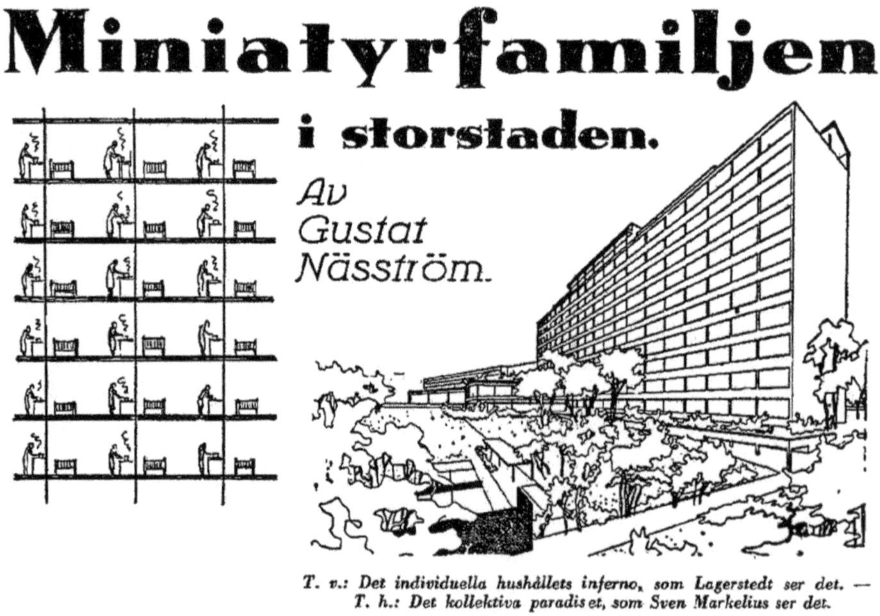

Figure 5.1 'The inferno of the individual household', *Stockholms-Tidningen*, 12 October 1932

(*storbarnkammare*), food was transported up through the elevator and heated in the tiny cooking nook. As it was, the residents were supposed to use the common restaurant on the first floor. The individual apartment constituted an individual space to withdraw to, but it was too small to isolate oneself for long periods from the community.

Standardization without uniformity, and collectivity and individuality – these were the two poles between which the ambivalence of integration and freedom was situated. The children were collectively socialized during the day; in the evening the parents were allowed to bring them into the apartment. Thereafter they were free to let the children spend the night with them or send them back to the children's section.

In this project, the house was to intersect as political metaphor (*folkhem*) and a real dwelling (collective house). The utopia of the *folkhem* was to be concretely realized with the collective house. The ideal of a conflict-free family community within the small space of the private home intertwined with the Social Democratic project of a conflict-free national community within the space of the *folkhem*. The idea was attractive but impracticable. Alva Myrdal participated in the building of only one of these houses, and herself moved with her husband into a functionalist villa. Although several collective houses were constructed until the 1980s, economically they barely sustained themselves. They symbolized the utopia of the *folkhem* more so than actually realizing it.[36]

Child rearing

Before Alva Myrdal was able to embark upon her career in international politics, she made a name for herself as an education expert. In 1935 she published her book *Stadsbarn: En bok om deras fostran i storbarnkammare* ('City children: A book about their upbringing in the collective nursery'). With her reflections, she inserted herself successfully into the discussion at the time about the upbringing of children:[37] she was a member of various organizations that dealt with child rearing and acted as an important instigator of the debate. Jointly with the chief architect of the large construction cooperative 'Hyresgästernas Sparkasse- och Byggnadsförening' (HSB) (Savings and Construction Association of Tenants), she planned the establishment of nurseries in all houses of the HSB.

The question of children's nurseries was a central issue for Alva Myrdal, since she posited the existence of a two-fold problem. She believed, for one thing, that children in modern industrial society had become a problem for adults. Financially a burden, they also prevented women's working life and independence – and thus their integration into modern industrial society. For another, adults had also become a problem for children. They clung to completely obsolete methods of child rearing and thus prevented the integration of children into modern industrial society. For Alva Myrdal, the maxim that mothers knew best what was good for children no longer held true. Mothers did not have to be taught to love their children, but to love them in an 'intelligent' way. Consequently, a professionalization of child rearing was called for, that is, the systematic training of parents and educators.[38]

Her plan propagated a positive pedagogical ideal. Education should be enjoyable for children, and it should be child-friendly. Boys and girls should be educated together. Girls should learn woodworking, boys sewing. Although engrained habits should be imparted to them, they should be allowed to engage in self-directed play. In principle they should be rewarded, and punished only in an emergency. If they were to be punished, then it should be with isolation, not blows. What made punishment harmful, according to Alva Myrdal, was the personal aspect, which created resentment and social unrest. Better than punishment was diverting children from 'undesirable activities'. Of course, normalization corresponded to this thoroughly modern approach. In nurseries and schools, the next generation was to be taught the fundamentals of a rational way of life, that is, the importance of daily hygiene, a healthy diet, and physical fitness. Teachers and (school) physicians were to note and correct every mental, social, and health-related deviation. Every child should be given a health card on which all observations were recorded, along with information about the parents and the living conditions at home. The goal was healthy bodies, healthy psyches, and healthy social behaviour.[39]

However, the educators were to create only a framework. It would be the children themselves who would educate one another. For in a group, Alva Myrdal maintained, even problem children learned to behave in conformity with the norms; if not, they would be – temporarily – excluded.

> If [the child] does not follow the rules of the game, it is not allowed to play. This is the basic theme of our entire social life, which the child teaches itself. ... The *social upbringing* to which the comrades expose one another is the most effective of all [methods of upbringing].[40]

Prohibitions and taboos would no longer rest on arbitrary, emotion-contaminated injunctions from parents; rather, they would be rooted in the collective and represent traffic rules that operated 'naturally'. Which child would violate rules it participated in creating? As a result, self-mastery and emotional control would be rehearsed almost by themselves. Children should be allowed to develop their personal strengths and interests in order to place them at the service of society.

Precisely because the goal was to increase vitality *and* productive capacity, education could not be a disciplinary factory system. Frowned upon was a collectivism that squeezed all children into the same mould. With these convictions, Alva Myrdal's notion of a democratic, impersonal education in the form of general rules becomes comprehensible. Nurseries and schools should empower children as early as possible to normalize themselves. 'The art is to prevail upon the child to adopt, from internal motivation, without consuming conflicts, the general rules for our conduct that are necessary in any case.'[41] Her vision went so far that children should be made receptive for a future family life already in preschool, by familiarizing them with the principles of child care through practical exercises (diapering babies and so on).[42]

Consumption

The regulation of consumption was also supposed to play an important role in social policy, something for which there were undoubtedly solid reasons of economic policy and everyday practicality. To be sure, the Swedish welfare state attracted international attention as early as the 1930s, but the Swedes were by no means living in prosperity. Many were wrestling with considerable material problems. That is why the Myrdals wanted to subject the manufacturing sector to state guidance. Industry should be prevailed upon to produce goods of daily consumption in large quantities and cheaply, but of good quality. The economy could be stimulated via mass production, and quality goods were to provide also poorer fellow citizens with the necessities. A small sector of luxury goods would be open to the wealthy.

There was, of course, more at stake: 'beauty for all' had been a common motto in Sweden since the early twentieth century. Tasteful objects and home furnishings were to elevate people morally and promote life within a harmonious community.[43] In this process good taste was not to be commanded from the top down. Beginning in the 1930s, the Swedish were to learn, in countless discussion circles, from the objects themselves how to set a tasteful table or furnish and decorate an apartment in a way that was pretty and reasonable. At the same time, the course literature, drawn up by experts, simultaneously assured that the freedom within an established framework was never exceeded. Divergent opinions and tastes were certainly sought after, but they had to be separately justified. The goal was to rationalize consumption so as to fit people into a rationalized, modern industrial society – while at the same time offering them better living conditions, because they were put in a position where they could recognize quality and acquire quality goods for a low price.[44]

Simplicity, clarity, and beauty were to characterize the interactions in all things. In this kind of environment, people would be transformed for the better. Of course the Myrdals – once again Alva chief among them – went another step beyond that. Through consumption, the contingency of an 'ambivalent modernity' (Z. Bauman) was to be tamed in a democratic way. For products to be mass-produced at reasonable prices, they had to be largely standardized. Standardization meant a reduction of the product palette to a number of basic types. At the same time, though, standardization allowed consumers a personal variation upon these basic types, which meant that there would never be uniformity.[45]

This example, in particular, illustrates vividly that the Myrdals were not primarily concerned with uniformity and oversight, but with the clarity of the social order through the clarity of forms – *without* destroying individuality. On the societal level, standardized types were supposed to reduce 'random' variety and thus confusion. But the variation of these types on the personal level, thanks to conscious choice, should allow for *some* variety and thus individuality – contingency permitted in a controlled manner was balanced by contingency reduced in a controlled manner. In this way, a framework and freedom were to be combined such that the boundary between 'order' and 'chaos' always remained marked, that is, recognizable, and therefore also defensible.

The individuals concerned were expected to practise consumption. Trained through systematic educational work (*konsumentupplysning*) in the capacity to distinguish junk from quality, the population was supposed to put industry under pressure. Once again the focus was on women, for it was they who shopped and had to put products to everyday use. They were styled – and not only by the Myrdals – into experts of everyday life.[46] Their expertise was to become the guiding standard in the production of consumer goods. This was more than mere lip service, since women were in fact in charge in several research institutions for household products. Their comments and judgements were able to influence industry through extensive tests.[47] Here, too, Alva Myrdal thus inserted herself into an already ongoing debate in order to propagate her overarching sociopolitical ideas; here, too, she provided impulses, especially in the 1940s.

Freedom and community

Alva and Gunnar Myrdal were never able to dominate the sociopolitical discussion of the process of the scientization of the social. But they were among the Swedish experts who had an especially strong presence with their ideas. They owed this to their immense productivity and to the strategy of making their voices heard in very divergent fields. They published scholarly monographs – and above all many books with great public resonance, works that are still familiar today. The books were not easy reading, but they found their way instantly into the citation machinery of Swedish journalism. With all journalistic media of the country at their disposal, they wrote countless articles in which they propagated their ideas. As members of the Social Democratic party, they gave campaign speeches; as members of the Riksdag and the government, that sought to exert direct influence on politics. As experts in the government commissions of inquiry (at times in charge of them), they devised concrete political approaches to solving problems. Added to this were their top posts at the UN, through which they consolidated their international reputation, which they had already begun to establish in the late 1930s; in turn, this increased their political weight in Sweden. That they later also made use of television almost goes without saying, and the two Nobel Prizes did their part. Although other Swedish experts had mastered the journalistic keyboard with similar virtuosity, hardly anyone was able to employ the entire range, from campaign speeches in the Swedish provinces to the Nobel Prize speech in front of an international audience.

While the Myrdals' world of ideas may seem shocking today, there are three things one must not fail to recognize. First, in spite of their public and political importance, they were able to implement their ideas only imperfectly. Especially Gunnar Myrdal failed in his attempt – as did so many intellectuals – to cross over the boundary separating scholarship and politics. He was hardly able to implement his ideas about economic policy as trade minister;[48] the couple's work on social policy ran into opposition in the political realm, because powerful ministers established very different priorities.

Secondly, they were never concerned merely with the collective and with controlling people. Especially Alva Myrdal, because of her own life experiences,[49] aimed emphatically at allowing women and children to realize their own interests. Emancipation *through* self-normalization was the paradoxical project.

Thirdly, these were not simply the utopian ideas of two popular (and controversial) intellectuals. Rather, they certainly met with agreement, for throughout 1930s Sweden one encounters techniques for rationalizing and objectifying everyday life through gymnastics, (social) hygiene, sensible recreation, settlement construction, or the cooperative organization of consumption. This was the unquestionably optimistic attempt to shape individuals into a collective that organized its life more rationally, sought to avoid society-eroding conflicts, and at the same time lived more healthily in order to protect the 'national body' from biological decay. That is also why the social policy of Swedish social democracy was not aimed solely at establishing the *folkhem*. It seemed, as already mentioned, that a 'healthy' people was the prerequisite for making society into a 'home' for everyone. For that reason, eugenic set pieces pervade wide segments of sociopolitical thinking. Extending all the way to the 'milk propaganda', a distinction was made between A and B people and individual health was related to the health of the nation.[50] The goal was always to use the technical means of modernity to design a space of normality, and to enable people, by virtue of an inner understanding, to shape this space. 'Freedom' was to find its place *within* the 'community', not endanger it.

Figure 5.2 shows, using agriculture as an example, how social engineers imagined the ideal movement from an inefficient past to a lucid future. Division into small units, low productivity, and the rudiments of a collective way of life characterize pre-modern agriculture (top). In the present, land consolidation makes possible a higher productivity, but the village community has been dissolved into individual farms, which is a cost-intensive construction method (middle). In the future: 'Small enterprises combined into large-scale production – share-based agriculture [*andelsjordbruk*], cooperative agriculture [*samjordbruk*] or an equally strong mechanization, high production. Concentrated residential buildings. Collective society' (below). Note the extremely tidy space, in which there no longer seems to be any room even for a pond and a brook.

Conclusion

One final example illustrates this paradoxical relationship between freedom, contingency, and framing. In the mid-1940s, another notable social expert, the leading city planner Uno Åhrén, also took aim at supposed problems of social integration. The development of Sweden did not fit with an orderly society. Some places were too large, others too small; their economic and population structure had become imbalanced. People built their houses wherever they wished, which reflected their existential disorientation. By contrast, the ideal *samhälle* – a term that describes here, alongside 'community' and 'society', an ideal communal structure – consists of 5000 residents (for industrial cities c. 20,000–50,000 residents) with a balanced demographic, age, gender, labour productivity, and economic structure. All buildings and zones (living, economic activity, leisure) are arranged in such a way that traffic movements can be reduced to a minimum. For the circulation of people

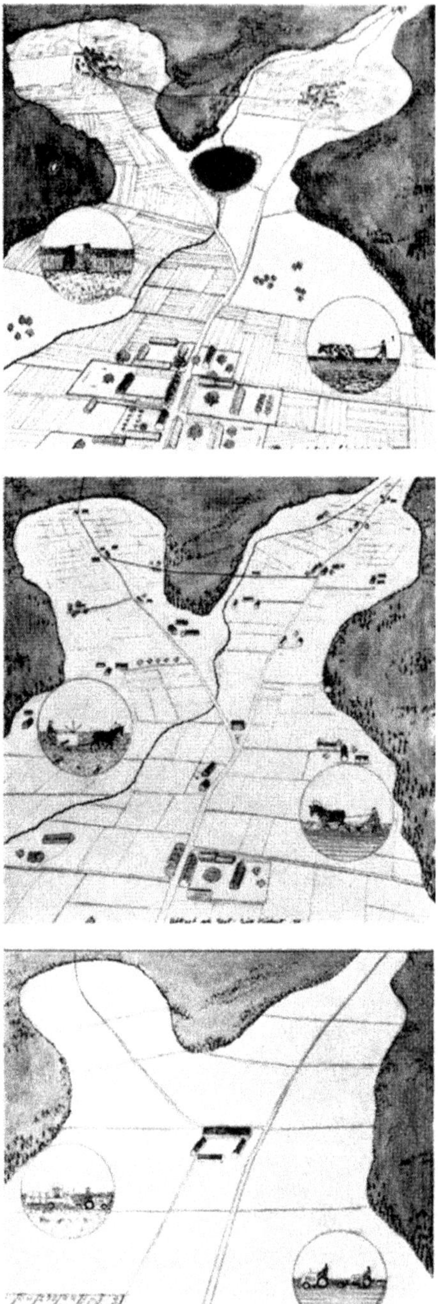

Figure 5.2 T. Åkesson (1944), 'Landbygdens bostäder', in *Bostadsförsörjning och sam-hällsplanering. Två radioföredrag av Uno Åhrén och Torvald Åkesson*, Stockholm: Svenska riksbyggen (© Riksbyggen Press, Stockholm)

seemed to Åhrén a waste of human capital and resources: spatially caused social mobility, accompanied by a change in values, that is, by social uncertainty. Only stable, manageable 'neighbourhoods'[51] allowed for the development of a democratic type of person characterized by personal initiative and a willingness to cooperate. Only there could children be reared in such a way that they would become capable of integration to the desired degree.

In the end, Åhrén's conception of a 'planned democracy' operated under the triad of 'enlightenment, cooperation, coercion': 'Voluntary conformity to the lines a social planned economy can draw should be the primary method, but, on the other hand, one should have legal possibilities of preventing what is clearly unsuitable from the societal perspective.' However, the instrument of coercion should be deployed only if the targets of the plan could not be reached through voluntary compliance.[52]

But the world of *social engineering* had no need for coercion. For experts like Alva and Gunnar Myrdal, coercion represented only the last resort of actions aimed at social reform. 'Reason' was their credo, and they believed that a system that had to enforce reason with coercion was doomed to failure. And so this social model was characterized by a peculiar, permanent back-and-forth between control and freedom, a controlled freedom, which constituted a normal arc of behaviour along with a corridor of permitted deviation. People *themselves* would eliminate the ambivalence of modernity in a flexible-cum-controlled manner by using their 'leeway' in a 'sensible' way. Through controlled movements that time and again level out at a balance of the various elements, they would cushion even the potentially uncontrolled dynamism of modernity – but not in a schematic way, since the normal curves, too, were never fixed once and for all.

As a result of a strict orientation towards reason and reality, experts were definitely able to adjust their models. For Alva Myrdal, the aspect of control in child rearing had completely lost importance many years later.[53] As early as 1950, architects dismissed the radicalized efficiency thinking of the 1930s.[54] Massively disciplining techniques were replaced with more elaborated 'technologies of the Self' (M. Foucault). The curve was continuously shifted, and *in this way*, to use Zygmunt Bauman's metaphor – as influential as it is askew – the 'weeds' in Sweden were to be eliminated by seeking to improve them into ornamental plants. Encapsulating the engagement with the 'ambivalence of modernity' in the notion of 'weeding' may apply to Germany or the USSR,[55] but does not even begin to do justice to the Swedish path to modernity.

Swedish history took a far more paradoxical course. It was a deeply *democratic* national community, which bet primarily on *inclusion* through social policy, and only in case it was absolutely necessary on exclusion through eugenic measures – which did not become *totalitarian*, but aimed at eliminating ambivalence and the creation of order through a *total* social structuring of the national body. Sweden demonstrated – and it is this that makes the country important for a comparative history of European modernity – that 'order' in a democratic, but in a Western sense decidedly anti-pluralistic state, could be achieved with means that could be suspiciously reminiscent, from a modern perspective, of National Socialism. This paradoxical engagement with ambivalent modernity and the scientization of the social is embodied in experts like Alva and Gunnar Myrdal.

Notes

1. L. Raphael (1996), 'Die Verwissenschaftlichung des Sozialen als methodische und konzeptionelle Herausforderung für eine Sozialgeschichte des 20. Jahrhunderts', *Geschichte und Gesellschaft* 22, pp. 165–93.
2. See M. Foucault (2007), *Security, Territory, and Population* (Basingstoke: Palgrave Macmillan) and idem (2008), *The Birth of Biopolitics* (Basingstoke: Palgrave Macmillan).
3. This distinction between a totalitarian and a total society was emphasized by B. Henningsen (1986), *Der Wohlfahrtsstaat Schweden* (Baden-Baden: Nomos), p. 372.
4. See K. Åmark (2005), *Hundra år av välfärdspolitik. Välfärdsstatens framväxt i Norge och Sverige* (Umeå: Boréa); R. Nilsson (2003), *Kontroll, makt och omsorg. Sociala problem och socialpolitik i Sverige 1780–1940* (Lund: Studentlitteratur); S. E. Olsson (1990), *Social Policy and the Welfare State in Sweden* (Lund: Arkiv förlag).
5. Per Albin Hansson, quoted in H. Dahlqvist (2002), 'Folkhemsbegreppet: Rudolf Kjellén vs Per Albin Hansson', *Historisk Tidskrift* 162, pp. 445–65, at 459. See A. Isaksson (1985–2000), *Per Albin*, 4 vols (Stockholm: Wahlström and Widstrand), vol. 3, pp. 172–92; V. Henze (1999), *Das schwedische Volksheim. Zur Struktur und Funktion eines politischen Ordnungsmodells* (Florence: European University Institute); H. Björck (2000), 'Till frågan om folkhemmets rötter. En språklig historia', *Lychnos*, pp. 139–70; N. Götz (2001), *Ungleiche Geschwister. Die Konstruktion von nationalsozialistischer Volksgemeinschaft und schwedischem Volksheim* (Baden-Baden: Nomos), pp. 190–280; H. Droste (2008), 'Das schwedische Volksheim – ein Erbe der frühneuzeitlichen Staatsbildung', in U. Schneider and L. Raphael (eds), *Dimensionen der Moderne. Festschrift für Christof Dipper* (Frankfurt am Main: Lang), pp. 129–48; S-O. Wallerstein and H. Mattsson (eds) (2010), *Swedish Modernism. Architecture, Consumption and the Welfare State* (London: Black Dog).
6. T. Tilton (1990), *The Political Theory of Swedish Social Democracy. Through the Welfare State to Socialism* (Oxford: Clarendon Press); S. O. Karlsson (2001), *Det intelligenta samhället. En omtolkning av socialdemokratins idéhistoria* (Stockholm: Carlssons Bokförlag), pp. 241–654; F. Sejersted (2005), *The Age of Social Democracy. Norway and Sweden in the Twentieth Century* (Princeton: Princeton University Press).
7. Paradigmatic: M. Childs (1936), *Sweden. The Middle Way* (New Haven: Yale University Press), and idem (1938), *This is Democracy. Collective Bargaining in Scandinavia* (New Haven: Yale University Press).
8. G. Adler-Karlsson (1967), *Functional Socialism. A Swedish Theory for Democratic Socialization* (Stockholm: Bokförlag Prisma).
9. For a detailed discussion, see C. Carlson (1990), *The Swedish Experiment in Family Politics. The Myrdals and the Interwar Population Crisis* (New Brunswick, NJ and London: Transaction); A-S. Kälvemark (1980), *More Children of Better Quality? Aspects on Swedish Population Policy in the 1930s* (Uppsala: Almqvist and Wicksell International); A-K. Hatje (1974), *Befolkningsfrågan och välfärden. Debatten om familjepolitik och nativitetsökning under 1930- och 1940-talen* (Stockholm: Allmänna Förlaget).
10. This compilation is in Statens Offentliga Utredningar 1946: 53 (1946), *Betänkande om befolkningspolitikens organisation m.m., avgivet av 1941 års befolkningsutredning* (Stockholm), pp. 41–3, 47–52.
11. See Nilsson (2003), *Kontroll, makt och omsorg*, pp. 219–92; E. Palmblad (2000), *Den disciplinerade reproduktionen. Abort- och steriliseringspolitikens dolda dagsordning* (Stockholm: Carlssons Bokförlag); I. Knobblock (1995), *Systemets långa arm. En studie av kvinnor, alkohol och kontroll i Sverige 1919–55* (Stockholm: Carlssons Bokförlag). What seems problematic is not the number of supervisory institutions, but their lack of coordination: Statens Offentliga Utredningar 1935: 49 (1935), *Betänkande med förslag rörande ändringar*

i vissa delar av hälsovårdsstadgan samt anordnande av förbättrad bostadsinspektion i städer och stadsliknande samhällen m.m. jämte därtill hörande utredningar, avgivet den 14 september 1935 av Bostadssociala utredningen (Stockholm), pp. 59*–79*, 79–89.

12. In detail: T. Etzemüller (2007), *Ein ewigwährender Untergang. Der apokalyptische Bevölkerungsdiskurs im 20. Jahrhundert* (Bielefeld: Transcript).
13. K. Johannisson (1991), 'Folkhälsa. Det svenska projektet från 1900 till 2. A världskriget', *Lychnos*, pp. 139–95, at 178.
14. In detail: G. Broberg and N. Roll-Hansen (eds) (2006), *Eugenics and the Welfare State. Sterilization Policy in Denmark, Sweden, Norway, and Finland*, second edition (East Lansing, MI: Michigan State University Press); M. Tydén (2002), *Från politik till praktik. De svenska steriliseringslagarna 1935–1975. Rapport till 1997 års steriliseringsutredning*, second edition (Stockholm: Almqvist and Wicksell International); M. Runcis (1998), *Steriliseringar i folkhemmet* (Stockholm: Ordfront).
15. Detailed statistics in G. Broberg and M. Tydén (2006), 'Eugenics in Sweden. Efficient Care', in Broberg and Roll-Hansen (eds), *Eugenics and the Welfare State*, pp. 77–149.
16. H. Lundborg (1921), *Rassenbiologische Übersichten und Perspektiven* (Jena: Fischer), p. 31; similarly J. V. Hultkrantz (1919), *Om rashygien. Dess förutsättningar, mål och medel* (Uppsala: Svenska sällskapet för rashygien), pp. 49–50.
17. Tydén (2002), *Från politik till praktik.* With this interpretation a plausible one in terms of institutional theory – of the sterilization practice as a bureaucratic automatism, Tydén has exonerated Swedish social democracy, which was accused (Runcis 1998, *Steriliseringar i folkhemmet*) of having overpowered segments of the Swedish population with the sterilization policy and excluding them from the welfare state.
18. Statens Offentliga Utredningar 2000: 20 (2000), *Steriliseringsfrågan i Sverige 1935–1975. Historisk belysning, kartläggning, intervjuer. Slutbetänkande av 1997 års steriliseringsutredning* (Stockholm).
19. F. Tönnies (1887), *Gemeinschaft und Gesellschaft. Abhandlung des Communismus und Socialismus als empirischer Culturformen* (Leipzig). English translation by J. Harris and M. Hollis (2001), *Community and Civil Society* (Cambridge: Cambridge University Press).
20. See M. Riedel (2004), 'Gesellschaft, bürgerliche' bzw. 'Gesellschaft, Gemeinschaft', in *Geschichtliche Grundbegriffe. Historisches Lexikon zur politisch-sozialen Sprache in Deutschland*, 8 vols (Stuttgart: Klett Cotta), vol. 2, pp. 719–862.
21. Three examples: G. H. von Koch (ed.) (1908), *Social Handbook* (Stockholm: Ljus); R. Kjellén (1917), *Der Staat als Lebensform* (Leipzig: Hirzel); U. Åhrén (1946), 'Ett planmässigt samhällsbyggande', in Statens Offentliga Utredningar 1945: 63 (1946), *Slutbetänkande, avgivet av Bostadssociala utredningen.* Del I: *Allmänna riktlinjer för den framtida bostadspolitiken. Förslag till låne- och bidragsformer* (Stockholm), pp. 580–640.
22. See B. Rothstein (1992), *Den korporativa staten. Intresseorganisationer och statsförvaltning i svensk politik* (Stockholm: Norstedts); M. Micheletti (1994), *Det civila samhället och staten. Medborgarsammanslutningarnas roll i svensk politik* (Stockholm: Fritzes).
23. Henningsen (1986), *Der Wohlfahrtsstaat Schweden*, p. 368.
24. The early letters have been edited by Y. Hirdman (ed.) (2003), *Alva & Gunnar Myrdal: De blå kuverten. Kärleksbreven juli 1919–augusti 1920* (Hedemora: Gidlunds Förlag). For an account of the marriage on the basis of these letters, see Hirdman (2008), *Alva Myrdal. The Passionate Mind* (Bloomington: Indiana University Press); H. Hederberg (2004), *Sanningen, inget annat än sanningen. Sex decennier ur Alva & Gunnar Myrdals liv* (Stockholm: Atlantis). A still excellent biography is S. Bok (1991), *Alva Myrdal. A Daughter's Memoir* (Reading, MA: Addison-Wesley).
25. G. Myrdal (1927), *Prisbildningsproblemet och föränderligheten* (Uppsala: Almqvist and Wicksell).
26. This is documented by the extensive collection of newspaper clippings among Myrdal's papers (Arbetarrörelsens arkiv och bibliotek, Stockholm, 405/5.1.1: 1–210).
27. For a detailed discussion, see T. Etzemüller (2010), *Die Romantik der Rationalität. Alva und Gunnar Myrdal – Social Engineering in Schweden* (Bielefeld: Transcript).

28. W. Tolman (1909), *Social Engineering. A Record of Things Done by American Industrialists Employing Upwards of One and One-half Million People* (New York: McGraw Publishing Company); C. Frederick (1913), *The New Housekeeping. Efficiency Studies in Home Management* (New York: Doubleday). For a critical analysis in the case of Sweden, see Y. Hirdman (1997), '"Social Planning Under Rational Control". Social Engineering in Sweden in the 1930s and 1940s', in P. Kettunen and H. Eskola (eds), *Models, Modernity and the Myrdals* (Helsinki: The Renvall Institute for Area and Cultural Studies), pp. 55–80.
29. See Raphael (1996), 'Die Verwissenschaftlichung des Sozialen'; T. Etzemüller (ed.) (2009), *Die Ordnung der Moderne. Social Engineering im 20. Jahrhundert* (Bielefeld: Transcript).
30. A. Myrdal and G. Myrdal (1934), *Kris i befolkningsfrågan* (Stockholm: Albert Bonniers Förlag); as a greatly expanded edition for the English language market, see A. Myrdal (1941), *Nation and Family. The Swedish Experiment in Democratic Family and Population Policy* (New York: Harper).
31. For a detailed account, see Etzemüller (2007), *Ein ewigwährender Untergang.*
32. See Myrdal and Myrdal (1934), *Kris i befolkningsfrågan*, pp. 217–26; A. Myrdal (1946), 'Kontanta barnbidrag kräver skärpt steriliseringslag?' *Tidskrift för barnavård och ungdomsskydd* 21, pp. 55–60; G. Myrdal (1940), *Population. A Problem for Democracy* (Cambridge, MA: Harvard University Press), pp. 174–213. It must be emphasized that eugenics had only comparatively marginal importance in the thinking of the Myrdals.
33. The most highly symbolic studies were the 'Housing Study' (*Bostadssociala utredningen*) from 1933 and the aforementioned population commissions of 1935 and 1941. A summary of their reception can be found in Carlson (1990), *The Swedish Experiment in Family Politics*, pp. 99–128.
34. E. Eriksson (2001), *Den moderna staden tar form. Arkitektur och debatt, 1910–1935* (Stockholm: Ordfront), pp. 340–497. See also K. Saarikangas (1997), 'The Policies of Modern Home. Organization of the Everyday in Swedish and Finnish Housing Design from the 1930s to the 1950s', in Kettunen and Eskola (eds), *Models, Modernity and the Myrdals*, pp. 80–108.
35. G. Näsström (1935), *Svenska Slöjdföreningens utställning Hem i kollektivhus* (Stockholm: Svenska slöjdföreningen). See also *Kollektivhuset i Stockholm. John Ericssonsgatan 6. Den moderna bostaden för moderna människor* (1937) (Stockholm: Bostadsrättsföreningen Fågelbärsträdet); A. Myrdal (1932), 'Kollektiv bostadsform', *Tiden* 24, pp. 601–8; S. Markelius (1932), 'Kollektivhuset. Ett centralt samhällsproblem', *Arkitektur och samhälle* 1, pp. 53–64.
36. Later, the history of the idea of the collective house was retraced, the small number of realized projects was praised, and the failure of the idea as a whole was lamented; see C. Caldenby and Å. Walldén (1979), *Kollektivhus. Sovjet och Sverige omkring 1930* (Stockholm: Statens råd för byggnadsforskning); D. U. Vestbro (1982), *Kollektivhus från enkökshus till bogemenskap* (Stockholm: Statens råd för byggnadsforskning). A good (rather more critical) first-hand account about the Myrdal collective house was written by S. Lamm and T. Steinfeld (2006), *Das Kollektivhaus. Utopie und Wirklichkeit eines Wohnexperiments* (Frankfurt am Main: S. Fischer).
37. See Å. Bergenheim (1994), *Barnet, libido och samhället. Om den svenska diskursen kring barns sexualitet 1930–1960* (Grängesberg: Höglunds Förlag); M. Börjesson and E. Palmblad (2003), *I problembarnens tid. Förnuftets moraliska ordning* (Stockholm: Carlssons Bokförlag); P. Lundquist Wanneberg (2004), 'Kroppens medborgarfostran. Kropp, klass och genus i skolans fysiska fostran 1919–1962' (Ph.D. thesis, Stockholm: University of Stockholm); K. Ohrlander (1992), *I barnens och nationens intresse. Socialliberal reformpolitik 1903–1930* (Stockholm: Almqvist and Wicksell International).
38. A. Myrdal (1935), *Stadsbarn. En bok om deras fostran i storbarnkammare* (Stockholm: Kooperativa förbundets bokförlag); see also idem (1936), *Riktiga leksaker* (Stockholm: Kooperativa förbundets bokförlag), and idem (1941), 'Familjen, fostran och den nya tiden', in *Hem och familj i skolans undervisning jämte några synpunkter på utbildningen*

för husligt arbete. Betänkande framlagt av Sveriges husmodersföreningars riksförbund (Stockholm), pp. 7–43. At that time she also established a 'Socio-pedagogical Seminar' to train educators.
39. This technique was emphatically propagated also by another educator, C. A. Ljunggren (1933), *Skolbarnens hälsouppfostran. Handbok för lärare och föräldrar* (Stockholm: Natur och kultur).
40. Myrdal (1935), *Stadsbarn*, pp. 94, 96 (emphasis in original).
41. Ibid., p. 87.
42. Myrdal (1941), 'Familjen, fostran och den nya tiden', p. 30. See also H. Bergman (2003), *Att fostra till föräldrarskap. Barnavårdsman, genuspolitik och välfärdsstat 1900–1950* (Stockholm: Acta Universitatis Stockholmiensis); A-K. Hatje (1999), *Från treklang till triangeldrama. Barnträdgården som ett kvinnligt samhällsprojekt under 1880–1940-talen* (Lund: Historiska Media).
43. Paradigmatic examples are two classics from the end of the nineteenth century: E. Key (1899), *Skönhet för alla. Fyra uppsatser* (Stockholm: Albert Bonniers Förlag), and C. Larsson (1899), *Ett hem. 24 målningar* (Stockholm: Albert Bonniers Förlag). See also P. Aléx (2003), *Konsumera rätt – ett svenskt ideal. Behov, hushållning och konsumption* (Lund: Studentlitteratur). 'Swedish Modern' is the term for a corresponding current of tradition within Swedish crafts. Three key documents of Swedish modernism have now been translated in L. Creagh, H. Kaberg and B. Miller Lane (eds) (2008) *Modern Swedish Design. Three Founding Texts* (New York: Museum of Modern Art).
44. See P. Aléx (1994), *Den rationella konsumenten. KF som folkuppfostrare 1899–1939* (Stockholm; Stehag: Brutus Östlings Bokförlag Symposion) and idem (2003), *Konsumera rätt.*
45. See A. Myrdal (1945), 'Folkvaror – kvalitetsvaror', *Vi* 32(8, 2), pp. 24f.; idem (1946), 'Orienterande inledning om tendenser och problem', *Runda bordet 1: Folkvaror 'Svensk Garantie'* (Stockholm), pp. 7–59.
46. An early example, noted by Elsa Törne in her 'solidarity books', was that hygiene, thrift, and order in the household would lead to a new solidarity among people: see E. Törne (1905), *Solidar. En livsfråga för hemmen* (Stockholm: Sandbergs bokhandel); idem (1906), *Solidar II. Renhet. En hafstång för den enskilde och samhället* (Stockholm: Sandbergs bokhandel); and idem (1907), *Solidar III. Hemmens lönereglering* (Stockholm: Sandbergs bokhandel).
47. See B. Lövgren (1993), *Hemarbete som politik. Diskussioner om hemarbete, Sverige 1930–40-talen, och tillkomsten av Hemmens Forskningsinstitut* (Stockholm: Almqvist and Wicksell International); B. Åkerman et al. (1984), *Kunskap för vår vardag. Utbildning och forskning för hemmen* (Stockholm: Förlaget Akademilitteratur AB).
48. For a detailed discussion, see Ö. Appelqvist (2000), *Bruten brygga. Gunnar Myrdal och Sveriges ekonomiska efterkrigspolitik 1943–1947* (Stockholm: Santérus Förlag).
49. See Bok (1991), *Alva Myrdal.*
50. The most important critical analyses of this process include Y. Hirdman (2000), *Att lägga livet till rätta – studier i svensk folkhemspolitit* (Stockholm: Carlssons Bokförlag); U. Olsson (1999), *Drömmen om den hälsosamma medborgaren. Folkuppfostran och hälsoupplysning i folkhemmet* (Stockholm: Carlssons Bokförlag); and Nilsson (2003), *Kontroll, makt och omsorg.*
51. The term 'neighbourhood community' was imported at the time from British debates and integrated into the concept of community – unlike in the 1970s, when 'neighbourhoods' were supposed to promote more strongly the 'emancipation' of individuals.
52. See U. Åhrén (1944), 'Stadsplanering och bostadsförsörjning', in *Bostadsförsörjning och samhällsplanering. Två radioföredrag av Uno Åhrén och Torvald Åkesson* (Stockholm: Svenska Riksbyggen), pp. 5–26, at 25; and idem (1946), 'Ett planmässigt samhällsbyggande'.

53. See the interview with Alva Myrdal in *Fackläraren* 7, 6 April 1981, pp. 22–4; similarly *Dagens Nyheter*, 21 April 1981, p. 42.
54. S. Markelius (1950), 'Människan i centrum?', *Plan* 4, pp. 51–7.
55. See Z. Bauman (2005), *Moderne und Ambivalenz. Das Ende der Eindeutigkeit* (Hamburg: Hamburger Edition); J. Baberowski and A. Doering-Manteuffel (2006), *Ordnung durch Terror. Gewaltexzesse und Vernichtung im nationalsozialistischen und stalinistischen Imperium* (Bonn: Dietz).

6
The Neurosciences and Criminology: How New Experts Have Moved into Public Policy and Debate

Peter Becker

Some people in the scientized society perform mediating roles, translating the experts' theories into ideas and images that the general public can grasp. Together with journalists, novelists are essential go-betweens and influencers in the uneasy triangular relationship between criminologists, public opinion, and the makers of penal policy. Personal interest stories in the newspapers and fictional plots about people who do not follow the rules of society – these intrigue their readers and can make ideas and narratives from the human sciences accessible. Such entertainment does not leave its consumers unaffected, but influences their engagement with real-world issues.[1]

It is a rare thing, however, to find criminological controversies thoroughly embedded into the plot of a crime novel and crime film. This is the case with William March's novel *The Bad Seed* (1954). It tells the story of the charming child-murderess Rhoda Penmark.[2] *The Bad Seed* was published at a time when expert discourse on social problems was shifting from a biological–eugenic defence against 'breakdown of society' to a behaviouristic-cum-sociological approach with a belief in the power of social engineering.[3] William March takes up the controversies that divided experts over why children can become offenders. Skilfully he interweaves them into the disturbing story of a charming little girl, who turns out to be a cold-blooded sociopath, consciously killing to achieve petty, childish ends.

I would like to use *The Bad Seed* to introduce the disputes the criminological experts had over the competing sociological and biological explanations of human behaviour. The Penmark case allows us to see the reaction to the debates of the 1950s, when a new configuration of criminological discourse appeared. It can be represented as a triangle consisting of criminal biology, sociological explanations of crime, and the new biological approach based on genetics and the neurosciences. At the end of the story, March introduces a brain surgeon who speaks not just about the physical condition of Rhoda Penmark's mother after surgery, but provides a new biological explanation of crime. He holds that it is not inheritance but genetic variation that lies at the root of deviance and violence.

The debate of the 1950s is important for a proper understanding of the second shift in biological explanations, which started in the 1980s, at a time, when, according to Lutz Raphael, the modernization paradigm lost its appeal. This second shift was not a turn back to the murky past of criminal biology with its eugenic connection,[4] but rather a turn towards a new form of expertise analyzing and drawing conclusions from the intricate relationship between the genetic and neurophysiological make-up of a person and that person's behaviour. It is the neuroscientists who claimed to have this expertise. They have emerged from their labs and enter debates on agency, violence, and education. Their success depends not least on the fact that they 'speak with several tongues', as sociologist Sabine Maasen pointedly argues. This provides them with a specific foothold on all three platforms of the triad of criminology, public opinion, and crime policy described earlier. As scientists, they claim superior authority in criminological debates because of their data; in the media they appear as public intellectuals and proclaim the need for reform of society and personal self-scrutiny based on conclusions drawn from their research;[5] and as entrepreneurs, they try to sell new technologies for conducting risk assessments to police and prison authorities.

Rhoda Penmark's legacy

In his novel, William March places Rhoda Penmark in an upper middle-class family, with a caring mother and a successful military officer as father. At first sight, Rhoda seems to fit well into this perfect domestic niche of the Cold War era. She is a highly self-disciplined child, successful at school and tidy at home. She responds to the emotional needs of her parents, especially those of her father, but does so in a calculating way that is not from the heart. Her mother notices this, and starts to have some doubts about her daughter's social skills and to worry about her highly manipulative engagement with adults. When she drops her daughter off at the summer party of her private school, she anxiously asks the teacher how well Rhoda gets on with the other pupils.

Slowly, Christine Penmark begins to discover the dark side of her daughter's character. She realizes indeed that the girl is a murderess. Pretending she is raising the issue out of pure curiosity, she asks acquaintances and family members with criminological knowledge of why and how children come to commit serious crime. Three different male characters provide explanations. March gives most space to the sociological argument. According to this, neglect from parents and bad environmental influences have the potential to turn innocent young children into monsters. This position is forcefully defended by Christine's father, a character March based on a well-known writer and journalist with a strong interest in forensic cases.

Another, less aired, explanation maintains that children like little Rhoda are simply *bad seeds* – that is, individuals whose criminal destiny is formed by inheritance. This argument is introduced by another writer and journalist, who represents the attitudes of a younger generation to whom the reasoning of eugenicists and physical anthropologists like Earnest Hooton still had an appeal. In Rhoda Penmark's case, this line of reasoning at first seems far-fetched. Both parents were raised in 'respectable' families. But all of a sudden we find that adoption is *very* relevant to the Penmark family, for Christine, the mother of murderous Rhoda, turns out to be an adopted child. And her own natural mother – what a surprise! – is revealed as having been an infamous serial killer, and in many respects similar to Rhoda: beautiful, charming on first impression, but in reality manipulative, and ready to pursue both rational and irrational goals without any inhibitions.

With this novel William March did more than contribute a best-seller to the genre of horror fiction. By introducing the different arguments about how pathological deviance arises, March situated the novel within a serious criminological discourse, where men reason about crime, basing their ideas either on 'sociology' or on biological science, and explain to women how crime is related to mothering – to women's bad parenting or to the way they gave birth. Following this highly loaded argument, a third voice has its say. This is the voice of the medical expert, who links statistical findings on genetic variation with its modulation of brain activities. This new voice has considerable appeal. It provides an answer to the puzzle of crime committed by children raised in a caring social environment, and it exculpates women from being at fault as 'defective' natural or social mothers.

The Bad Seed announces a sea change in criminological discourse in the years after the Second World War. This transformation cannot be understood as a simple takeover by scientists, moving into a field previously commanded by participants without a background in laboratory research. Nor can it be seen as a bid from a newly formed expert group for public and political recognition, as Lutz Raphael argues in his Introduction. Rather, we are confronted with futile defensive strategies criminal biologists adopted to withstand in the contest against sociological theories, a contest that reached its climax in the 1950s. As the criminological debate in *The Bad Seed* indicates, the shift in authority towards sociology did not silence biologically minded authors. It merely reduced their influence on the general public and on policy-makers.

The second biological wave in the 1980s – foreshadowed by an extension of genetic and evolutionary thinking to social behaviour – can be viewed as arising from the appearance of new scientific 'experts' on the criminological scene, joining sociologists, psychologists, and psychoanalysts. Neuroscientists brought new explanations for violence, its perpetrators and their social environment into scholarly debate, and into the political field and that of forensic practice. The sociology-minded experts now found themselves on the defensive and responded by incorporating some biological reasoning into their own analytical framework.[6] Both the shift towards sociological explanations in the 1960s and the second biological wave of the 1980s raise questions about the transformations of expertise. They are different from the ones related to the shift in criminological reasoning during the late nineteenth and early twentieth centuries, when the scientization of the social and the medicalization of penal law were important themes.[7]

My exploration of the second biological wave, on which this chapter will focus, is indebted to insights from the work of Michel Foucault, in particular his reflections on shifting sites of authority from which particular groups are able to speak about social problems. In his analysis, the 'authorities of delimitation' are based on the recognition of certain fields of knowledge and their institutionalization by public opinion, the law, and government, and by the function that the discourse carries out in a field of non-discursive practices.[8] The link to penal policy-making and to institutional practices in the forensic field is particularly relevant in the case of criminology. Neuroscientists would not have so much presence in the public sphere if their input did not answer demands from the media for spectacular discoveries and from political actors for new technologies to use in the management of crime problems. This we can learn from the remarks René Sève, director of the Centre for Strategic Analysis, made in 2009 when he opened a workshop on neurosciences in judicial procedures:

> The main objective, in all these different areas, was to enhance our ability to predict what public measures may be needed. In controlled schemes, we want to pool and make use of all the findings and ideas the various human sciences can bring – especially sociology, economics, psychology and the neurosciences, a research field in full growth.[9]

To understand the success neuroscientists have had in moving from the lab into politics, we need to move beyond mere critical assessment of criminal–biological textbooks and research in neuroscientific laboratories. Thinking along the lines of science/policy interaction, we should regard the issue of *recognition* by the public as a crucial feature. What made the insights and arguments of neuroscientists sufficiently attractive to gain public and political acceptance?

The triangular relationship of neuroscientists, public opinion, and crime policy requires additional analytical input to be systematically assessed. A useful lead can be taken from actor–network theory, which looks at the agency of human and non-human actors, breaks up aggregate institutional actors, and searches for the building of alliances.[10] Following the methodology of actor–network theory, we can take a new look at the second biological wave. The appearance of neuroscientists on the scene did not just happen as the result of political, social, and cultural change (though the relevance of this background in providing space for their emergence cannot be denied). Rather, it is best seen as the result of a variety of strategies deployed by different actors: researchers, lobbyists, journalists, brain scanners, political actors, and the brains themselves. An important actor has been the new visualization method of functional magnetic resonance imaging (fMRI), which measures and colourfully documents changes in blood flow related to neural activity in the brain.[11]

When the whole cast – neuroscientists, visualization technologies, lab design, journalists, officials in funding agencies, and politicians – is seen from the perspective of actor–network theory, the strategies the experts have used to form alliances become intelligible (rather than conjectural). The alliances can be studied as the result of successful boundary work of neuroscientists with media actors, funding authorities and scholars from the social sciences and the humanities. This generated recognition and political capital.

Towards a 'neurocentric age'?

> We are looking for genetic, biological, environmental, medical causes of violence. The sociologists have had their day and things have only gotten worse. It's time for the biologists, the geneticists, the neuroscientists to take over.
>
> Dr Palmquist[12]

I begin my exploration of the second biological wave with the help of another literary source – Richard Dooling's novel *Brain Storm*, published in 1999. The book shifts the perspective to an actor viewpoint. We are invited to follow the beautiful and witty neuroscientist Dr Palmquist in her quest for recognition in legal circles. Her claim to have complete neurological explanations to satisfy criminological discourse and its related political programmes is fictitious but not fictional, as even a cursory look at the contributions of neuroscientists to political debates about violence and education shows.

Within the multi-faceted, global research community of neurosciences[13] the specialists involved in crime studies[14] are an obvious minority, but a highly visible one. Their aspirations are not limited to effecting a shift in theories, research tools, and narratives. They have much bolder claims: they aim at joining, and eventually replacing, social and political scientists as advisors in political decision-making processes. This reaches beyond a change of disciplinary labels and a redirection of communication networks between academic institutions and policy-makers. It involves a radical redefinition of criminal subjects and the political options related to them. Political actors are receptive to these advances: they are looking for new solutions to the problem of crime prevention – as the statement quoted from René Sève illustrates. Some of the political actors share the perception of Dr Palmquist, that *things have gotten worse* under the previous approach – meaning that experts from sociology have failed to deliver the solutions to the control and prevention of crime they promised.

The neuroscientists' presence in the political field has been made possible by heavy investment in brain sciences, by technological advances enabling the brain to be viewed in close-up, and by breakthroughs in the understanding of cognition and neural functioning. Neuroscientists seem to be the heralds of a new scientific age, about to solve the last riddle of mankind: reconstructing (in both senses) the actual working of the brain. Their scientific discoveries and the popular imagination related to them have begun to change the ways in which we make sense of ourselves and of society. Jake Dunagan, research director at the American Institute for the Future, believes we are entering a 'neurocentric age'.[15] Colin Blakemore, who directs the steering group of the Brain Waves project, feels this too:

> Neuroscience and neurotechnology reach far and wide into disparate fields. The array of 'neuro' disciplines lend themselves to applications in diverse areas of public policy such as health, education, law, and security. More broadly, progress in neuroscience is going to raise questions about personality, identity, responsibility, and liberty, as well as associated social and ethical issues.[16]

The onward march of the neurosciences, however, is likely to result in a series of questionable changes in the field of penal policy. If neuroscientific tools and concepts are used as 'black boxes' for the classification of individuals, it would reduce the possibility of open debate over the rationale behind long-term exclusion of individual offenders. The political role of the neurosciences is certainly not just the outcome of their infallibility or success (giving them a right to automatic recognition). To understand this recognition, and therefore be able to look critically at penal decisions, we need to consider three intertwined processes.

The first process is well expressed in the claim of the fictional Dr Palmquist. It is related to the recognition of neuroscientists as producers of knowledge that can guide social, educational, and penal policies. This recognition implies that laboratory studies on rodents can produce results of significance for the understanding of human social interaction. The neuroscientists themselves are convinced that patterns seen in mouse experiments throw light on aggressive human behaviour,[17] and open a way to overcome the failure of existing intervention programmes. The point is strongly emphasized in Debra Niehoff's fine book, *The Biology of Violence*. As an American neurobiologist and science writer, Niehoff bemoans the fact that researchers looked for so long in the wrong places as they sought an answer to the problem of violence. Why did they not look in that 'thoroughly logical place: the brain'? After completing an exploration of the brain as the site of violence and aggression, she presents some political recommendations in her concluding chapter, inviting governments to stop wasting money on retribution (lengthy prison sentences) and on failed welfare schemes. Instead, she suggests, they should draft 'well-designed, carefully targeted intervention programs', in which welfare and neuroceutical interventions are restricted to subjects known to be at risk because of their genetic, neurochemical, and neuroanatomical characteristics.[18]

The 'neurocentric age' is set to bring wider repercussions yet. A systematic study of society and its subjects may become entirely framed within neuroscience. This is the second process we need to look at. According to Jake Dunagan, the impact of this expansion of neuroscience 'is being felt in other disciplines, and new approaches to epistemology, research methods, and long-standing philosophical dilemmas are being generated and debated'.[19] The reframing goes far beyond the use of new biological metaphors based on the 'body social' analogy (which the American sociologist Kurt W. Back, writing in the early 1970s saw as the 'transmission belt' between the physical and the human sciences).[20] The fields incorporated into neuroscience have now reached the status of a new *Leitwissenschaft*,[21] and their methods are increasingly implicated in the research design of other disciplines. New labels for new fields of knowledge keep appearing with the prefix 'neuro', clearly showing their basis.

We have to consider a third concomitant process linked to the coming of the 'neurocentric age'. Utopian visions such as equal opportunities, absence of violence and deviance, and better integration of cultural and ethnic minorities are increasingly being relocated from frameworks acknowledging the complex interplay between institutions, programmes, and subjects to the subjects themselves and their performance in 'markets'. Individuals suffering from an unbalanced brain physiology, it is said, should be given access to neuroceuticals. The result should be a society peopled by subjects who take responsibility for their brains and can live up to their true potentials.

The British sociologist Nikolas Rose has linked these wider cultural transformations to the emergence of a new kind of selfhood, which he fittingly labels the 'neurochemical self'. This self has the capability of maximizing his or her potential with the support of smart drugs operating on the molecular level of the brain.[22]

Criminology and the 'neurochemical self'

> If we want to make real progress in reducing the level of violence in our society, we have to stop reacting and start thinking. ... We don't lack for resources, programs, people, or character. What we need is a whole new perspective. ... If any entity deserves a greater voice in the ongoing debate about violence, it is the brain.[23]

This passage from Debra Niehoff's book prompts us to carry reflection on the neurocentric wave somewhat further. It introduces the brain itself as an actor – one that needs attention not just in the lab but in the offices of policy-makers too. What does this imply for Niehoff? When one reads further in her argument and browses through the paratexts clustered around her book, it becomes obvious that she is not outing herself as a follower of actor–network theory, where giving greater voice to the brain would imply an analytical, not a political shift in perspective. In her argument, the brain is not given this kind of agency even though she characterizes it as an 'organic historian'.[24] For her, the brain is a neurochemical organism – its fundamental processes being understood as communication of signals between neurons through neurotransmitters. The brain responds to stimuli from the outer world on the basis of past experiences and their internalized correlates (hence Niehoff's analogy between the brain and the historian). This concept of the brain emerged in the 1970s, replacing a cybernetic model based on an analogy between brain functioning and electronic computing devices.[25]

The neurochemical brain appears as key interface between the world of genes and proteins inside the body, the body's social and physical environment, and evolutionary and developmental processes. Even though it is regarded as a 'computational organ that works on physical principles',[26] it differs from the kind of technical interface known from the digital world. The main difference is its plasticity and malleability. It does not just run programs but actively creates them and adapts itself permanently to meet the challenges resulting from physical and environmental stimuli. This has implications for an understanding of violence. It subverts the criminal–biological conceptualizations of human behaviour, which considered the brain to be a given, subject to the influences of inheritance and environment only during its formation. Intervention was directed at limiting the spreading of *bad seeds* through eugenic measures and at improving environmental conditions for pregnant women. Individuals who were destined to violence and crime could be isolated from the community, and the only 'cure' was through the radical surgical intervention of lobotomy.

Aberrant behaviour is – as to the neuroscientists – explained on the basis of a homeostatic notion of normality: 'undesirable habits and thoughts, here, are the outcome of processes that may be pathological in their outcomes but are not pathological in themselves'.[27] The serotonergic system, its receptors, and transmitters do not in themselves destine a person to excessive forms of violence. Only if the quantity of a neurotransmitter in the central nervous system falls below a threshold level is the probability of impulsive violent behaviour radically increased.[28] The functioning of neurotransmitters and the reasons for their quantitative fluctuation is assessed in lab studies with animals and in clinical studies with human subjects. The results of the research trials are presented in highly technical language accessible only to a small group of specialists.[29]

The concepts and technical language of the neurosciences do provide a 'voice' for the brain. It needs to be heard! To create the societal impact Niehoff and others are aiming at, neuroscientific results need to be 'translated'. This means translation not just in linguistic terms; the task requires finding a way of *socializing* the insights, visions, and devices of the neurosciences. If they can explain and help regulate criminal behaviour, they need to be made available to a lay audience. It is the co-production of politically relevant knowledge and tools rather than more neuroscientific research in laboratories and clinical facilities that could make the dreams of Niehoff and the fictional Dr Palmquist come true.[30]

Before looking in more detail at the presence of neuroscientists in penal policy and forensic practice, I would like to look closer at their *stories*. What kind of insights do neuroscientists' ears and eyes gather from the numerical whispering and mumbling of the brain? At the very centre stand genetic polymorphisms – that is, variations in the genetic modulation of the serotenergic system through coding of relevant DNA sequences. These polymorphisms work together with negative social environment impulses in the creation of violence. Genetic polymorphisms were already the analytical basis of the argument the neurosurgeon put to his listeners in *The Bad Seed*.

Studying the genetic modulation of violence is a highly complex task, as it involves the search for causal links between the coding of individual DNA sequences (below the level of genes) with the working of brain physiology. A tendency to violence is part of our evolutionary legacy, which we share with primates and other mammals. So research into the neurochemical processes giving rise to this kind of behaviour can be carried out with animals (rodents are made to suffer for the betterment of the human race). It should come as no surprise that the redefinition of violence as the result of a complicated interaction between genetic modulation, brain physiology, and environmental factors privileges neuroscientists as the expert group in charge of understanding and controlling violence. They are the only experts with the competence and the tools to identify and to realign derailed brain chemistry through environmental stimuli, medication, and relearning programmes.[31]

Through a focus on the brain, the physical self reemerges as a crucial point of reference in scholarly, public, and political debates about violence as a social problem. This is not to say that the physical self was absent before. It was a different kind of self, though, which was reflected and acted upon. Inside the neurochemical self, Niehoff and her colleagues ignore the psychodynamic aspects accompanying traumatic experiences – unless they find corresponding modifications in the neurochemical data. Biography, environment, and life experiences are thus reduced merely to molecular traces they leave in the physical brain. This points to the ontological blindness characteristic of this approach.[32] And it foreshadows a new style of thought about social problems and their solution.[33]

The neuroscientific argument sweeps away all the complications involved in assessing multiple causative factors (whose relative impact must always remain uncertain). Neuroscientists promise to establish a direct causal link between specific forms of malfunctioning and the emergence of socially maladaptive behaviour. The malfunctioning they can detect in the brain is linked to the specific states of neurotransmitters, to certain levels of activity in specific brain areas, or to neuroanatomical abnormalities caused by disease or accident. Differentiated assessments of the serotonin system, research into the genetic programming of neurochemical processes, and fMRI studies on brain functioning in different groups of people are all aimed at localizing the source of violent behaviour in functional pathologies of the cerebrum.

This redefinition of violence comes at a high price. Neuroscientists have lost sight of a variety of analytical dimensions. The *International Handbook of Violence Research* (2002) provides a comprehensive survey of the existing literature on violence. Perpetrators and victims are discussed in various contexts: the performative aspect of committing violent acts and their relation to social and group identities, the bodies of victims as targets of interpersonal physical violence, the politically legitimated use of violence against the bodies of opponents – to name just a few.[34] These 'bodies' differ in many respects from the biological entities examined in neuroscientific studies: they are situated in political, cultural, and social contexts, which bestow richer meanings on them.

The neurosciences and penal policies

> The particular appeal of neuropsychological evidence in the criminal context is the expert's ability to bring quantified, normative data on brain–behavior relationships to bear in support of what have traditionally been professional opinions based on mental status examinations and clinical interview techniques.[35]

In his almost visionary statement (from a paper of 1992), Daniel Martell foresaw the move neuroscientists would make from the lab to the courtroom. He referred to the American situation, which is somewhat peculiar because of its tradition of case law and the presence of strong lobbying activity from the DANA Foundation. The DANA Foundation, founded in 1950, is a private philanthropic organization, which in 1985 identified the neurosciences as one of its funding priorities. In the early 1990s (the beginning of the so-called decade of the brain), it persuaded leading neuroscientists to establish an outreach programme under the title of the DANA Alliance for Brain Initiatives. This had the objective of 'translating the advances in brain research to the public, the ultimate beneficiary of these advances"[36] The foundation started a series of interdisciplinary seminars in the 2000s to foster collaboration between neurosciences on the one hand and representatives from law and education on the other. Thanks to its lobbying activities, a substantial grant from the McArthur Foundation was secured to establish The Law and Neuroscience Project to systematically integrate neuroscientific expertise into penal policies and legal decisions.[37]

Outreach projects from the DANA and McArthur Foundations targeted trial judges to make them understand the validity of neuroscientific evidence. The presence of neuroscientists as expert witnesses 'has changed the contours of the playing field, and no matter which side of the divide we find ourselves on, we must acknowledge that reality', the legal expert Michael Perlin, director of the International Mental Disability Law Reform Project, points out when referring to the increasing presence of neuroscientists in criminal proceedings in the United States since the 1990s.[38] In the meantime, even in Italy, neuroscientists have been called to testify about the mental state of a defendant. The Court of Appeals in Trieste invited a molecular scientist and a cognitive scientist to run a series of tests on a defendant based on the latest advances in neuroscience and genetic screening. They were to find 'significant genetic polymorphisms to modulate responses to environmental variables including, in particular, those concerning, in this case, the exposure to stressful events and the reaction to the same type of impulsive behavior'.[39] Based on their results, the Court of Appeal granted a reduction of sentence, and this started a heated debate in Italy about what evidence was admissible in court.

The two American law experts Michael Pardo and Dennis Patterson summarize what is at stake:

> At stake are issues involving voluntary conduct (actus reus), mental states (mens rea), diminished capacity, insanity, theories of punishment, and the death penalty. ... Scholars have made strong claims about the ways neuroscience can and ought to, or potentially will be able to, aid our understanding of these fundamental doctrinal issues. These claims typically confuse properties of people with properties of brains.[40]

The issues reach into legal philosophy, but are not restricted to it.[41] Prediction is equally at the heart of a future 'neurolaw'.[42] Neuroscientists are convinced that their technologies will offer better assessments about the future dangerousness of individuals.

The historian cannot help being struck by parallels to the claims of criminal biology in the early twentieth century. The strongest impact of the first biological wave in criminology was not in the codification of penal law but in the development and implementation of technical tools for assessing prisoners.[43] These tools met the demand of a prison system in transformation, where a revival of the will to reeducate required a sound prognosis of the future moral and social performance of prisoners. This had existed before, but the introduction of criminal–biological questionnaires provided the kind of 'objectivization' that neuroscientific screening and scanning technologies impose. There was a complete disregard for what an experienced practitioner could see with his/her eyes – the *'practical gaze'*[44] – and there is an 'obsession with figures', as a French neuroscientist puts it.[45]

The historical precedent of criminal biology is a warning. It requires us to look beyond the supply side – all these solutions proposed by neuroscientists – and to consider the demand side – the need of the penal system to assess and classify offenders. How far do neuroscientists respond to this demand? To answer the question, I would like to turn to the French professor of medicine and biology, Jean-Claude Ameisen, who considers *prediction* to be one of the inroads neuroscientists can make into the penal system – an inroad still fraught with ethical questions:

> The relation between the neurosciences and justice raises plenty of other ethical questions. One of these is related to prediction. The law as it stands allows us to hold certain people indefinitely in preventive detention: they have committed a crime and have served their sentence, but – in terms of probability – it is predicted that they could still be dangerous. This is a worrying situation, recalling the one portrayed in the film *Minority Report*: locking up a person for a crime he or she hasn't (yet?) committed. This continued incarceration after the sentence has been served constitutes an application of risk avoidance that borders on extreme inhumanity.[46]

Ameisen refers to personality assessments in the French prison system in order to take 'dangerous' offenders out of circulation for an extended time: they came in with a law passed on 25 February 2000. Every prediction of dangerousness presents a moral problem: it is making a drastic intervention in the life chances of an individual based solely on previous behaviour, along with 'expert' opinions and a medical–psychological examination. Standardized procedures guide experts 'in systematically collecting, reviewing, weighing, integrating, and combining information'.[47] The status of the resulting predictions is probabilistic – the norm in risk management in modern and post-industrial societies.[48] The predictions try to limit negative impact from the general contingency of human behaviour by directing intervention against those identified (through institutional and clinical evidence) as prone to 'antisocial behaviour'.[49]

What added value comes from the neuroscientists in this process? Serge Stoléru, a French psychiatrist with research interests in the field of neurolaw, raises this question explicitly.[50] The protagonists of the coming *neurolaw* would certainly respond in an affirmative manner. Two influential authors, Terry Moffitt and Avashalom Caspi, are quick to point out the enormous leap in analytical precision in neuroscientific assessments. They claim that neuroscientists are able to lift the study of antisocial behaviour from the 'risk-factor stage' to an understanding of causal processes at an individual and group level. Such prescience, as they argue, could significantly improve the success rate of political intervention. So far, 'valuable resources have been wasted because intervention programs have proceeded on the basis of risk factors, without sufficient research to understand causal processes'.[51]

Moffitt and Caspi aspire to overcome this limitation by studying the correlation between genetic and environmental influences, not on the general level of adoption and twin studies but with a new empirical design: they use child maltreatment as a measured environmental risk and correlate it with MAOA polymorphism in an identified gene.[52] Monoamine oxidase A (MAO-A) is an enzyme that breaks down serotonin and other neurotransmitters. The amount of this enzyme in the brain is modulated genetically and strongly influences the serotonergic system. With their research design Moffitt and Caspi offer fascinating and frequently quoted insights. They claim to show how environmental factors and genetic dispositions can work together to produce higher risks of antisocial behaviour. Although their conclusions are more differentiated than those offered by previous sociological models, their findings are far from presenting the promised causal processes at the individual level. A more realistic appraisal comes from Sébastian Tassy, a French neuroscientist, who emphasizes that any prediction about dangerousness is necessarily restricted to populations and cannot be done on the level of single individuals.[53]

The political gain from neuroscientific studies on violence is the development of intervention strategies based on improved prediction of deviant behaviour by linking conclusively neurochemical and anatomical brain figurations to a propensity for deviant behaviour. The German neuroscientist Hans Markowitsch uses the Moffitt–Caspi study to propose a systematic screening of children to identify those at risk of having future criminal careers.[54] In a similar vein, neuroscientists have used advanced visualization techniques to establish a connection between brain activity and sociopathic personality traits, which could be used in the classification of offenders.[55]

But Markowitsch and his peers ignore the complexity of neuroscientific predictions when they dream about a new neuroscientific utopia bringing a super-efficient prevention of crime.[56] Even when a strong correlation between deviant behaviour and a particular neurochemical and neurophysiological pattern can be found, a direct causal link to criminal acts is impossible to establish.[57] Moreover, the results from brain imaging are very difficult to read, as Sève reminds us. He calls for prudence in the use of this evidence for scientific and (even more) for institutional and political purposes.[58] Tassy, another member of the more critically minded French neuroscientific community, points to three major problems: precisely localizing sites for deviant behaviour in the brain; finding a convincing definition of 'normal' behaviour; and establishing what are the 'normal' neuroanatomical and neurochemical correlates.[59]

The neurosciences do not limit their input in the political debate to the field of prediction. Because the brain remains malleable, there is an opportunity for manipulation. In a way unknown to criminologists of the first wave, undesirable behaviour can now be redressed by highly focused interventions into the brain. As Nikolas Rose put it: 'what is learned may be unlearned, or may be replaced by a new and less damaging set of learned ways of thinking and acting, an array of competences and skills of life management that are more desirable to others, and indeed to the dependent person themselves'.[60]

The current biological wave is based on a complex understanding of the brain, its interaction with genetic modulation and the social and physical environment, and the correlation of neurochemical patterns with personality and social behaviour. This complexity promises individualized interventions like the prescription of neuroceutical drugs to redress neurochemical imbalances. The hope is to reintegrate violent offenders into society.[61] The plasticity of the brain thus opens up a radically different perspective on the correlation between specific brain features and deviant behaviour. However, as the French neurobiologist Catherine Vidal points out, any particularities found in the brains of criminal offenders could just as plausibly be the *result* of the acts committed as the cause.[62]

In the style of Cesare Lombroso, neuroscientists plan to use their technologies for detection purposes too. Hanns Gross, the famous *fin-de-siècle* Austrian crime expert, would have been delighted to see such a strong renewed interest in criminal psychology. Even more delightful to the old master from Graz would be the recurrence of an ontological concept of truth and deception (with a hope of finding their specific neurological correlates).[63] He would have liked also the way neuroscientists assess the statements of defendants, victims, and witnesses appraising their truth claims. For this purpose, neuroscientists and their start-up companies in the United States market their new lie-detection technology for use in criminal law procedures, for litigation purposes and for police investigations.[64]

Heavy investment in the security sector drives the development of neuroscience-based technologies for the screening of large populations passing through airports and train stations. They are meant 'to glean intentionality of a suspect through a combination of agent interpretation and neuroimaging, body language, temperature, and other physical data'.[65] It is again a French neuroscientist, Hervé Chneiweiss, who shows up the mistaken reasoning. At its best this technology can only show if a person is telling *his own personal truth*. Hans Markowitsch argues otherwise. He claims that the brain knows better than the person about the truth of its representations. (This tells us a lot about his epistemological approach.) Accessing memories, desires, and thoughts should then be – if possible – directly through the brain and not through a social communication process.[66] Chneiweiss strongly opposes this ignorance of cultural context:

> The way we use the brain is inextricably bound up with the social context we're in. Whether we like it or not, the brain is steeped in the cultural concepts and social prejudices we all carry with us.[67]

We might be living in a world, where the Brain Overclaim Syndrome, wittily described by Stephen J. Morse, becomes epidemic in media and intellectual circles. Propelled by the strong influence of brain imaging studies on the popular imagination,[68] an increasing number of scholars and influential lay people have started imagining a better, more efficient legal and penal system based on *neurolaw*. Commentators predict pressure on the legal system to use the 'ever more sophisticated understandings of normal and abnormal behaviour' produced by the biological and behavioural sciences. The legal system cannot turn a blind eye to such advances. So there is much debate to come. The moral, political, and ethical issues need to be discussed openly. Legal procedures may have to adapt, but the judiciary must also resist premature implementation of unverified neuroscientific technologies even if urged by the media.[69]

Media-generated authority

> I'm struck by the sheer speed with which the neurosciences have infiltrated not just the social sciences but everyday life. ... Our society can't tolerate uncertainty any more: we keep asking more and more from the neurosciences, whether it's about creating our images or about genetics. ... How can you expect society to resist when neuroscience has supplied things that grab people so?[70]

Didier Sicard, a specialist in medical ethics, here points to the importance public opinion will have on how far the neurosciences get taken up in politics. How can societies withstand the claims and subtle suggestions of neuroscientists wanting to frame policies according to their own ideas, if the media are bestowing authority on them?[71] This question is all the more pertinent, as stories and references to neurosciences and neuroscientists have increased rapidly in the media over the last 15 and more years.

The impressive rise in media coverage has gone largely unnoticed by scholars from the social sciences and cultural studies. The few writers who have systematically paid attention to it have focused on the *feuilleton* and the debate over the question of free will. I cannot offer a comprehensive explanation for the rising star of the neurosciences in the media. Fundamental political changes such as the growth of a neo-liberal society with its specific form of governance and subjectivity, and the replacement of social sciences and political philosophy by biology and economics as leading methods of enquiry are among the reasons.

In the remaining part of this chapter, I will develop a point regarding the role of the media in the neuroscientists' move from the lab into politics. My approach is inspired by Bruno Latour.[72] At first sight, Latour's argument seems to sum up the processes through which media-derived authority is generated. Influential gatekeepers and agenda-setting journalists decide to pick up the ball of the neuroscientists and start running with it. Upon closer inspection, however, we realize that clear-cut distinctions between brain researchers and journalists do not exist. To be sure, a journalist does not produce research results: he is part of the network that links the laboratories to funding agents, politics, scholars from other disciplines, themes to be explored, and a broader public. In doing so, the journalist helps turn findings into accepted 'facts'. At the same time, the journalist's activities might result in questioning of findings and reopening the 'black box'.

The increased attention in the media does not answer the question of how the boundary between science and media has been crossed. To answer this in a tentative form, I will concentrate on the more limited corpus of media reports that deal with the neuroscientific contribution to the debate on violence. The German *Frankfurter Allgemeine Zeitung*, the Austrian *Presse*, and the Swiss *Neue Züricher Zeitung* are papers that pay particular attention to neurosciences. All of them are high-quality papers with an extensive readership but without a mass circulation. In these papers, coverage of neuroscientific arguments on violence comes under the sections *Feuilleton*, *Culture*, and *Science*. These are sections where journalists are given sufficient space to develop an argument in reasonable depth. The articles seldom assess scientific advances critically – either in a positive or negative way. Usually news of neuroscientific research is presented more or less as a matter of fact. Critical voices persist, but are increasingly marginalized by positive and neutral reports. This finding provides additional evidence for the coming of a neurocentric age, for we find it simply taken for granted that neuroscientist experts should speak about violence as a social and political problem.

What narrative strategies do the authors of media reports use to make the neurosciences and their explanations of crime attractive to their audiences? The articles translate neuroscientific findings for the general reader in two ways. The first way is to make the language intelligible. The terminology of neuroscientific research papers is highly specialized and not at all attuned to the reading habits of newspaper audiences. The second way is a consistent attempt to link neuroscientific research to political issues and public concerns. This process of translation feeds back into the scientific arena, as it directs funds, and thus shifts attention to, particular research questions. Reports in mass-circulation newspapers use neuroscientific research on topics like violence to promote new solutions to social problems. There is a focus on preventative detention and other preventative schemes such as screening programmes,[73] and also on the use of neuroscientific tools for the prognosis of recidivism and for the assessment of the *mens rea* of people appearing in court charged with serious crime.[74]

When journalists write about the neurosciences and their contributions to the solution of social problems, the positioning of the discipline and its representatives is of crucial importance. Are neuroscientists introduced as team players – siding up with social scientists and psychologists in a joint effort to contain violence? Or do they appear as new actors on an existing stage? A close reading of the media articles reveals three different strategies. The first strategy presents the neurosciences as part of a broad range of disciplines all addressing the same issue: identifying and controlling criminals. Here neuroscientists are invoked as the ultimate authorities.[75] The second strategy plays off the irrefutable scientific pronouncements of neuroscientists against the 'soft', unrealistic approach of liberal social scientists and penal reformers.[76] The third strategy simply discounts previous efforts by social scientists and psychologists. It presents the biology of violence as an entirely new field with promising results for future prevention and control.[77]

Conclusion: Towards a brave new neuro-world?

In this chapter, I used the novel *The Bad Seed* and its cinema adaptation as a starting-point for surveying a striking contemporary phenomenon: the neuroscientists' move from the lab into social and penal politics. In the mid-1950s, when the book and the film came out, brain research was just beginning to be used in criminological debates – and rather cautiously. Criminal biologists welcomed the contribution of neuroscientists as support in their increasingly desperate contest against sociological theories of crime. Between the 1950s and the 1980s, when a second biological wave emerged in criminology, genetic/neuroscientific researchers detached themselves from the old criminal biology. Their investigations into crime and violence were based on a different concept of the brain, which was now understood as malleable and flexible, and they had new ideas about how it interacted with the environment.

The argument I go on to present focuses as much on the neuroscientists' production of criminological data as on the demand for their expertise from actors in politics and the media. Following the reasoning of Nikolas Rose, I link the increasing recognition of neuroscientists as public intellectuals and as experts in social and crime policy to broader social, political, and cultural changes. In particular, I trace the emergence of a new form of subjectivity, which Rose aptly calls the *neurochemical self*. Neuroscientists have risen to positions of prominence in criminology and regularly appear as expert witnesses in US courts. Fully to understand this rise, one has to consider their success in the design of new diagnostic and therapeutic strategies in medicine: they cast new light on Alzheimer's and on Parkinson's disease, for example. Trumpeted by the publicity campaign of the *decade of the brain*, these discoveries bestowed credibility on neuroscience and conferred on it an authority that became extended to a broad range of other issues linked to the brain.

The media played a crucial role in conferring this authority. It presented leading figures from the neurosciences as public intellectuals with a mandate to talk about philosophical questions concerning free will and political issues like the reform of primary education and the control of representations of violence in the media – and also, of course, about criminology and penal policy. Neuroscientists cannot claim their new status of public intellectuals everywhere. In Germany, Britain, and the United States their presence in the media has significantly risen; but France and Italy have not seen a similar development.

If we look at neuroscientists' arguments, we can identify an attempt to become the 'obligatory passage point' (Latour) in debates involving human cognition and social interaction. This claim has been bolstered by a systematic confusion of 'properties of people with properties of brains' (Pardo/Patterson). Redefinition of 'what people are' licenses a far-reaching change in research design for the study of human behaviour. Since neurochemical brain processes are now privileged over people, studies of laboratory animals to establish 'mouse models' (Lesch/Merschdorf) for violent behaviour are often unquestioningly accepted. They are then tested on the human population.

The first criminal biological wave in the 1880s used metaphors, concepts, and research strategies that were widely accessible: they focused on genealogy and consideration of visible parts of the body. The second biological wave, overtaking criminology in the 1980s, has created a much wider gap between researchers and the general public. The gap has been bridged by a broad variety of actors: researchers, journalists, funding agencies, political actors, criminologists, and officials in the penal law system. Boundary work is of the utmost importance. An important instrument bringing neuroscience to the lay audience is fMRI, which produces colourful images of brain activities people feel they understand. These images are present in the media and are also increasingly introduced into the courtroom as evidence for the prosecution or defence.

The boundary work of neuroscientists, journalists, and practitioners in the penal system is strongly supported by lobbying agencies like the DANA Foundation, with its outreach programmes and interdisciplinary seminars. In these workshops, the complex reality of the brain and its functioning is explained in accessible chunks targeted audiences can take in. The objective is to convey new scientific insights to parts of the public, political, and institutional domain where they can be put into practice. New theoretical and technical instruments for the assessment of defendants, new tools for the prognosis of recidivism, new devices for the identification of false statements are all in the toolbox offered to the penal system.

Looking at this development from a science studies perspective, we can only recommend caution and a degree of scepticism. Neither the findings nor the technical solutions proposed are neutral. As Michael Perlin insists, they change the 'contours of the playing field, and no matter which side of the divide we find ourselves on, we must acknowledge that reality'. I have tried to show the truth of this statement for the criminological debate, which has only just started.

Notes

1. On crime films, see N. Rafter (2000), *Shots in the Mirror. Crime Films and Society* (New York: Oxford University Press), pp. 7–9.
2. W. March (1954), *The Bad Seed* (New York: Rinehart & Co.); the cinema adaptation was directed by Mervyn LeRoy and produced by Warner Brothers. It was released in 1956.
3. A critical reflection on the link between criminology and eugenics can be found in R. F. Wetzell (2000), *Inventing the Criminal. A History of German Criminology, 1880–1945* (Chapel Hill: University of North Carolina Press), Chapter 7.
4. Cf. O. D. Jones and T. R. Goldsmith (2005), 'Law and Behavioural Biology', *Columbia Law Review* 105(2), pp. 405–502, at 428 and 485f.
5. See N. Rose (2007), *The Politics of Life Itself. Biomedicine, Power, and Subjectivity in the Twenty-First Century* (Princeton: Princeton University Press), pp. 222f.; S. Maasen (2006), 'Hirnforscher als Neurosoziologen? Eine Debatte zum Freien Willen im Feuilleton', in J. Reichertz and N. Zaboura (eds), *Akteur Gehirn – oder das vermeintliche Ende des handelnden Subjekts. Eine Kontroverse* (Wiesbaden: VS), pp. 287–303, at 288.
6. A good case in point is the edited volume: P-O. H. Wikström and R. J. Sampson (eds) (2006), *The Explanation of Crime. Context, Mechanisms and Development* (Cambridge: Cambridge University Press).

7. See L. Raphael (1996), 'Die Verwissenschaftlichung des Sozialen als methodische und konzeptionelle Herausforderung für eine Sozialgeschichte des 20. Jahrhunderts', *Geschichte und Gesellschaft* 22(2), pp. 165–93.
8. M. Foucault (2002), *Archeology of Knowledge* (London: Routledge), pp. 46, 75.
9. R. Sève (2009), 'Ouverture', in Centre d'analyse stratégique (ed.), *Perspectives scientifiques et légales sur l'utilisation des sciences du cerveau dans le cadre des procédures judiciaires* (Paris: Centre d'analyse stratégique), pp. 5–6.
10. B. Latour (2005), *Reassembling the Social. An Introduction to Actor–Network Theory* (Oxford: Oxford University Press).
11. Cf. S. J. Morse (2006), 'Brain Overclaim Syndrome and Criminal Responsibility. A Diagnostic Note', *Ohio State Journal of Criminal Law* 3, pp. 397–412, at 403.
12. R. Dooling (1999), *Brain Storm* (New York: Picador), p. 234.
13. See M. Hagner (2006), *Der Geist bei der Arbeit. Historische Untersuchungen zur Hirnforschung* (Göttingen: Wallstein), p. 26.
14. See M. Enserink (2000), 'The Search for the Mark of Cain', *Science*, n.s. 289, 28 July, pp. 575–9, at 575.
15. J. F. Dunagan (2010), 'Politics for the Neurocentric Age', *Journal of Future Studies* 15, pp. 51–70, at 51.
16. C. Blakemore (2011), 'The *Brain Waves* Project', in Royal Society (ed.), *Brain Waves Module 1: Neuroscience, Society and Policy* (London: The Royal Society), pp. 1–2, at 1.
17. See K. P. Lesch and U. Merschdorf (2000), 'Impulsivity, Aggression, and Serotonin. A Molecular Psychobiological Perspective', *Behavioral Sciences and the Law* 18, pp. 581–604, at 581.
18. D. Niehoff (1999), *The Biology of Violence. How Understanding the Brain, Behavior, and Environment can Break the Vicious Circle of Aggression* (New York: Free Press), pp. ix and 267; see also J. P. Wright, S. G. Tibbetts, and L. E. Daigle (2008) *Criminals in the Making. Criminality across Life Course* (Los Angeles: Sage), pp. 253–60.
19. Dunagan (2010), 'Politics for the Neurocentric Age', p. 55.
20. K. W. Back (1971), 'Biological Models of Social Change', *American Sociological Review* 36(4), pp. 660–7, at 660.
21. See W. Frühwald (1997), 'Ein Ende ahnen, neuen Beginn erfahren', *Forschung & Lehre* 12, p. 618.
22. Rose (2007), *The Politics of Life Itself*, pp. 222f.
23. Niehoff (1999), *The Biology of Violence*, p. 260.
24. See interview with D. Niehoff at http://umbral2.uprrp.edu/files/The%20Biology%20of%20Violence.pdf [accessed 14 February 2011].
25. Hagner (2006), *Der Geist bei der Arbeit*, pp. 195–214; see also the chapter on the cybernetic brain in idem (2004), *Geniale Gehirne. Zur Geschichte der Elitengehirnforschung* (Göttingen: Wallstein), pp. 288–96.
26. Jones and Goldsmith (2005), 'Law and Behavioral Biology', p. 422.
27. N. Rose (2003), 'The Neurochemical Self and its Anomalies', in R. V. Ericson and A. Doyle (eds), *Risk and Morality* (Toronto: University of Toronto Press), pp. 407–37, at 418.
28. See the review of studies on the association between violence and the serotonergic system: M. Krakowski (2003), 'Violence and Serotonin. Influence of Impulse Control, Affect Regulation, and Social Functioning', *The Journal of Neuropsychiatry and Clinical Neuroscience* 15, pp. 294–305.
29. Lesch and Merschdorf (2000), 'Impulsivity, Aggression, and Serotonin', p. 583.
30. See S. Maasen (2010), 'Science Policy', in P. Becker, R. Kreissl, and S. Maasen 'Neurolaw Ahead', unpublished manuscript.
31. See H. J. Markowitsch and W. Siefer (2007), *Tatort Gehirn. Auf der Suche nach dem Ursprung des Verbrechens* (Frankfurt am Main: Campus), Chapters 6 and 7.
32. See M. Hagner (1996), 'Der Geist bei der Arbeit. Überlegungen zur Visualisierung cerebraler Prozesse', in C. Borck (ed.), *Anatomien medizinischen Wissens* (Frankfurt am Main: Fischer), pp. 259–86, at 278.

33. P. Becker (2009), 'New Monsters on the Block? On the Return of Biological Explanations of Crime and Violence', in M. S. Hering Torres (ed.), *Cuerpos Anómalos* (Bogotá: Universidad Nacional de Colombia), pp. 265–99, at 267–9.
34. W. Heitmeyer and J. Hagan (eds) (2003), *International Handbook of Violence Research* (Doordrecht: Kluwer Academic Publishers).
35. D. A. Martell (1992), 'Forensic Neuropsychology and the Criminal Law', *Law and Human Behavior* 16(3), pp. 313–36, at 315.
36. See the webpage of the DANA Alliance: http://www.dana.org/danaalliances/about/ourhistory.aspx [accessed 4 March 2011].
37. See M. S. Gazzaniga (2008), 'The Law and Neuroscience', *Neuron* 60(3), pp. 412–15; and, applied to the treatment of psychopaths, S. J. Morse (2008), 'Psychopathy and Criminal Responsibility', *Neuroethics* 1(3), pp. 205–12; see also http://www.lawneuro.org/ [accessed 16 February 2011].
38. M. L. Perlin (2009), '"And I See Through Your Brain". Access to Experts, Competency to Consent, and the Impact of Antipsychotic Medications in Neuroimaging Cases in the Criminal Trial Process', *Stanford Technology Law Review* 4, par. 2 [online].
39. Court of Assizes of Appeal in Trieste, 1 October 2009, in *Riviste penale* 1 (2010), p. 74, quoted from E. Musumeci (2010), 'Cesare Lombroso and Neuroscientists. A Failed Patricide', unpublished conference paper at the workshop *Was Lombroso Right? The Historical Legacy of Neuroscience*, Vienna 2010, p. 8.
40. M. S. Pardo and D. Patterson (2010), 'Philosophical Foundations of Law and Neuroscience', *University of Illinois Law Review* 4, pp. 1211–50, at 1231.
41. See J. Greene and J. Cohen (2004), 'For the Law, Neuroscience Changes Nothing and Everything', in S. Zeki and O. R. Goodenough (eds), *Law and the Brain* (London: Royal Society), pp. 1775–85.
42. C. Byk (2009), 'Les difficultés légales et éthiques liées à l'utilisation des neurosciences', in Centre d'analyse stratégique (ed.), *Perspectives scientifiques et légales sur l'utilisation des sciences du cerveau dans le cadre des procédures judiciaires* (Paris: Centre d'analyse stratégique), pp. 50–6, at 54.
43. On the institutionalization of criminal biology, see most recently T. Kailer (2011), *Vermessung des Verbrechers. Die Kriminalbiologische Untersuchung in Bayern, 1923–1945* (Bielefeld: Transcript).
44. See P. Becker (2001), 'Objective Distance and Intimate Knowledge. The Rhetoric of Criminological Narratives', in P. Becker and W. Clark (eds), *Little Tools of Knowledge. Historical Essays on Academic and Bureaucratic Practices* (Ann Arbor: University of Michigan Press), pp. 197–235, at 206–13.
45. See Kailer (2011), *Vermessung des Verbrechers*, pp. 147–91; H. Chneiweiss (2009), 'Les neurosciences, nouvelle branche de la médecine légale?', in Centre d'analyse stratégique (ed.), *Perspectives scientifiques et légales*, pp. 45–50, at 46f.
46. J-C. Ameisen (2009), 'Neuroscience et éthique', in Centre d'analyse stratégique (ed.), *Perspectives scientifiques et légales*, pp. 56–63, at 62. See also Byk (2009), 'Les difficultés légales et éthiques liées à l'utilisation des neurosciences', especially p. 55.
47. V. de Vogel (2005), *Structured Risk Assessment of (Sexual) Violence in Forensic Clinical Practice. The HCR-20 and SVR-20 in Dutch Forensic Psychiatric Patients* (Amsterdam: Dutch University Press), p. 141.
48. See M. J. Hassett and D. G. Stewart (1999), *Probability for Risk Management* (Winsted, CT: Actex).
49. On the link between risk management and the sophisticated analysis of the role of time and future in contemporary culture, see J. Arnoldi (2009), *Risk. An Introduction* (Cambridge: Polity Press), p. 146.
50. S. Stoléru (2009), 'Communication During the General Debate', in Centre d'analyse stratégique (ed.), *Perspectives scientifiques et légales*, p. 64.
51. T. Moffitt and A. Caspi (2006), 'Evidence from Behavioral Genetics for Environmental Contributions to Antisocial Conduct', in Wikström and Sampson (eds), *The Explanation of Crime*, pp. 108–52, at 109.

52. Ibid., pp. 109 and 142.
53. S. Tassy (2009), 'Contribution to the Debate', in Centre d'analyse stratégique (ed.), *Perspectives scientifiques et légales*, p. 29.
54. Markowitsch and Siefer (2007), *Tatort Gehirn*, pp. 229–31.
55. Ibid., pp. 237f.
56. Similar utopian visions have underpinned criminological debates since the late eighteenth century. See Becker (2002), *Verderbnis*, pp. 16f.
57. Byk (2009), 'Les difficultés légales et éthiques', pp. 50–6, at 54; O. Oullier (2009), 'Presentation', in Centre d'analyse stratégique (ed.), *Perspectives scientifiques et légales*, pp. 7–8, at 8.
58. Sève (2009), 'Ouverture', p. 6.
59. S. Tassy (2009), 'La relativité du concept de comportement normal', in Centre d'analyse stratégique (ed.), *Perspectives scientifiques et légales*, pp. 19–23, at 19. On prediction, see also J. Dunagan (2004), 'Neuro-Futures. The Brain, Politics, and Power', *Journal of Future Studies* 9(2), pp. 1–18, at 13.
60. Rose (2003), 'The Neurochemical Self and its Anomalies', p. 418.
61. See Markowitsch and Siefer (2007), *Tatort Gehirn*, pp. 206f.
62. C. Vidal (2009), 'Contribution to the Debate', in Centre d'analyse stratégique (ed.), *Perspectives scientifiques et légales*, pp. 32f.
63. M. S. Pardo (2006), 'Neuroscience Evidence, Legal Culture, and Criminal Procedure', *American Journal of Criminal Law* 33(3), pp. 301–37, at 302f.
64. Ibid., pp. 304 and 322.
65. Dunagan (2010), 'Politics for the Neurocentric Age', pp. 59f.
66. Cf. Markowitsch and Siefer (2007), *Tatort Gehirn*, pp. 105–8.
67. H. Chneiweiss (2009), 'Les neurosciences, nouvelle branche de la médecine légale?' in Centre d'analyse stratégique (ed.), *Perspectives scientifiques et légales*, pp. 45–50, at 49.
68. Morse (2006), 'Brain Overclaim Syndrome', p. 403.
69. Ibid., p. 412; Jones and Goldsmith (2005), 'Law and Behavioral Biology', p. 421.
70. D. Sicard (2008), Opening remark to the discussion 'Les Enjeux ethiques, philosophiques, cliniques psychologiques, sociaux, juridiques et economiques', in A. Claeys and J-S. Vialatte (org.), *Exploration du Cerveau, Neurosciences. Avancées scientifiques, enjeux éthiques* (Paris: Office parlamentaire d'evaluation des choix scientifiques et technologiques), pp. 65–6.
71. On media generated authority, see S. Herbst (2003), 'Political Authority in a Mediated Age', *Theory and Society* 32(4), pp. 481–503.
72. Cf. B. Latour (1987), *Science in Action. How to Follow Scientists and Engineers through Society* (Cambridge, MA: Harvard University Press), p. 104.
73. *Oberösterreichische Nachrichten*, 4 October 2006; *Frankfurter Allgemeine Nachrichten*, 19 September 2003; *Süddeutsche Zeitung*, 13 January 2004; *Profil*, 23 September 2002; *Frankfurter Allgemeine Sonntagszeitung*, 5 November 2006; *Focus*, 8 October 2007.
74. *Süddeutsche Zeitung*, 18 October 2005; *Spiegel*, 30 July 2007; *Focus*, 8 October 2007.
75. A. Miller (2001), guest article in the *Frankfurter Allgemeine Zeitung*, 6 October.
76. Article (2008), 'Kein Ort für Jungs', *Badische Zeitung*, 8 November.
77. J-P. Changeux (1998), *Neue Züricher Zeitung*, 17 June.

Part II
Diagnosis and Therapy

7

The Psychological Sciences and the 'Scientization' and 'Engineering' of Society in Twentieth-Century Britain

Mathew Thomson

It is tempting to see the increasing influence of psychological sciences in the twentieth century as an exemplary case in a history of the 'scientization' and 'engineering' of society. In Britain, just as in the Netherlands and the United States, and indeed across the Western world, the psychological sciences and their technologies, standards, and languages have come to play an evident role in shaping the way modern life is governed, understood, and conducted. This is not in dispute. However, the nature (and thus degree) of this influence, as well as the route to this influence, is complex. As such, rather than offering yet another case study of the inexorable advance towards a psychological and therapeutic society, this chapter highlights reasons for a more critical and probing attitude towards processes of (psychological) scientization and engineering of society.

Elsewhere in this volume, Harry Oosterhuis demonstrates the increasing influence, from the mid-nineteenth to the twenty-first century, of ideas and practices of mental health in the Netherlands. In other contributions, we see the advance of theories and applications of psychology in the management of social life in European contexts and in the United States, ranging from a role in the organization and delivery of education, the management of workforces in industry, and in gauging and manipulating public opinion whether in the interests of managing democracy or increasing sales. This is in line with an emerging body of scholarship on the growing importance of psychological science and its applications in these areas of modern life and government.[1] This scholarship also highlights several areas of influence not covered in detail in this volume, perhaps most notably in the organization of the armed forces and the management of fear, but also in the development of penal welfare in reshaping ideas about responsibility and thus challenging existing attitudes towards punishment.[2] What makes the case for the influence of this branch of the social sciences even more compelling, and perhaps even unique, is that such psychological thought and practice was also taken up at a popular level, and thus we find an emerging historiography on how psychological governance overcame the problems of access and resistance that beset other attempts to intervene on individual subjects, and instead became embedded in everyday life and internalized within the individual psyche. In an era in which confidence in the onward march of the state has waned, such an alternative route towards government of the modern self is emerging as a rich vein for historical enquiry.

When it comes to the case of Britain, we already have a powerful case for such a story of advancing influence for the '*psy* disciplines' in the works of sociologist Nikolas Rose.[3] Drawing on Foucault, Rose describes this process in the language of governmentality. He demonstrates how psychological science provided languages and technologies that would shape the modern self, initially driven by the development of applications in the health services, school, workplace, and army, but extending particularly in the second half of the century to shape the way individuals understood and managed their own capacities.[4] Rose follows Foucault in seeing governmentality as the 'common ground of all our modern forms of political rationality' and crucially one that focused on the target of population, and he sees the psychological sciences as important in linking subjective and intersubjective life into the realm of such governmentality.[5]

Here, two features were crucial. First, governmentality is dependent upon knowledge of populations, and the psychological sciences would help to identity and isolate subjectivity as an object for government: 'to make its features notable, speakable, writable, to account for them according to certain explanatory schemes'. Secondly, it required processes that could transcribe such features of the population into calculable form: into maps, diagrams, and, crucially, numbers. As such, it involves historians exploring the role of administrative techniques, bureaucratic processes, and the machinery of forms and tests that have become such an integral part of modern life.[6] However, Rose also acknowledges the limitation to such authority: a consequence both of ongoing dispute over truth claims within psychological science, and the constraints on power over individual subjectivity within a liberal democracy. As a consequence, it has been the ability of psychological science to embed its techniques within everyday life and to extend its truth claims to the general public, and thus to create systems of self-regulation, that have been crucial to this modern system of psychological government.[7]

Rose's work has become the standard reference point for any historian wanting to point to the increased influence of the interrelated complex of psychology, psychiatry, and psychoanalysis in modern Britain. He presents us with a compelling and seductive equation between the creation of new psychological languages, norms, and techniques, and the making of subjectivity within a liberal democracy. On the other hand, it is worth recognizing that he does so without too much concern for some of the questions that the historian who turns to the subject more closely might worry about. Such historians may, for instance, be left with the task of examining and testing the extent of this psychological influence, its nature in practice, the interplay with other systems of knowledge, professional interests, and existing values, and the possibility that changes may instead have been driven by more fundamental underlying social and economic structural forces.

The language of 'engineering society' and 'scientizing society', adopted by this volume, is not prominent within the existing historiography on the influence of the psychological sciences in modern Britain, or for that matter on the more general influence of the social sciences in the country.[8] Nevertheless, the subject of enquiry at the heart of this volume is one that is attracting increasing attention within the historiography on modern Britain. Firstly, and specifically, there has been the emergence of a historiography that parallels Rose's project and which begins to add the sort of historical nuance that one would expect from more detailed historical research.[9] Secondly, and more generally, there is currently a turn towards paying greater attention to the impact of the social sciences in shaping the categories of analysis that are central in the writing of history of this period: a move towards seeing these sciences as actors in the history of the period rather than just tools for describing it. This has included work on the history of social anthropology in the making of ideas of Englishness,[10] and, most recently, it has put that most central of categories in British history – class – under the microscope, directing us to a history of the role of social science in shaping the evidence of the surveys about social life that have been the bread and butter of social history, and also in shaping the way that contemporaries understood themselves.[11]

Though the terms 'engineering society' and 'scientizing society' are not used in such work, they would not be out of place. For the purposes of this chapter, and without strong precedent, the terms may at times be used rather imprecisely and interchangeably, both describing a broad process in which the psychological sciences came to govern, shape, and give meaning to aspects of modern life and society. On the other hand, though closely linked, there are some potential differences between the two terms that seem worth distinguishing at this point, and the chapter will attempt to differentiate accordingly. The process of 'engineering' will be used to refer more specifically to practical interventions at a group or individual level, shaped by and contributing to a broader agenda to organize and shape society; while 'scientization', which is the more awkward and unusual term in English,[12] will refer to the process in which science (or social science) gains influence in society, and the degree to which the resulting practices are in turn genuinely reflective of this scientific orientation (or, for instance, were in fact adapted in processes of translation and accommodation away from this initial form, thus putting the process of scientization in question). In the end, the chapter devotes more consideration to scientization than engineering – in part because the framing of the question is more novel in the British context; in part because analysis of the problems of assessing processes of engineering demands a type of complex analysis of multiple systems and the limits of implementation, which is difficult, other than in a cursory fashion, in such a survey.

As already noted, it would be very possible to write this account of the psychological sciences in Britain and the engineering and scientization of society as one of onward expansion and permeation; indeed, it is difficult to see how any other basic trajectory would be conceivable. For instance, the picture of the advancing reach of psychological sciences in one area – that of medicine – and the expanding domain of 'mental health', carefully set out in Harry Oosterhuis's study of the Netherlands in this volume, could largely be replicated for Britain. Of course, there are some intriguing cultural differences that could be explored, and very probably some significant differences of timing and degree. Impressionistically, it does seem to be the case that the embrace of mental health in the Netherlands in the final third of the century may have been more enthusiastic than in Britain, with the Netherlands perhaps catching up with, and overtaking, British development in this period. Likewise, the British always recognized themselves as lagging well behind the pace and scale of development in the United States, and this seems justified.[13] But we will not be in a place to compare, and then hopefully explain, such variations until these studies of national contexts move beyond offering us a patchwork of discrete developments and pronouncements to provide us with a clearer sense of the scale and nature of development. For instance, in the area of mental health, how many psychiatrists, psychologists, psychotherapists, and psychiatric social workers are we talking about; how many people were treated by each of these, for how long, and in what ways; and, in the realm of popularization, how many people bought popular advice books or belonged to self-help organizations? Even then, the fact that these phenomena are so culturally inflected and mediated by differences in language and law means that the comparative historian will face a very considerable challenge.

The purpose of this chapter instead is to provide an analysis that may encourage more critical reflection of the processes at stake. Of course, the period saw the advance of psychological science and its application, but this is only part of the story. An analysis of the British story aims to highlight the gap between aspirations and reality, processes of resistance, accommodation, and adaptation, and thus the need to temper our perspective on the engineering and scientization of society. It leaves us consequently with a picture of scientization, if we are to use this term, which was a bottom-up as well as a top-down process. The reader may take from this the idea of some kind of British exceptionality. This would be a mistake; there may be something to such a thesis of course, but the aim instead is to highlight processes that complicate the history of the engineering and scientization of society in all national settings.

British psychological sciences and the scientization of society

Before proceeding any further, it is necessary to reflect upon and discuss the definition of the object of enquiry at the centre of this analysis: the psychological sciences. This is one of those categories that the historian applies to the past, rather than one that has clearly defined boundaries and identity in that past, and this needs to be highlighted but also justified. It is a category that brings together the histories of psychiatry, psychology, psychotherapy, and psychoanalysis. Each of these has its own complex history. Each sits in tension with the others. And in truth each could be subject to its own paring exercise, revealing a host of competing subfields and professional rivalries. However, one way of looking back at these histories is to regard the boundaries as manufactured by the politics of disciplinary and professional formation rather than as in any sense natural and fixed. This is the approach taken in this chapter. To do so also highlights the fact that the '*psy*' disciplines and professions were for much if not all of this period in a state of evolution, sometimes a state of emergence, and often a state of considerable internecine warfare and dispute, and thus one of insecurity over their scientific status: it is not a picture of a coherent or unified system.[14] For this fundamental reason, scientization of society remained both a vital aspiration for a field so keen to prove itself, but also something that was difficult to gain sanction for in practice.

Even at the start of the twentieth century, there is a good case for arguing that psychiatry, for instance, was not yet a fully formed and thus fully authoritative profession or medical science when it came to Britain.[15] Much of the most important work in extending the reach of psychological medicine in this period, which aimed at tackling borderline conditions and preventing mental illness in the community, had in fact relatively little to do with the still heavily stigmatized institutions for the mentally ill.[16] Here we see longstanding efforts at addressing conditions of the nerves developing into psychotherapeutic work on the neuroses and psychoneuroses, which was more often the work of elite private practitioners catering to the problems of the rich worried well. We see charities looking to address the problems of personality and social adjustment that threatened the social order. And we see a bohemian fringe who turned to psychotherapy as part of a broader taste for a new age lifestyle.[17] These new orientations towards the borderline, towards prevention of mental illness and promotion of mental health, and towards psychotherapy would, as often as not, be viewed initially as little more than quackery or vulgar populism by medical scientists.[18] Even within medicine, those who turned to psychological approaches did so in part out of dissatisfaction with the alienating style of an overly materialist medicine: psychology appealed as a way to humanize medicine, and thus to make it in some respects less rather than more scientific.[19] Psychotherapy also appealed to churches looking to modernize their pastoral and confessional work, and to doctors who saw space for Christian values in the psychodynamic armoury.[20] There were even strong intersections with mysticism and the occult.[21] And perhaps most significant of all, the fragility of the boundaries of expertise on the one hand, and the enthusiasm for the liberating and self-improving potential of a psychology that opened up a hidden dimension to the self on the other, meant that the floodgates were open to a host of populists who translated science into vernacular practical advice, and in doing so challenged and threatened to undermine the scientific reputation of the field.[22]

Of course, the boundaries of what counted as scientific have changed – this was a period in which several of Britain's leading scientists were closely involved in the field of psychic research or in efforts to bring together science and religion, for instance – but the lingering association with the esoteric, the unorthodox, and the popular should be acknowledged as complicating our assumptions about the advance of such approaches, too simplistically, as processes of scientization; at the very least, it points to the need to reorientate our perspective from seeing scientization as a top-down and monolithic process, to one that could emerge from below in heterodox form.[23] This applies particularly to the early part of our period, but suspicion of the scientific status of psychological medicine, and dispute between its different branches, which both reflected and contributed to this status of scientific insecurity, persisted in a way that marked out this area of medicine from the more secure verities of an increasingly laboratory-, drug-, and technology-driven scientific medical world after the Second World War. Psychoanalysis, in particular, an increasingly dominant presence in the psychotherapeutic field, was subject to attack on grounds of its scientific basis.[24] Even towards the end of our period, as psychiatry turned first to drugs as a tranquillizing panacea for the stresses of modern life and then to neuroscience, there remained a good deal of scepticism about whether questions of mind, emotion, and human relationships could be turned into a science of chemicals and brains, and thus a new era of the neurochemical self.[25]

As with psychiatry, psychology was an embryonic field at the start of the twentieth century, and one that still had to prove itself as science. It sat at the cusp of the fields of philosophy and physiology, and in the nineteenth century had been more closely allied to the former than the latter.[26] It lagged well behind the experimental tradition that was emerging in Germany and the United States, a fact reflected in the lack of laboratories and university positions: at the start of the twentieth century, the only (small) research laboratories were in University College London and Cambridge; and it was not until 1919 that Britain had its first chair in psychology.[27] A new ambition and sense of identity was heralded by the formation of the British Psychological Society in 1901 and the *British Psychological Journal* in 1904; membership of the Society did grow slowly across the interwar period to reach 800 at the start of the 1940s.[28] However, much of this expansion depended on the ability of the field to develop as an applied rather than theoretical discipline. As Nikolas Rose has argued, in the interwar period, the discipline expanded less through its scientific authority than through an ability to measure human capacity that proved useful in an increasingly complex society looking to organize and manage the populations of systems such as education, employment, social welfare, and the armed forces.[29]

Another difficulty with the concept of 'scientization' when it comes to psychology is that the applied focus on the problems of society, and the work of sifting through and organizing the mass of data that emerged in the psychological calibration of the stresses and strains of modern systems of government, meant that, as a theoretical field, psychology's scientific identity would in the long term be torn between that of a pure, laboratory-based, experimental, and physiologically orientated science and that of a social science. Here, it had to compete against the claims of other disciplines that came to the fore at much the same time. Initially, its links to physiology and to a Darwinian paradigm, and its insights thereby on causation, may have helped it in developing a scientific identity among the emergent social sciences: it was offering something more than the mere observational and social survey tradition that was so strong in British sociological work.[30] And it is significant that people coming from a psychological background became influential in others of the new social sciences of this period: early issues of the flagship journal *The Sociological Review* regularly gave a voice to psychologists, and at the country's leading academic venue for the study of sociology, the London School of Economics, there was a distinctive psychological hue to the work of leading figures such as Graham Wallas, Leonard Hobhouse, and Morris Ginsberg. But in the longer term, as these psychological models were exposed as propositions more often than verifiable truths, psychology found itself struggling to extend its claims as a science of individual behaviour to one that could explain the behaviour of groups and societies; or do so in ways that trumped the explanations centring on social structure and economics that came to dominate the postwar social sciences.

A final, ongoing problem was that the field developed as one of myriad competing schools of thought: behaviourist, psychoanalytic, physiological, and statistical, for instance. This lack of theoretical unity, like the division between a role in the laboratory or as an applied tool of management, and like the division between its orientation as a science or social science, all complicate our thinking about the role of psychology in the scientizing of society: hence the attraction here of talking about the psychological sciences rather a single science or system. Such lack of unity contributed to a relatively low status and to dispute over the scientific basis of certain psychological claims.

The insecure scientific status of the psychological sciences is important in modifying conclusions about scientization, but there is a paradox here: at times, such a situation provided opportunity and incentive for experts in this field to be more, rather than less, ambitious about their claims as scientists of society. This is particularly apparent in the first decades of the century. Here, there was a conjuncture of two developments. The first was that psychology turned to the social. This was partly spurred on by the contact with other cultures in the intersecting nascent field of British anthropology, encouraging a generation of influential British psychologists and their pupils to look beyond the individual to see all individuals as set in a field, and all psychology as necessarily social. And it was partly a product of the way a post-Darwinian, capacious instinct theory provided scope for speculation about individual subjectivity having inbuilt inclinations towards certain patterns of social behaviour – the 'herd' instinct, in one of the era's most notorious instances of such thinking – and having social values embedded in the individual psyche as a necessary brake on the primitive power of raw instinct.

The second development was an exponential shift in the responsibilities and expected future responsibilities of the state, associated politically in Britain with the rise of the Labour Party and in particular the 'New Liberalism', which turned social behaviour into an increasingly pressing issue of statecraft.[31] This was given ongoing momentum through the practical problems of managing fear and the resulting anxieties about human nature associated with the First World War and then the social, economic, and political difficulties associated with the depression of the interwar years.[32] It was also significant that the British social sciences during this period were relatively poorly developed – still in a 'gentlemanly', ethnographically orientated era, in the words of one recent analysis – and thus that there were opportunities for psychology to make some surprising inroads.[33] For these reasons, despite its insecure scientific status, the social orientation and resulting ambition of psychology was marked in the interwar period, reaching a peak in the crisis of confidence over human nature and the opportunities for planning in the Second World War and its immediate aftermath. Psychology in this period offered not just science but values.[34]

Without the stricter limits that come with disciplinary orthodoxy, Wilfred Trotter – surgeon by background rather than psychologist – provided one of the most influential early instances of this turn to the psychology of society in his 1919 study, *Instincts of the Herd in Peace and War*, a manifesto for psychological expertise in understanding the challenges of both domestic and international politics.[35] A similar message came from William McDougall, who probably has best claim to being the leading British psychologist of this generation.[36] And in this instance we see the dangers for scientific status of such claims for a role in statecraft, with McDougall's reputation within the discipline never fully recovering from his reactionary pronouncements on how psychological insight needed to be listened to in defending against racial decline.[37]

In the interwar era, this social orientation persisted.[38] Spurred on by a war-borne anxiety about loss and aggression, but also exposed to the problems of behaviour in expanding systems of education, penal welfare, and child guidance, psychologists began to offer up psychological explanations for the social problems of the day.[39] In the long term, the argument that the infant child needed the constant affection of a mother, which was already developing in this interwar context and found powerful further validation in the problems encountered in wartime evacuation of children, provides us with the most important legacy of how such a psycho-social orientation could have a deep impact on society: postwar 'Bowlbyism' taking the blame, even if in truth only a contributory factor, for the return of working women to the home at the end of the war and the difficulty of achieving equality in the postwar decades.[40]

New opportunities emerged with increased state activism, pessimism regarding human nature, and a recognition of the importance of morale, in the Second World War.[41] Under such circumstances, it was tempting to extend a language of psycho-social pathology to diagnoses of national character and political health – mental health being an essential prerequisite for a democratic, tolerant, and peaceable society. As with Trotter's earlier foray into such a line of analysis in the context of the First World War, it was difficult for such experts to remain objective in the context of wartime patriotism, particularly since there was also considerable excitement about the opportunities for influence and for establishing a vital role for the field. Indeed, British psychological scientists were at the very centre of efforts to extend this equation between mental and political health to the international stage via agencies such as the World Health Organization, UNESCO and, in particular, the World Federation for Mental Health, which looked to science to help build a healthy postwar world order. For a while, it seemed that mental health could be a vital *mantra* and goal in this era.

If ever there was a moment of opportunity for the psychological sciences to scientize society, the mid-century seemed to be it. Even in the more pragmatic and grounded context of managing wartime morale and then debate about postwar reconstruction at home, for a time it seemed that the insights of psychological science would have to be given far more serious consideration than in the past.

Practical achievements and real influence, however, proved a pale shadow of the often grandiose rhetoric, and hard-nosed politicians and bureaucrats had little sympathy for some of the wilder psychoanalytic speculation. This is why the influence was so marked on the international stage with a group like the World Federation for Mental Health, whose early business it was to offer expert visions and inspiration rather than practical solutions.

At home, there were real achievements, but these were in the domain of extending psychiatric and psychological provision, rather than in offering broader blueprints for social organization. Those outliers who still offered grand pronouncements about the relationship between mental health and national character now found the crude leaps of faith, generalizations, and lack of evidence base come under scientific attack, even ridicule.[42] Even the *idea* of mental health attracted criticism from a social science looking to rid itself of such loose, value-laden tools.[43] The social orientation of psychological science found a focus instead in efforts to construct a more secure and sophisticated science, and as such its gaze narrowed to a focus on particular problems and sites, and to the complexities of a field of social influences on individual subjectivity.[44] One advocate of a more ambitious role for mental health as a tool in the shaping of society reflected how far the postwar ambition had slipped by the start of the 1970s.[45] Instead sociologists, with their studies of social mobility and the breakdown of traditional community structures, or the discoverers of new forms of poverty and social deprivation set the agenda. There was nothing emerging from Britain to compare with a series of influential, sweeping psycho-social diagnoses of postwar American culture from figures like David Riesman and Christopher Lasch.

Perhaps the closest the psychological sciences came to an influential message about the broader organization of society was when they adopted the language of diagnosis, sickness, and health. As already noted, eugenic concern about national fitness, which reached a peak in the first decades of the twentieth century, was significant in this regard. Though, here, when psychiatrists and psychologists warned of the dangers of the feeble-minded to the eugenic stock and to the social and moral fabric of the nation, they were validating and reinforcing social, moral, and economic anxieties that already existed; and it is revealing that when critics set out to attack eugenic policies, they often targeted the scientific basis, recognizing ongoing public suspicion of such claims, and were invariably able to back this up with a lack of consensus among the experts.[46] By the interwar period, cultural pessimism, which had initially fuelled the eugenic vision of eliminating the unfit, and which rapidly ran up against the limits of British traditions of liberty of the subject, turned instead towards a discourse of mental health and hygiene. This could incorporate the ongoing eugenic concern, but it could marry it with a more optimistic and broad-ranging strain of psycho-social diagnosis and resulting arguments about the need to reshape society accordingly. Here, the psychological sciences were deployed in calculations about the economic and social cost to the nation of a very broadly defined mental ill health. Social phenomena such as unemployment, delinquency, sickness or morale were presented either as being caused by underlying psychological circumstances or as having psychological consequences.[47] However, once again the rhetoric was far more impressive than the actual achievements. It was easy for politicians and the media to pay lip service to the importance of such mental hygiene, and this concern about the cost of mental health to the nation has resurfaced in recent years in the new language of happiness and well being, and perhaps in recognition that social problems are invariably also mental ones. Apart from small steps in supporting the extension of mental health services, however, there was little sign of willingness to address the radical underlying message that the very planning of society needed to embrace this psychological economics.[48]

In summary, it is undoubtedly a significant part of the history of the influence of the psychological sciences that they have presented opportunities for a 'scientization' of society: as a vehicle for audit, diagnosis of ills, or the key to understanding the population, and thus as a vital tool of national and international statecraft. But rarely if ever were such grand visions realized. One of the key reasons for this lies in the fragile scientific authority in this area. And because of this, scientization tended to be most successful when it was supported, often for additional reasons well beyond the realm of scientific argument, by government, social bodies, and the public. In other words, scientization – if we are to use this concept – was a process driven as much from outside as within the bodies who represented themselves as its main authorities.

The psychological sciences and the engineering of society

As noted in the introductory section, one of the tasks this chapter has set itself is that of exploring the differentiation of a process of 'engineering' society from one of 'scientization' – the former an issue of practical achievements, the latter a more intangible and diffuse issue concerning the influence of theory and its status as science. The chapter now turns to the engineering of society. Again, it takes for granted the significance of an expansion of influence over the course of the century and elects instead to highlight the way in which such a narrative needs, nevertheless, to acknowledge a series of significant limitations and qualifications to this line of argument. It focuses in particular on the role of psychological science as an applied discipline in relation to education, healthcare, and the world of work.

The role in education via mental testing was probably the most significant of all these instances of the psychological sciences providing applied tools that would become part of the fabric of society. Here, the demands of a system of universal state elementary education, which had come into formation after 1870, overcame the still insecure status of psychology at the start of the twentieth century. However, it was need on grounds of efficiency, social differentiation, and an ideology of meritocracy (aiming to organize a system of mass education in terms of ability) that drove testing forward – not, primarily, the authority of psychology. There was ongoing suspicion of the latter (and hence, in fact, only a tentative adoption of mental testing despite the advance of selection in the British context).[49] In other circumstances, such as the workplace, where the subjects being tested had more of a voice than the child, the advance of psychological testing was even less impressive. As Adrian Wooldridge has suggested, the success of mental testing owed as much to a broader psychological message of the importance of child-centredness and shaping education to the needs of the individual child, as it did to what Nikolas Rose has described as a 'psycho-eugenic' form of social engineering (excluding mental defectives, and turning the education system into a filter to promote a more eugenically fit population).[50] Indeed, close analysis of the debates leading to parliamentary acceptance of legislation to confine 'mental defectives' and to set up separate special schools for backward and 'feeble-minded' children in 1913 and 1914 indicates the efforts made by supporters of the legislation to obscure the role of experts (and that of either eugenics or psychology) and to emphasize instead the importance of social arguments about the need for care and control.[51] In the same vein, though it attracts less attention than the mental test, the use of psychology to bolster the case for a progressive, child-centred 'learning-by-doing' and a play-centred pedagogy was just as important a development. Here, teachers, often eager students of psychology via new training programmes, provided the type of key intermediary role performed elsewhere by social workers: they extended and translated (and in the process most likely diluted) the role of psychological theory and practice in the engineering of the population.[52]

This points towards very different ends from that of a psycho-eugenic engineering. On the other hand, these new teaching methods would likewise struggle against entrenched academic and disciplinary values and the harsh realities of the under-resourced state classroom. In the end it still seems fair to conclude that nowhere was psychological engineering more significant than in the way the theory and techniques of psychology shaped both the structure of education and the delivery of teaching in the classroom. The shaping of structure is seen, for instance, in the tripartite system of grammar, technical, and secondary modern schools, devised to mirror different types of mind among secondary school children and to facilitate a ladder of opportunity based on intellectual merit via testing. Yet psychological theory could just as readily be drawn upon by critics of these structures and modes of pedagogy. And ultimately we need to recognize that policy and practice were driven by political rationality and (as one analysis drawing on the words of the figurehead behind the famous 1944 Education Act put it) 'the art of the possible'.[53]

In the workplace, it would always prove very difficult to satisfy the competing interests of workers and employers, and to demonstrate the worth of Taylorite efficiency engineering, though British industrial psychologists attempted to do so by presenting their role as one of humanization rather than 'engineering' – the latter term striking the wrong tone for British tastes.[54] In the context of the Second World War, the productionist demands of the state would boost the role of a human-relations psychology in management.[55] However, when the labour supply increased in peacetime circumstances, employers and the state were less inclined to invest finance and authority in industrial psychology. In the long term, economic bargaining, not psycho-social engineering, remained the crucial determinant of the British workplace and its ailing industrial relations and efficiency in the second half of the century.[56]

In the workplace, and even the school, and then in the expanding healthcare system, ambition for the role of applied psychology was also held back by the sheer lack of trained experts, resources, and will, given first the economic cuts of the interwar period and then the expanding demands on the state after the Second World War. Claims, by the 1930s, that mental health was at the root of a third of all days missed in sickness might have been taken quite seriously well beyond the mental hygiene movement itself; but the resources to address a problem of the 'sick society' and to extend psychotherapeutic efforts substantially beyond the seriously mentally ill did not exist.[57] Here, the role of psychiatric social workers in translating ideas about mental health into everyday social policy certainly deserves far more consideration than it has attracted to date; though we also see calls by the 1970s for a shift away from psychological to more social models – a move from analysis based on pathology and adjustment to one centring more on inequality and access to state resources.[58]

The postwar decades also saw the development of a psychiatry that explored the epidemiological relationship between environment and mental health, highlighting in particular some of the ways in which modern urban life added to the stresses of the population; but there is little indication that this work was able to make the next step of changing policy and shaping a healthier environment.[59] Psychological therapy for the unhappy (and not yet seriously debilitated) remained largely something for those who could afford it, rather than something that the state was able or willing to use as a tool of engineering. This remained the case after the Second World War, despite calls by a figure such as Michael Balint for a national mental health service and subsequently for psychological therapy as an integral element of general practice.[60] Even as late at the mid-1970s, unless you were rich you would be lucky to find a psychotherapist, there being just 550 of them in practice (compared to 5500 by 1999).[61] In that sense, the widespread reality (rather than the vision) of a therapeutic society is in fact a very recent phenomenon when it comes to Britain. Despite renewed calls for the state to extend the availability of mental health services and psychotherapy within the National Health Service, past history suggests that ambition will continue to surpass reality by a long way.

In fact, throughout the period under discussion, the main influence of psychology within everyday life was less likely to be imposed from above than actively embraced, consumed, and appropriated from below. In the process, knowledge, practice, and the norms of psychological engineering tended to be translated to serve popular tastes. Such popularization often entailed a rejection of professional claims to a unique authority and shifted the focus from the pathological to the cultivation of the self. It also invariably involved an accommodation with existing values and desires. There has been a tendency to regard popularization of such knowledge and practice as an extension of professional power, and thus of an engineering of society – indeed, to suggest that the internalization of psychological norms, expectations, and definitions at an individual level saw engineering of society turn into an even more powerful governance of the soul. However, this has yet to move beyond assumptions based on popularity of texts to sensitive studies of reception at a social, let alone an individual, level.[62]

In the British context, popularization was just as important a phenomenon before as it was after psychology itself had assumed a reasonable degree of professional and scientific authority. And, particularly in some of its early manifestations, the enthusiasm for psychology included an element of hostility towards professionals because of perceived efforts at policing the popularization of psychological thought and practice. This, of course, questions the idea that the increasing permeation of everyday life and popular understanding of the self was simply a top-down process. An indication of this is the way in which there remained a popular appetite for a psychology that left the door open for, or reframed, a spiritual component to self-development.[63] There was also little popular sympathy for the crude mental testing that has gained so much attention in studies of the advance of applied psychology, or for what was regarded as the determinism of behaviourism or the pathological introspection of psychoanalysis. And there was a common critique of the expert and his attempt to police and monopolize the field, with an emphasis instead on practical routes towards taking control of one's own psychological potential – all attractive at a time when the resources for treatment and therapy were so limited.[64]

Conclusion

This chapter has attempted to problematize the idea that the psychological sciences in twentieth-century Britain provided an authoritative system and set of tools for the scientization and engineering of society. In doing so, it proposed certain modifications to Rose's influential model of governmentality in this area. It suggested that the influence of the psychological sciences was perhaps both more and less than often assumed: *more* through acknowledging the considerable excitement about the way these sciences opened up a new map of society – a history, albeit in large part of ambition (and fluctuating ambition according to circumstances) rather than practical results; more, too, through broadening our notion of popularization to move beyond the diffusion downwards of psychological norms; but *less*, probably, if we look at the relative underdevelopment (in comparative perspective) as an applied discipline, judged in terms of therapeutic resources and the numbers of psychologists working as engineers of society in areas such as education, the workplace, and the clinic; and less too if we recognize that influence depended on processes of translation and accommodation with existing values and political realities.

By all means, we can see the diffusion of psychological thinking and practice to a popular level, and its potential internalization, as strengthening the case for its role in 'engineering' and 'scientizing' society. But in doing so, if we collapse into a single whole one form of psychological engineering and scientization coming from above that was so obviously often in tension with another coming from below, we are in danger of doing an injustice to the richness of this history.

We are now beginning to have a far clearer picture (albeit complex and nuanced) of the shifting discourses available for psycho-social engineering and scientization, and of the way in which broader political, ideological, and socioeconomic circumstances may have affected the viability of turning ideas into practice – though we need to go much further when it comes to the postwar era in particular. What also remains to be seen, however, is whether we can move from an understanding of discourses, social practices, and institutional techniques to an understanding of how social engineering does, or does not, map on to individual subjectivity.

Notes

1. These issues are explored in greater depth in M. Thomson (2006), *Psychological Subjects. Identity, Health and Culture in Twentieth-Century Britain* (Oxford: Oxford University Press).
2. Indeed, these issues have attracted some of the most extensive attention in the historiography on Britain, where coverage has been uneven. There has been a particular interest in shell-shock in the First World War, and more recently in the role of psychological thought and practice in relation to homosexuality. For instance, M. Stone (1985), 'Shell-Shock and the Psychologists', in W. Bynum, R. Porter, and M. Shepherd (eds), *The Anatomy of Madness*, vol. 2 (London: Athlone), pp. 242–71; B. Shephard (2000), *A War of Nerves. Soldiers and Psychiatrists, 1914–1994* (London: Jonathan Cape); C. Waters (1998), 'Havelock Ellis, Sigmund Freud and the State. Discourses of Homosexual Desire in Interwar Britain', in L. Doan and L. Bland (eds), *Cultural Sexology. Labelling Bodies and Desires, 1890–1940* (Cambridge: Polity), pp. 165–79; I. Crozier (2000), 'Taking Prisoners. Havelock Ellis, Sigmund Freud and the Construction of Homosexuality, 1897–1951', *Social History of Medicine* 13, pp. 447–66.
3. N. Rose (1985), *The Psychological Complex. Psychology, Politics, and Society in England, 1869–1939* (London: Routledge); idem (1989), *Governing the Soul. The Shaping of the Private Self* (London: Routledge); idem (1996), *Inventing Ourselves. Psychology, Power and Personhood* (Cambridge: Cambridge University Press).
4. The first process is the particular emphasis of Rose's *Psychological Complex*. The second is the focus of the latter sections of *Governing the Soul*, which takes the story from the first to the second half of the twentieth century.
5. Rose (1989), *Governing the Soul*, p. 5.
6. Ibid., p. 6.
7. Ibid., pp. 10–11.
8. Though Rose gets close to this language in N. Rose (1997), 'Assembling the Modern Self', in R. Porter (ed.), *Rewriting the Self. Histories from the Renaissance to the Present* (London: Routledge), pp. 224–48.
9. Particularly in the area of the history of what Rose calls 'psycho-eugenics' – the history of the feeble-minded or mentally deficient: M. Thomson (1998), *The Problem of Mental Deficiency. Eugenics, Democracy and Social Policy in Britain, 1870–1959* (Oxford: Oxford University Press); and P. Dale (2003), 'Implementing the 1913 Mental Deficiency Act. Competing Priorities and Resource Constraint Evident in the South West of England before 1948', *Social History of Medicine* 16, pp. 403–18. On the development of mental health care beyond the asylum, see L. Westwood (2001), 'A Quiet Revolution in Brighton. Dr Helen Boyle's Pioneering Approach to Mental Health Care, 1899–1939', *Social History of Medicine* 14, pp. 439–57. And in relation to expansion of concern regarding mental hygiene, see J. Toms (2005), 'Mental Hygiene to Civil Rights. MIND and the Problematic of Personhood, c. 1900 to c. 1980' (Ph.D. thesis, University of London).

10. P. Mandler (2009a), 'One World, Many Cultures. Margaret Mead and the Limits to Cold War Anthropology', *History Workshop Journal* 68, pp. 449–72; idem (2009b), 'Margaret Mead amongst the Natives of Great Britain', *Past and Present* 204, pp. 195–233; idem (forthcoming), 'Being His Own Rabbit. Geoffrey Gorer and English Culture', in C. Griffiths, J. Nott, and W. Whyte (eds), *Cultures, Classes and Politics. Essays on British History for Ross McKibbin* (Oxford: Oxford University Press).
11. M. Savage (2010), *Identities and Social Change in Britain since 1940* (Oxford: Oxford University Press).
12. In the GHI conference that inspired this volume, the keynote speaker, Lutz Raphael, in fact used the alternative term 'scientification'. The American historian of social science Dorothy Ross has used the term 'scientism', which is easier in English but may have implications about a relationship to the scientific disciplines that shift its meaning: D. Ross (1993), 'An Historian's View of American Social Science', *Journal of the History of the Behavioural Sciences* 29, pp. 99–112. This essay uses 'scientization' for consistency with the title of the volume.
13. H. Pols (1997), 'Managing the Mind. The Culture of American Mental Hygiene, 1910–1950' (Ph.D. dissertation, University of Pennsylvania); idem (2010), '"Beyond the Clinical Frontiers". The American Mental Hygiene Movement, 1910–1945', in V. Roelcke, P. Weindling, and L. Westwood (eds), *International Relations in Psychiatry. Britain, Germany, and the United States to World War II* (Rochester, NY: Rochester University Press), pp. 111–33; N. Hale (1971), *Freud and the Americans. The Beginnings of Psychoanalysis in the United States, 1876–1917* (Oxford: Oxford University Press); idem (1995), *The Rise and Crisis of Psychoanalysis in the United States* (Oxford: Oxford University Press); J. Pfister and N. Schnog (eds) (1997), *Inventing the Psychological. Toward a Cultural History of Emotional Life in America* (New Haven and London: Yale University Press).
14. R. Smith (1998), 'Does the History of Psychology have a Subject?', *History of the Human Sciences* 1, pp. 147–77; G. Richards (1987), 'Of What is History of Psychology a History?', *British Journal for the History of Science* 20, pp. 201–11.
15. M. Thomson (1996), '"Though Ever the Subject of Psychological Medicine". Psychiatrists and the Colony Solution for Mental Defectives', in G. Berrios and H. Freeman (eds), *150 Years of British Psychiatry*, vol. 2 (London: Athlone).
16. R. Porter (1996), 'Two Cheers for Psychiatry! The Social History of Mental Disorder in Twentieth Century Britain', in Berrios and Freeman (eds), *150 Years of British Psychiatry* 2, pp. 383–406.
17. J. Oppenheim (1991), *Shattered Nerves. Doctors, Patients and Depression in Victorian England* (New York: Oxford University Press); Thomson (2006), *Psychological Subjects*, Chapters 1 and 3.
18. T. Turner (1996), 'James Crichton Browne and the Anti-Psycho-Analysts', in Berrios and Freeman (eds), *150 Years of British Psychiatry* 2, pp. 144–55.
19. C. Lawrence (1998), 'Still Incommunicable. Clinical Holists and Medical Knowledge in Interwar Britain', in C. Lawrence and G. Weisz (eds), *Greater than their Parts. Holism in Biomedicine, 1920–1950* (Oxford: Oxford University Press), pp. 94–111.
20. G. Richards (2000), 'Psychology and the Churches in Britain, 1919–1939. Symptoms of Conversion', *History of the Human Sciences* 13, 57–84.
21. A. Owen (2004), *The Place of Enchantment. British Occultism and the Culture of the Modern* (Chicago and London: Chicago University Press).
22. Thomson (2006), *Psychological Subjects*, Chapter 1.
23. P. Bowler (2001), *Reconciling Science and Religion. The Debate in Early Twentieth-Century Britain* (Chicago: Chicago University Press).
24. H. Eysenck (1957), *Sense and Nonsense in Psychology* (Harmondsworth: Penguin); idem (1966), *The Effect of Psychotherapy* (New York: International Science Press); and H. Eysenck and G. Wilson (1973), *The Experimental Study of Freudian Theories* (London: Methuen).

A similar assault came from the polemicist of a biological psychiatry: W. Sargant (1967), *The Unquiet Mind. The Autobiography of a Physician in Psychological Medicine* (London: Heinemann).

25. On the turn towards neuroscience, see N. Rose (2007), *The Politics of Life Itself. Biomedicine, Power, and Subjectivity in the Twenty-First Century* (Princeton: Princeton University Press). For a critical review, see R. Cooter and C. Stein (2010), 'Cracking Biopower', *History of the Human Sciences* 23, pp. 109–28.
26. L. Hearnshaw (1964), *A Short History of British Psychology, 1840–1940* (London: Methuen).
27. Thomson (2006), *Psychological Subjects*, p. 59.
28. A. D. Lovie (2001), 'Three Steps to Heaven. How the British Psychological Society Attained its Place in the Sun', in G. C. Bunn, A. D. Lovie, and G. D. Richards (eds), *Psychology in Britain. Historical Essays and Personal Reflections* (Leicester: BPS Books), pp. 95–114.
29. Rose (1985), *The Psychological Complex*.
30. M. Bulmer (1985), 'The Development of Sociology and Empirical Research in Britain', in M. Bulmer (ed.), *Essays on the History of British Sociological Research* (Cambridge: Cambridge University Press), pp. 3–36; A. H. Halsey (2004), *A History of Sociology in Britain* (Oxford: Oxford University Press).
31. A small number of intellectuals in both parties recognized the potential importance of the new psychology: M. Freeden (1986), *Liberalism Divided. A Study in British Political Thought, 1914–1939* (Oxford: Oxford University Press), pp. 228–46; J. Nuttall (2003), 'Psychological Socialist, Militant Moderate. Evan Durbin and the Politics of Synthesis', *Labour History Review* 68, pp. 235–52.
32. S. Alexander (2000), 'Men's Fears and Women's Work. Responses to Unemployment in London between the Wars', *Gender and History* 12, pp. 401–25; M. Roiser (2001), 'Social Psychology and Social Concern in 1930s Britain', in G. Bunn, S. Lovie, and G. Richards (eds), *Psychology in Britain. Historical Essays and Personal Reflections* (Leicester: BPS Books), pp. 169–87; J. Bourke (2001), 'Psychology at War, 1914–1945', in Bunn, Lovie, and Richards (eds), *Psychology in Britain*, pp. 133–49; J. C. Lerner and N. Newcombe (1982), 'Britain between the Wars. The Historical Context of Bowlby's Theory of Attachment', *Psychiatry* 45, pp. 1–12; R. Hayward (2007), 'Desperate Housewives and Model Amoebae. The Invention of Suburban Neurosis in Inter-War Britain', in M. Jackson (ed.), *Health and the Modern Home* (London: Routledge), pp. 42–62; idem (2009), 'Enduring Emotions. James L. Halliday and the Invention of the Psychosocial', *Isis* 100, pp. 827–38.
33. Indeed, this 'gentlemanly' quality is deemed as extending into the first decade after the war, and the importance of psychology is still evident in a survey of material in *The British Journal of Sociology* from the period 1950–9: Savage (2010), *Identities and Social Change*, pp. 103–7. This interpretation contrasts with analysis of the increasing 'scientism' of the American social sciences in the first half of the century: D. Ross (1993), 'An Historian's View of American Social Science', *Journal of the History of the Behavioral Sciences* 29, pp. 99–112; idem (1999), *The Origins of American Social Science* (Cambridge: Cambridge University Press).
34. R. Smith (2003), 'Biology and Human Values. C. S. Sherrington, Julian Huxley, and the Vision of Progress', *Past and Present* 178, pp. 210–42.
35. W. Trotter (1919), *Instincts of the Herd in Peace and War* (London: Ernest Benn).
36. W. McDougall (1920a), *Anthropology and History* (Oxford: Oxford University Press); idem (1920b), *The Group Mind* (Cambridge: Cambridge University Press); idem (1921a), *Is America Safe for Democracy?* (New York: Scribner); idem (1921b), *National Welfare and National Decay* (London: Methuen).
37. R. Soffer (1989), 'The New Elitism. Social Psychology in Prewar England', *Journal of British Studies* 8, pp. 111–40; idem (1978), *Ethics and Society in England. The Revolution in the Social Sciences, 1870–1914* (Berkeley: University of California Press).

38. For instance, J. Hadfield (1923), *Psychology and Morals. An Analysis of Character* (London: Methuen); I. Suttie (1935), *The Origins of Love and Hate* (London: Kegan Paul, Trench, Trubner); E. Glover (1946), *War, Sadism and Pacifism* (London: G. Allen and Unwin).
39. The area of child guidance has attracted particular interest from historians: D. Thom (1992), 'Wishes, Anxieties, Play and Gestures. Child Guidance in Interwar Britain', in R. Cooter (ed.), *In the Name of the Child* (London: Routledge), pp. 200–19; J. Stewart (2006), 'Child Guidance in Inter-War Scotland. International Context and Domestic Concern', *Bulletin of the History of Medicine* 80, pp. 513–39.
40. D. Riley (1983), *War in the Nursery. Theories of the Child and Mother* (London: Virago). Most closely associated with the writings of psychologist John Bowlby, particularly via his influential WHO report of 1951, republished in popular form as *Child Care and the Growth of Love* (1953), 'Bowlbyism' directed attention to the importance of maternal attachment – as important as the effect of vitamins – in the first five years of the child's life: J. Holmes (1993), *John Bowlby and Attachment Theory* (London: Routledge).
41. Thomson (2006), *Psychological Subjects*, Chapter 7.
42. This was the case with the criticism of Geoffrey Gorer's notorious postwar study of Russian national character and its links to the use of swaddling in child-rearing: Mandler (forthcoming), 'Being his Own Rabbit'.
43. B. Wootton (1959), *Social Science and Social Pathology* (London: George Allen and Unwin).
44. This is evident for instance in the content of the postwar journal *Human Relations*, which was associated with the Tavistock Clinic.
45. J. D. Sutherland (1971), *Towards Community Mental Health* (London: Tavistock), p. vii.
46. E. J. Larson (1991), 'The Rhetoric of Eugenics. Expert Authority and the Mental Deficiency Bill', *British Journal of the History of Science* 24, pp. 45–60.
47. J. L. Halliday (1948), *Psychosocial Medicine. A Study of the Sick Society* (London: William Heinemann). On British mental hygiene, see M. Thomson (2010), 'Mental Hygiene in Britain during the First Half of the Twentieth Century. The Limits of International Influence', in V. Roelcke, P. Weindling, and L. Westwood (eds), *International Relations in Psychiatry. Britain, Germany and the United States to World War II* (Rochester, NY: Rochester University Press), pp. 134–55.
48. In the first decade of the twentieth century, expenditure on mental health was still being condemned as scandalous – just 12% of the NHS total despite mental health accounting for half of all disability. More fundamentally, GDP was being recast as a 'hopeless measure of welfare' and attention instead turned once again to the measurement of well being: R. Layard (2003), *Happiness. Has Social Science a Clue?* (London: Centre for Economic Performance), pp. 2, 42; idem (2005) *Happiness. Lessons for a New Science* (London: Allen Lane).
49. See, for instance, G. Sutherland (1994) *Ability, Merit, and Measurement. Mental Testing and English Education, 1880–1940* (Oxford: Oxford University Press).
50. A. Wooldridge (1994), *Measuring the Mind. Education and Psychology in England, c. 1860–1990* (Cambridge: Cambridge University Press). Rose (1985), *The Psychological Complex*.
51. Thomson (1998), *The Problem of Mental Deficiency*; Larson (1991), 'The Rhetoric of Eugenics. Expert Authority and the Mental Deficiency Bill'.
52. Thomson (2006), *Psychological Subjects*, Chapter 4.
53. D. Thom (1986), 'The 1944 Education Act. The Art of the Possible?' in H. Smith (ed.), *War and Social Change* (Manchester: Manchester University Press), pp. 101–28.
54. C. S. Myers (1927), 'The Efficiency Engineer and the Industrial Psychologist', *Journal of the National Institute for Industrial Psychology* 3, pp. 168–72. For a fuller discussion, see Thomson (2006), *Psychological Subjects*, Chapter 5.
55. Rose (1989) *Governing the Soul*, pp. 55–118.

56. J. Cronin (1984), *Labour and Society in Britain, 1918–1979* (London: Batsford).
57. Halliday (1948), *Psychosocial Medicine.*
58. E. Younghusband (1978), *Social Work in Britain, 1950–1975* (London: George Allen and Unwin).
59. H. Freeman (1984), *Mental Health and the Environment* (London: Churchill Livingstone); M. Shepherd (ed.) (1983), *The Psychosocial Matrix of Psychiatry* (London: Tavistock).
60. C. P. Blacker (1946), *Neurosis and the Mental Health Services* (London: Oxford University Press); M. Balint (1964), *The Doctor, the Patient and the Illness* (London: Pitman Medical). Born in Hungary, Balint immigrated to Britain in 1939. Based at the Tavistock Clinic after the war, he was involved in the group approach to psychotherapy, first in marriage guidance and, in the 1950s, in relation to general practice: Thomson (2006), *Psychological Subjects*, pp. 204–5.
61. P. Halmos (1981), *The Faith of the Counsellors* (London: Constable), p. 46; F. Furedi (2004), *Therapy Culture. Cultivating Vulnerability in an Uncertain Age* (London: Routledge), p. 10.
62. Some recent cultural history makes a contribution in this direction: for instance, M. Roper (2005), 'Between Manliness and Masculinity. The "War Generation" and the Psychology of Fear in Britain, 1914–1950', *Journal of British Studies* 44, pp. 343–62; J. Hinton (2010), *Nine Wartime Lives. Mass-Observation and the Making of the Modern Self* (Oxford: Oxford University Press).
63. See, for instance, G. Richards (2000), 'Psychology and the Churches in Britain, 1919–1939. Symptoms of Conversion', *History of the Human Sciences* 13, pp. 57–84.
64. For more on this recasting of popular psychology: Thomson (2006), *Psychological Subjects*, especially Chapter 1; and idem (2001) 'The Popular, the Practical and the Professional. Psychological Identities in Britain, 1901–1950', in Bunn, Lovie, and Richards (eds), *Psychology in Britain*, pp. 115–32.

8
Mental Health as Civic Virtue: Psychological Definitions of Citizenship in the Netherlands, 1900–1985

Harry Oosterhuis

This chapter discusses how, in the Netherlands from around 1900 until the mid-1980s, the idea of 'citizenship' acquired new definitions in the context of developing 'mental hygiene' and outpatient mental health care. Formulating views about the position of individuals in modern society and their potential for self-development, psychiatrists and other mental health workers linked mental health with ideals of democratic citizenship. Thus, they were involved in the liberal-democratic project of promoting not only productive, responsible, and adaptive citizens, but also autonomous, self-conscious, and emancipated members of an open society.

Since the 1980s, in many Western countries, 'citizenship' has become a fashionable concept throughout the political spectrum, and is seen as a means of facing a number of developments: the crisis of the welfare state; ongoing individualization and a presumed loss of social cohesion; growing ethnic and religious diversity; declining trust in parliamentary democracy; and globalization. Shifting the emphasis from social rights and welfare benefits to obligations and responsibilities, the new ideal focuses on the intrinsic qualities of the practice of citizenship. It refers to an ensemble of social abilities and public virtues: independence, self-reliance, self-control, sound judgement, reasonableness, open-mindedness, the capacity to discern and respect the rights and opinions of other individuals and social groups, tolerance without indifference, political participation and 'civil' behaviour in the public sphere. The 'citizenship' argument is that the resilience of modern democracy and the vigour of a pluralist society depend on the appropriate attitudes of its citizens.[1]

In modern liberal democracies, the concept of citizenship is generally used to describe what draws individuals together into a political community and what keeps that allegiance stable and meaningful to its participants. This overall definition can be made more specific by pointing to three basic constituents of citizenship.[2] First, citizenship refers to the more or less enduring allegiance between the individual and the political community and/or civil society – the social space of free association and relational networks that is separate from the state, the market, and private life. The domain of citizenship can be distinguished from the private sphere of intimacy and emotion as well as from a market swayed by economic interests. Secondly, citizenship is about (political and civil) participation on the bases of a sense of individual autonomy and a sense of public commitment. Citizenship presupposes a capacity for self-direction and is not easily reconciled with subordination and dependence. Thirdly, citizenship has to do, on the one hand, with rights and entitlements (granted and guaranteed by the state) and, on the other, with responsibilities and obligations (towards the state and the community).

Present-day calls for a revitalization of the intrinsic quality of democratic citizenship are far from unique, at least in the Netherlands, the country that is the focus of this chapter. From the late nineteenth century, citizenship took on a broad meaning, not just in terms of political rights and duties, but also in the context of material, social, psychological, and moral conditions that individuals should meet in order to develop themselves and also be able to act according to those rights and duties responsibly. The connection between mental health and citizenship that I am exploring here was embedded in changing sociopolitical frames of thought and debate on self-development that not only reflected, but also constituted social realities of mental health and citizenship. On the basis of the three different ideals of self-development that I identify, my account is divided into three periods: 1900–50 (adaptive self-development); 1950–65 (guided self-development); and 1965–85 (spontaneous self-development).[3] In the conclusion I will also briefly discuss developments in the last two decades.

My starting point is the supposition that both citizenship and mental health are ambiguous and complicated multi-layered concepts open to a variety of definitions and interpretations. Their specific meanings are historically contingent and cannot be established without considering the sociopolitical contexts in which they were employed. Mental health and citizenship are also essentially contested concepts: their definition and also their application and realization are not self-evident, but open to continuous debate, negotiation, and strife. As empirical and descriptive as well as prescriptive and normative categories, they are the focus of varied ways of understanding problems. These diverse interpretations reflect as well as constitute different social and political interests.

Its very vagueness and flexibility made 'mental health' a powerful metaphor in discourses about citizenship. From the 1920s on, the mental hygiene movement and the psychiatric outpatient sector successfully established themselves in the Netherlands by appealing to various social groups and connecting various social domains. The notion of *geestelijke volksgezondheid* (public mental health), which grouped together a number of diverse problems in and between people, fulfilled a major strategic function in linking the social application of the human sciences to broader sociopolitical issues. *Geestelijke volksgezondheid* related to both the individual and society, and established a connection between the private and public spheres. *Volksgezondheid* (public health) evoked associations with medicine and hygiene, while *geestelijk* – which means 'spiritual' as well as 'mental' – referred to religious, moral, and cultural values as well as matters of the psyche. Thus it was possible to establish an explicit connection with the strong charitable tradition in the Netherlands and the nineteenth-century liberal-bourgeois civilization offensive, whose moral-didactic ethos was adopted and continued by both Christian and social democratic groups in the first half of the twentieth century. The ideal of mental health tied in with the need to articulate a belief in public morals and a utopian message, not only among liberals and Christians, but also, especially in the 1960s and 1970s, among socialist and other leftist groups who strongly believed in social engineering.

Adaptive self-development, 1900–50

The leading members of the Dutch Society of Psychiatry, founded in 1871, were liberal and positivist-minded physicians who believed science and social responsibility to be crucial for progress. Despite their focus on scientific medicine, they not only pointed to biological causes of insanity and nervous disorders, but also blamed the spread of mental disorders on certain harmful behaviours and social–cultural influences, such as pauperism, poor hygiene, immorality, alcoholism, sexual debauchery, bad upbringing, and the hectic pace of urban life. The assumed danger of degeneration and the increase in the number of new clinical diagnoses such as neurasthenia, moral insanity, and criminal psychopathy provided psychiatrists with arguments to expand their domain of intervention from mental asylums to society at large. From around 1900, several of them – including the private neurological practitioner F. J. Soesman in The Hague, the Amsterdam professor of psychiatry K. H. Bouman, the Amsterdam social psychiatrists F. S. Mijers and A. Querido, the Groningen professor of psychiatry W. M. van der Scheer, the Catholic psychiatrist C. T. Kortenhorst and the Protestant psychiatrist J. van der Spek – thus aligned themselves with social hygiene, in which the effort to prevent people from falling ill through a reform of their living conditions and way of life was centre stage. To counter the debasing influences of modern society that were supposed to undermine people's minds and nerves, they pointed to the importance of self-control, will power, moral awareness, and a sense of discipline.[4]

The growing involvement in social hygiene by psychiatrists fitted into a more general effort to tackle the 'social issue' by integrating the lower classes into the nation in a peaceful way. If classic liberalism started from the self-reliance and diligence of individuals and sought to minimize state intervention, reform-minded social-liberals acknowledged that opportunities for self-development depended not only on individual talents and will power, but also on changing economic and social circumstances and the general risks of life. An extension of the role of the state in society and also collective social insurance were thought necessary to protect the socially weak from disaster and to create conditions in which those lagging behind might improve their social position: in due time, this would render them eligible for full political citizenship.[5]

The emergence of mass society and the gradual process of political democratization, which in 1919 led to universal suffrage, caused mounting concerns among the upper and middle classes over the dominance of irrational emotions and drives among the lower classes, which could lead to unruliness and social disintegration. Divergent behaviours, ranging from drinking, gambling, and other forms of 'low entertainment' to idleness and money squandering, and from sexual licentiousness to child abandonment and crime, became the targets for interference and intervention by both voluntary organizations and the state.

The question behind this was whether all people had the necessary rational and moral qualities to meet the demands of modern society and to act as responsible citizens. What mattered was not just the resolution of social wrongs and misfortunes like poverty, illness, backwardness, and exploitation, but equally that everyone should achieve a virtuous life and a sense of social responsibility. As the democratization of society progressed, it was deemed all the more essential to raise the lower orders morally and to inculcate in them a civil sense of responsibility and decency. In demands for a national education of the people, 'character formation' was a basic component. Its programme was mainly geared towards social integration on the basis of middle-class values, which were now less exclusively tied to the social position of the bourgeoisie and were assuming a more general significance as civic virtues that applied to all members of society. Central notions were self-control and social responsibility. The aim was a proper balance between individual independence and community spirit as well as long-term personal and collective well being. An industrious and productive existence, a sense of order, duty, and family values acted as cornerstones for the democratized middle-class ideal of citizenship.[6] Apart from politicians, social reformers, and moral entrepreneurs, proponents of this social–moral activism were found especially in civil society among the professional groups that were gaining influence and self-awareness. These included physicians, teachers, youth leaders, social workers, and later on, mental health workers. Although the state supported such activism, it did not itself play a leading role.

Between the First World War and the early 1940s, the groundwork was laid for the psycho-hygienic movement and a national network of outpatient mental health care provisions, which mainly developed independently from asylums and mental hospitals. The individuals involved included psychiatrists as well as physicians, teachers, psychologists, lawyers, social workers, and also clergymen. (Not all of them were professionally trained mental health workers.) The domain claimed by these psycho-hygienists was wide: it stretched from the social care of mentally ill, mentally disabled, and psychopathic individuals to marriage, family life, sexuality, education, work and leisure, alcoholism, and crime. The psycho-hygienic ideal took material form in a growing number of counselling centres for alcoholics (23 in 1936), social-psychiatric pre- and aftercare services for the mentally ill and retarded (18 in 1936), child guidance clinics (six in 1936), centres for marriage and family problems (the first one was established in 1941), and public institutes for psychotherapy (the first one being founded in 1940). Although eugenics was a topic of discussion among psycho-hygienists, in these facilities they mainly adopted social, moral-didactic, pastoral and psycho-dynamic approaches rather than biomedical ones.[7]

The underlying reasoning of the psycho-hygienists was rooted in a broadly shared cultural pessimism about the harmful effects of rapid changes in society, as well as in an optimistic belief in the potential of scientific knowledge to solve these problems. The psycho-hygienists viewed modernization, with its attendant social fragmentation and disorientation of individuals, as a major cause of the increase in mental and nervous problems. A rising number of people would have trouble keeping up with the complexity and high-paced lifestyle of industrialized and urbanized society. This growth could be contained, psycho-hygienists argued, by taking preventive measures, such as treating the early stages of mental and behavioural problems so as to prevent them from becoming worse. Fearing cultural decay and social disintegration, the psycho-hygienists repeatedly stressed the significance of spiritual values and a sense of community.

Basically, the psycho-hygienic doctrine fitted in with efforts to 'civilize' the lower classes. In the nineteenth century, such efforts had been promoted by the liberal bourgeoisie, but after the turn of the century they became entangled with confessional social activism (Catholic and Protestant) as well as with social democratic politics aimed at furthering the emancipation and national integration of their constituencies. These efforts suggested an optimistic belief in the perfectibility of mankind, even though the vision was frequently couched in a more or less high-minded moral-didactic paternalism. With their particular understanding of public mental health, psycho-hygienists closely aligned themselves with the paradigm of an orderly mass society that was based on the unconditional adaptation of the individual to middle-class norms and values. In their view, responsible citizenship required the development of mental qualities associated with 'character': self-control, will power, a devotion to duty, and, above all, social adaptation and community spirit.[8]

After the Second World War, outpatient mental health care facilities expanded rapidly in numbers and in their average staffing. In the years between 1949 and 1965 social-psychiatric pre- and aftercare services for the mentally ill grew in number from 24 to 35; child guidance clinics multiplied from 12 to 73, centres for marriage and family problems from eight to 67, and public psychotherapy institutes from one to three.[9] Increasingly these facilities were staffed by professional mental health care workers: psychiatrists, psychologists, and (psychiatric) social workers. The growth was much advanced by worries about social disruption in the wake of the German Occupation and the subsequent liberation by Allied forces. The war and Nazism epitomized the cultural pessimism of the psycho-hygienists, and in the postwar years their doctrine won increased support among politicians and in other significant circles. Various forms of misconduct and presumed lack of ethical standards – idleness, juvenile mischief, trading on the black market, lack of respect for authority, and also family break-up, growing divorce rates, and sexual licence – were considered serious threats both to the moral fibre of the nation and to public mental health.

The leitmotif of this widespread anxiety was the observation that uncontrollable drives had gained the upper hand. It was widely felt that in order to rebuild the devastated country, achieve unity and hold off the new threat of communism, the people's moral resilience should be strengthened. Again, an insistence on self-discipline and sense of duty served to underline the importance of responsible citizenship in a democratic mass society as well as in the emerging welfare state.[10] In their striving for a recovery in the mental health of the Dutch people, psycho-hygienists displayed a great sense of mission while claiming an even broader professional domain than before. Through the use of medical–biological metaphors – society viewed as a body, the family as a vital organ, the individual as a cell, social wrongs as pathologies, and problem groups as sources of infection – social and moral problems were represented as public mental health issues. The notion of mental health was turned into a comprehensive concept linked to the prevention of war, opposition to fascist and communist totalitarianism, and the realization of a better world.[11]

Guided self-development, 1950–65

From the 1950s on, the development of mental health care was part of the emerging welfare state, and it was strongly influenced by the specific ways in which experts in the field interpreted social transformations. When, around 1950, moral panic about the disruptive effects of the war had faded, the experts began to focus especially on the potentially harmful influences of ongoing modernization. The 1950s brought a new and vigorous economic dynamic, based on a great confidence in science and technology. Large-scale urbanization, industrialization, and infrastructural innovation had far-reaching effects on people's social relationships and on everyday life. Spatial and social mobility rose sharply, enabling increasing numbers of individuals to evade the paternalism and social control of small communities, the church, and their families. Steadily increasing prosperity provided more material security, while class differences and other hierarchical relationships gradually lost their edge. Increasingly, the new dynamic of everyday life was at odds with the still prevailing middle-class and Christian norms and values, with their clearly defined dos and don'ts.[12]

In the statements psycho-hygienists articulated about these developments, tones of cultural pessimism reverberated. In essence, they felt that the mental and moral development of man was falling out of step with ongoing economic and technological progress.[13] Like other intellectuals, they argued that modernization was putting society under the domination of a one-sided, instrumental rationality, which jeopardized moral and spiritual values. Their critique focused on the so-called mass-man, the embodiment of all the evils that accompanied modernity. The mass-man was lonely and uprooted, had no fixed norms and values, and no longer had any ties with religion, tradition, and community. He let his life be dictated by his unconscious drives and emotions, and showed no regard whatsoever for moral authority. His inner emptiness was shown in his flight into material consumption, popular entertainment, and sexual gratification. This rudderless man, critics argued, undermined social solidarity and democratic citizenship. They argued in favour of social planning and a normative education of the people, so as to prevent democratic mass society from degenerating either into anarchy or into dictatorship. Although rationalization was regarded as one of the main causes of the cultural crisis, there was at the same time great confidence in the possibility of reforming society with the help of the social and human sciences. This is why sociologists and psycho-hygienists believed they had a major task to fulfil.[14]

Initially, mental health workers stressed the need for a fixed collective morality and the social adaptation of the individual. They looked for solutions in moral-pedagogical measures. In the course of the 1950s, however, their defensive stance towards modernization yielded to a more accommodating approach. More and more they acknowledged that restrictions and coercion affected only the outer behaviour of people while leaving their inner selves untouched. What in the late 1940s was seen as lack of moral strength was, in the 1950s, increasingly explained in psychological and relational terms. It was now believed that personality flaws, developmental disorders, and unconscious conflicts caused by a defective education and poorly functioning families constituted the underlying causes of deprivation and misbehaviour. The older moral-didactic discourse about the need to develop 'character' was beginning to give way to a new, more psychological one about the development of 'personality'.

The belief that socioeconomic progress was inevitable brought a new perspective on the task of mental health care. The results of preventive psychiatric treatment on Allied soldiers during the war, the psycho-dynamic model, and new American psycho-social methods such as social casework and counselling raised expectations of how psychiatry and the behavioural sciences could shape people's mental make-up. Inspired by the World Federation of Mental Health, mental health experts stressed that it was not only the prevention and treatment of mental disorders that mattered, but that it was also crucial to improve mental health in general and thus ensure maximal opportunities for all citizens to develop themselves in a wholesome way. Aside from offering support to people failing to keep pace with rapid modernization, mental health workers should also enhance the mental attitude and abilities that individuals needed to function properly in a changing society. Thus the pursuit of more dynamic and flexible adaptation took the place of frantic attempts at restoring morality and community spirit. It was now believed that the new social conditions required a redirection of norms and values, and that people should be granted more individual responsibility for self-development.[15]

Leading psycho-hygienists – in particular the Catholic psychiatrist C. J. B. J. Trimbos, the Protestant psychiatrists S. P. J. Dercksen and A. L. Janse de Jonge, the psychiatrist A. Querido, the psychiatric social worker E. C. Lekkerkerker, the pedagogue M. J. Langeveld and the Catholic psychologists F. J. J. Buytendijk and H. M. M. Fortmann – presented themselves as guides who could prepare individuals for the dynamism of modern life by encouraging a change in their mentalities and behavioural patterns. Inspired by phenomenological psychology and personalism, both of which stressed personality formation, they now identified 'maturity', 'inner freedom', and 'self-responsible self-determination' as the basic constituents of mental health. Such mental qualities were the opposite not only of passive obedience, but also of impulsive behaviour: they entailed inner regulation, which would guarantee that people could do without external constraint to lead a responsible life. It became the individual's task to develop into a 'personality' and to achieve a certain measure of inner autonomy and stability in relation to the fluid outside world. The mentally healthy were not those who uncritically subjected themselves to rules, but those who were independent, pursued optimal self-development and, at the same time, adapted to social modernization in an aware and considered way.[16]

If individuals were to decide on their own how to shape their lives, scrupulous self-examination was needed to ensure that their intentions were well thought out and based on good grounds. People were trusted to follow their own convictions, but were also expected to do so in line with the social requirements of a morally responsible mode of life as articulated by mental health professionals and other experts. Individuals could only develop their personalities in a meaningful way if, of their own accord, they were able to internalize social norms and values into an autonomous self. By fostering constant reflection among individuals on their conduct and motivation, mental health care would contribute to creating the conditions for participation in civil society and political involvement, and thus for maintaining and deepening democracy.[17]

The ideal of citizenship promoted in the 1950s and early 1960s by mental health experts can be characterized as 'guided self-development', and this was geared towards socioeconomic modernization, a process that called for a functional individualization, meaning flexibility and mobility. Self-identity, which used to be a product of given and more or less stable social categories such as class, religion, and family background, was increasingly being turned into a product of personal qualities and preferences. This individualization was understood as an inescapable effect of modernity, but in an effort to avoid social disintegration, psycho-hygienists considered it essential to offer guidance and add normative standards as a counterbalance to the individual's growing freedom.

Spontaneous self-development, 1965–85

Psycho-hygienists believed in controlled modernization and guided self-development under the supervision of a morally inspired and professionally trained elite. Their patronizing approach was characteristic of the postwar period of reconstruction, but from the mid-1960s it came under attack. In the ensuing decade the Netherlands changed from a rather conservative, Christian nation to one of the most liberal and permissive countries in the Western world.[18] Secularization, together with growing prosperity and an expanding welfare state, caused more and more people to break away from established traditions and hierarchical relationships in an effort to enhance their independence and individuality. Various protest movements voiced their concern for democratization, liberation, and self-determination – and voiced it loudly. The control of emotions and the adaptation of the individual to society were no longer considered signs of responsibility, but were thought of as repression of the authentic self.

The ideal of spontaneous self-realization, which extolled self-expression, paved the way for an assertive individualism that, together with the democratization movement, rocked the foundations of Dutch society and the mental health care system as well. If, beforehand, individuals had been expected to comply with the social order, now society itself had to change to facilitate the optimal self-development of people and the ultimate fulfilment of democratic citizenship. In 1848 a liberal constitution had provided the Dutch people with basic civil rights; the introduction of universal suffrage in 1919 had made them citizens in the political sense; and the postwar welfare state guaranteed their material security. Now, some psycho-hygienists argued, the time was ripe for taking the next step in this continuing process of emancipation: the meeting of immaterial needs in order to advance personal well being for all.[19]

Embracing some of the basic tenets of the protest movements and ideas from anti-psychiatry, responding too to more and more assertive clients, mental health workers increasingly voiced self-criticism. A growing number of them were trained in psychology, pedagogy, and the new discipline of *andragologie* (adult educational theory) as well as in sociology and social work. Training took place not only in universities, but also in the rapidly expanding academies for social work. Inspired by a leftist political agenda and an activist stance, they demanded that attention be given to the social causes of mental distress. Therapeutic treatment of individuals that aimed at adapting them to society became subject to debate. Instead, people needed to be liberated from the coercive 'social structures' that caused intolerable personal situations and restricted spontaneous self-development. The realization of this objective appeared to be more dependent on welfare work and political activism than on psychiatry and mental healthcare.

However, while institutional and medical psychiatry were both forced on to the defensive, psycho-social and especially psychotherapeutic services increased more than ever in size and prestige during the 1970s. The very dissatisfaction people had with medical psychiatry, as articulated by the anti-psychiatry movement, prompted new pleas for alternative and better mental health care, with, for instance, therapeutic communities and more outpatient facilities in society at large.[20] The growth of this care provision was facilitated by the welfare state: more and more collective social funds guaranteed the broad accessibility of mental health care. The prevailing trend between 1965 and 1985 was one of substantial increase and scaling-up of services aimed at a broad spectrum of psycho-social disorders and problems, and catering for steadily growing numbers of clients.[21]

The 1970s saw the heyday of psychotherapy, which, aside from the work of psychiatrists, was practised more and more by psychologists and social workers in a growing number of public psychotherapeutic institutes. Although socially critical mental health care workers blamed the evils of society for psychological problems, psychotherapy focused on the individual's inner self. A growing number of people began to consider it more or less self-evident that they should seek psychotherapeutic help for all sorts of discomforts or personality flaws that bothered them, matters previously not regarded as mental health problems at all. Both therapists and clients viewed themselves more or less as a cultural avant-garde: psychotherapy would liberate individuals from unnecessary inhibitions and limitations, and provide them with opportunities for personal growth and improvement in the quality of their lives.[22]

While engaging in heated debates on the political implications of their work, mental health workers widened their professional domain to include welfare work, a sector that in the 1970s experienced enormous growth. Together with social workers, they undertook the task of encouraging clients to become aware of their true needs and to develop their authentic self. As the psychiatrist J. A. Weijel explained, personal unhappiness was not to be viewed as an individual fate, but as a social evil that could be remedied by implementing welfare policies and by championing civil rights.[23] Weijel's view was echoed in a report on mental health entitled *Verbeter de mensen, verander de wereld* ('Reform people, change the world') published by a think tank associated with the Social Democratic party. Mental health care was a form of politics, the authors claimed: democratization, social activism and, if necessary, protest against 'the prevailing norms and rules of conduct' were essential preconditions for the realization of individual self-development and social welfare.[24]

Motivated by such ideas, mental health workers presented themselves as inspired advocates of personal liberation in the areas of religion, relationships, sexuality, education, work and drugs, as well as in issues concerning the emancipation of women, youngsters, the lower classes, and disadvantaged groups like homosexuals and ethnic minorities. As some of them emphasized, countering prejudice and advancing tolerance was part of a broader effort to improve the quality of social relations and to 'democratize happiness'.[25] Psychiatrists and other psycho-hygienic experts played a crucial role in public debates, and some of them put controversial and sensitive issues on the social agenda.

Already, in the 1950s and early 1960s, some leading psychiatrists (Trimbos, Janse de Jonge, Querido, C. van Emde Boas, H. Musaph, F. J. Tolsma, and M. Zeegers) had strongly contributed to a change in the moral climate in the areas of family, marriage, and sexuality. In the 1960s and 1970s, psychiatrists promoted a new outlook in other fields: recognition of the mental suffering of war victims and other traumatized people (E. de Windt, P. Th. Hugenholtz, J. Bastiaans, D. van Tol, and H. Keilson); the self-determination of patients and a depenalization of euthanasia (R. H. van den Hoofdakker, J. H. van den Berg, N. Speijer, F. van Ree, Trimbos, and Van Tol); abortion (Van Emde Boas, B. E. Chabot, C. Th. van Schaik, A. van Dantzig, G. A. Ladee, and Trimbos); and drugs (P. C. Kuiper, P. J. Geerlings, Ladee, D. Zuithoff, P. A. H. Baan, and Van Ree).[26]

In this way, they contributed to the implementation of practices that were considerably more liberal than the prevailing norm, certainly from an international perspective. In doing so, they drew on the 1960s culture of liberation and democratization, but also followed in the footsteps of the reform-minded psycho-hygienists of the 1950s. By raising issues that had earlier been largely kept quiet, they sought to break taboos and put an end to hypocrisy, thus paving the way for more openness, understanding, and tolerance. To achieve this, they explained, the requirements were a sense of responsibility, conscientious positioning, a sincere exchange of arguments, and a willingness of people to listen to each other. As Van den Hoofdakker wrote: 'in a world of emancipated and independent human beings', there was only one way to overcome outmoded ideas and habits, and that was by 'talking, talking, talking'.[27]

Rules and laws should not be rigidly applied, but discussed and sensibly interpreted. Emphasizing the 'debatability' of an issue – which in the Netherlands became a major norm, providing the basis for policies of controlled toleration – was essentially the opposite of being non-committal or outright permissive. What mattered was to counter the hidden abuse of specific liberties and to channel and control these liberties carefully, in good faith, and in open-minded deliberation. Making sensitive issues debatable was inextricably bound up with belief in a fully democratized society. Only self-reflective and socially involved citizens empathized with others, managed not to shy away from unpleasant truths, regulated their emotions, were capable of rational consideration and – through negotiation and mutual understanding – could arrive at balanced decisions. From a psychological perspective, this psycho-hygienic ideal of citizenship made great demands on people's competence.

Discussion: Democratization and psychologization

The link between the democratization and psychologization of citizenship, illustrated here by following the development of the mental health discourse in the Netherlands, can be viewed as part of a more general, long-term historical process in the Western world. In traditional systems of political domination, which placed people in subjection by coercion and force whether they willed it or not, people's inner selves were of little relevance. The need to 'form' individuals and make them internalize certain values and behaviour patterns became greater the more society was democratized. If in the nineteenth century citizens were largely appraised on external aspects (such as economic autonomy, male sex, tax duty, and education), in the twentieth century, when the adult suffrage was established and the welfare state softened the contradiction between formal political rights and social–economic inequality, the formation of an appropriate mental attitude gained prominence. Democratic citizenship presupposes a sense of public commitment based on individual autonomy, self-determination, and self-direction. It was in democratic societies, with their claim of a social order founded on the autonomous consent of individuals, that the inner motivations of citizens and their control of drives and emotions were considered of crucial importance for the quality of the public domain.

Against this backdrop the Dutch developments are hardly unique. In Britain, for example, from the 1920s on, mental health provided a paradigm for articulating in psychological terms a secular ideal of self-development as the groundwork for responsible democratic citizenship. In the United States the mental hygiene movement linked the ills of modern society with malaise in individuals, and mental health experts used theories of personality development to show how they could contribute to the formation of robust and self-reliant democratic subjects. In Germany during a striving for fundamental reforms in psychiatry in the 1960s and 1970s, the Nazi past was critically used as a spectre, and it was especially at this time that mental health care acquired a strong political dimension. Against the complicity of psychiatry in the atrocities of the Third Reich, a democratic counter-vision of mental health care emerged, based on a conception of citizenship that stressed political awareness, independence of mind, and the social rights of the infirm and indigent.[28]

However, in all these countries what was often missing was an extensive network of public outpatient facilities to underpin the rhetoric about mental health and citizenship with actual care-providing practices. In the Netherlands, models of psychological self-development were not mere abstract theories: in outpatient mental health care practices these ideals really materialized. From the 1950s on, in various facilities, clients were encouraged to be self-reflective about their conduct and motivations – initially in private life, but then with respect to their attitudes in the public domain as well.[29] The psycho-hygienic movement and outpatient services were more long-lasting, more broadly spread, and more generously funded in the Netherlands than in Britain, Germany or the United States.

Other striking elements of the Dutch outpatient sector were its broad orientation and its strong psycho-social approach. It incorporated various blends of psychotherapy, pedagogical technique, and social work intervention rather than sticking to a purely biomedical focus. In other European countries the institutional and public mental health sectors were more exclusively geared towards psychiatric patients, while there was also a closer link with the domain of (poly)clinical psychiatry.[30]

In the Netherlands, which used to be rather conservative and Christian, the cultural revolution of the 1960s was more sweeping than in other countries because it coincided with rapid secularization. In few countries was direct control and coercion from above curtailed so radically as in the Netherlands.[31] The ensuing moral or spiritual vacuum was partially filled by a psychological ethos; from the 1960s on, mental health care – several forms of psychotherapy in particular – expanded at an unprecedented rate. The strongly developed democratization of public and everyday life replaced hierarchy and social coercion with individual emancipation and informal manners. To find the proper balance between assertiveness and compliance, people needed self-knowledge, subtle social skills, empathy, and an inner, self-directed regulation of emotions and actions.

Thus interactions between people and the ways in which they assessed each other became increasingly determined by psychological insight. With their emphasis on self-reflection and raising sensitive issues, mental health experts articulated new values and offered a clear alternative to the old strict morality of dos and don'ts. Moreover, a strong inclination towards psychologization has gone hand in hand with the flourishing of a service economy. Reinforcement of this trend has come from the way in which, in the Dutch culture of consensus, social and ethical issues are often handled by experts. Their expertise is frequently called in because their supposedly objective professional stance neutralizes social conflicts over sensitive issues. In the articulation and implementation of policies on euthanasia, sexuality, birth control, abortion, and drugs, for example, professionals such as physicians, psychiatrists, psychologists, and social workers have had a large say. They have contributed generally to the formulation of practical solutions that are both pragmatic and well considered and that focus on individual conditions and motivations.

The Dutch social elites believed that social and political democratization called for an 'inner mission' to educate the masses in the responsible exercise of rights and duties. A democratic social order could only be maintained if individuals were capable of using their basic liberties in a self-controlled way. What mattered – according to mental health experts in particular – was an internalization of democratic values through the development of personality. Ironically, the pursuit of individual self-determination went hand in hand with gentle but persistent pressure on people to open their inner selves to scrutiny by others (for example, by mental health workers) and to account for their urges and motivations. This perspective would confirm Michel Foucault's interpretation of modernity.

Many works on the history of psychiatry and mental health care are more or less dominated by the Foucauldian perspective.[32] Foucault and his followers argue that the interlocking of power and knowledge conditioned medicine and the human sciences. The modern ideal of (mental) health and well being was historically constituted when scientific experts and the helping professions differentiated and segregated the 'abnormal' from the normal, first in institutions like prisons, hospitals, asylums, and schools, and later also through the various agencies of social work and mental health care. The medicalization and psychologization of social relations turned deviance into a form of abnormality or pathology and resulted in the disciplining of bodies and minds. By stigmatizing deviance as sickly deviation, physicians and mental health professionals, exponents of an anonymous 'biopower', controlled people and imposed social order and conformity to liberal society.[33] From a similar angle, professionalization theories have focused on strategies of the (mental) health professions to expand their field of action by 'medicalizing' and 'psychologizing' individual as well as social problems. They were able to do so because the state had granted them professional competence and autonomy in matters of health and well being.[34]

Recent work on the prominent role of professional expertise in modern liberal democratic societies has elaborated Foucault's notion of 'governmentality', and it has thereby refined and qualified the disciplinary perspective.[35] The growing role of professionalism has been explained against the background of tension between, on the one hand, an increasingly felt need to respond to what the elites framed as social problems and disturbances in modern mass society, and, on the other, the liberal (and in the Netherlands also the Christian Democratic) reluctance to allow state intervention in social and private life. Since liberal democracy was based on the principle that civil liberties and rights should be respected and advanced, social policies were often not based on direct state interference, but rather took the indirect form of 'governing at a distance' with the help of professional expertise outside the state apparatus. The lack of democracy associated with professionalism was compensated by the professional ethos, which presupposed personal integrity, scientific competence, technocratic efficiency, and disinterested dedication to the public good. At the same time, professional regimes and strategies did indeed reflect liberal-democratic values. Their operations were not restricted to putting constraints upon people or disciplining them; they also worked by gaining people's cooperation and by encouraging and guiding their self-control, self-direction, and self-advancement. Thus the helping professions were involved in the creation of democratic citizens who were able to exercise their own (controlled and regulated) freedom.

We have seen that the Dutch national government generally kept a low profile in the organization and implementation of mental health care and the articulation of civic virtues. At least until the late 1960s, when it began to play a more active role, the state mainly left these issues to voluntary organizations or to local government. Models for self-development and citizenship were hardly imposed by the state from above at all, but were instead developed and enunciated by leading groups in civil society itself. By delegating the execution of social policies to more or less independent helping professions and their administrative networks – state regulated and state subsidized, but not state controlled – interventionist strategies were removed from the disputed terrain of politics. It was the professionals who applied putatively neutral scientific knowledge to lay down what was normal, virtuous, healthy, and efficient. They used various methods of social work, counselling, and psychotherapy – in the Netherlands, education, persuasion, disciplining and inducement for the most part until the mid-1950s; incitement, motivation, encouragement, and management more and more from the mid-1950s on. Through these means, political concerns about social disorder, criminality, depravity, abnormality, and ill health could be translated into the neutral language of expertise, with the promise that such problems would subsequently be dealt with by socio-technical means. Targeting individuals and groups who supposedly did not behave in their own self-interest or who seemed to be indifferent to their own development, Dutch mental health professionals applied themselves to creating responsible citizens who would be capable of regulating themselves and of coping with a kind of controlled freedom.

This indeed fits in with the Foucauldian notion of governmentality. However, whereas Foucault views the quest for self-knowledge and self-control as the consequence of the unstoppable, all-embracing wielding of knowledge and power by scientific and professional regimes, I maintain that other social processes should be taken into account as well – in particular social levelling, democratization, emancipation, and informalization (the relaxing of codes of behaviour and the widening of alternatives for individuals in their emotional lives). My perspective partly dovetails with that of Norbert Elias and some of his followers.[36] Both the Foucauldian and the Eliasian approaches focus on the interplay of social coercion and rational (self-)control as the fundamental characteristic of modern liberal-democratic society. The first stresses the disciplining of the abnormal by scientific knowledge and professional power, and the rise of a normalizing governmentality in which expertise plays a crucial role, while the second emphasizes growing self-restraint as a consequence of broader social changes.

In Elias's theory of the civilizing process, a growing preoccupation with the inner self and with mental health can be explained in the context of increasing individualization and self-control. While Foucault stresses that modern psychological subjects and their social relations are produced by prescriptive discourses and techniques of scientific experts, the Elisian view implies that the theories of the human sciences and the practices of the helping professions are simply reflecting a more fundamental civilizing process. The essence of this process is the growing social pressure towards self-control or what Elias spells out as the shifting balance from *Fremdzwang* to *Selbstzwang*, which results from functional specialization (especially an increasing division of labour), social differentiation, lengthening chains of interdependence between more and more people, and shifting hierarchies with equalizing balances of power between individuals and social groups (classes, sexes and generations).

Following the Elisian approach, I would argue that mental health care played a major part in the articulation of the psychic dimension of personal as well as public life, but that the spread of a psychological *habitus* among the Dutch population from the 1950s on also stemmed from more general social developments that heightened demands on personal self-regulation and management of emotions. Psychologization, a change of mentality characterized by a growth in individuality and the ability to internalize, was connected with the democratization and informalization of social relationships, the shift from social coercion to self-regulation, and an increasingly subjective way of fashioning personal identity. The conduct of people was increasingly viewed, and experienced, as a reflection of personal wishes, motives, and feelings. It became more and more common for people to talk about themselves or others in psychological terms and to refer to their moods or feelings as a way of legitimizing their behaviour.

The role the human sciences and the helping professions had was to advance and to articulate psychological interpretations of human behaviour and motives, which to a large extent had already evolved in Dutch society, first among intellectual middle-class groups and later among other sections of the population. Although the clients of the mental health experts, and other lay people as well, sometimes adopted the specialists' language and modes of interpretation to deal with their problems and anxieties – an example of what the Dutch sociologist Abram de Swaan has characterized as 'proto-professionalization' and the British sociologist Anthony Giddens calls the 'reflexivity of modernity' – the helping professions had not constructed the social experiences from which such problems and anxieties emanated.[37]

Conclusion

The modernization of Dutch society and evolving views of democratic citizenship provided a sociopolitical context for the pursuit of a mental health programme in which cultural pessimism coexisted with an optimistic belief in social progress. In the development of the programme, it is possible to identify a turning point in the mid-1950s. In this decade a defensive response to the modernization process gave way to a much more accommodating stance. In attitudes to citizenship there was a shift from unconditional adaptation to the existing system of values and norms ('character') to a belief in individual self-development ('personality'). People's personal experience and inner motivations took centre stage, and therapeutic treatment and social integration were definitively favoured over coercion and exclusion. In the years 1950–65, by building on the ideal of guided self-development, mental health care aligned itself with social modernization and the development of a consumer society: individuals were expected to shape their own personalities, develop autonomy, be open to renewal, and, in a responsible way, achieve self-realization. In the late 1960s and the 1970s, mental health workers embraced spontaneous self-development as a core value, thus legitimizing the need for democratization and personal liberation, also, to a certain extent, endorsing consumerism and hedonism.

The mid-1980s were another turning-point. Earlier confidence that individuals could be motivated to integrate as full citizens of an egalitarian and democratic society through 'soft' person-centred psycho-social support and nudges towards self-guidance began to subside. From the early 1980s on, mental health experts were far less vocal in their articulation of civic virtues. In this respect, there was no linear evolution or continuous expansion of the psychologizing perspective. The decline of psychological definitions of citizenship was connected with the transformation of the mental health care system as well as with the growing impact of neo-liberalism in the Netherlands, together with a hardening of the sociopolitical climate.

In the wake of the crisis of the welfare state, the focus of mental health care shifted in the 1980s. The government, and also some psychiatrists, repeatedly argued that outpatient facilities were one-sidedly geared to clients with minor psychological afflictions, and that this would cause a boundless increase in the demand for mental health care. Instead, the treatment of acute and chronic psychiatric patients should become a priority, a strategy that would also keep the number of admissions to mental hospitals as low as possible. Public funding, of psychotherapy in particular, was criticized as over-generous. Psychotherapists were accused of making good money by serving a privileged clientele with mild psycho-social problems while neglecting patients with serious disorders. The effort to develop community care for these latter patients became a governmental priority. Also, calls for efficiency, rationalization, a partial re-medicalization of psychiatry and careful assessment of costs and benefits replaced the emancipatory ideals of the 1960s and 1970s. In the late 1990s, to improve cooperation between psychiatric hospitals and outpatient services, the government pressured the two to merge in comprehensive mental health organizations that offered intramural as well as extramural care. All of this marked a break with the historically developed constellation of mental health care facilities, which since the 1930s had been divided between clinical psychiatry for serious mental disorders and an outpatient sector with a strong psycho-social orientation for a wide spectrum of milder problems.[38]

The strategies of community psychiatric care echoed some of the democratic ideals of the 1960s and 1970s, such as the need to increase the personal autonomy of psychiatric patients and respect their civil rights. The aim was that they should regain their proper status as citizens, an aim that since the 1970s had been championed by the critical patients' movement. The degree to which they would be able to realize themselves as more or less independent citizens relied in part on a programme of mental health care that guaranteed that they were not isolated from the rest of society. In practice, however, the citizen status of psychiatric patients met with obstacles all the time and came under challenge in the 1990s. Critics pointed out that the emphasis on autonomy and social participation, meant to create a meaningful mode of existence, entirely ignored what in effect constituted the essence of mental illness: a limited power of self-determination and the loss of basic, taken-for-granted patterns of social interaction. Fragile as they were, psychiatric patients were in need of protection and a quiet, simple life shielded from the dynamic of society. As long as the defining qualities of citizenship were agency and social participation (especially having regular work), the mentally ill and handicapped were at best consigned to the category of marginal citizens.

The positive value given to self-determination was questioned too because it allowed mentally ill people with serious behavioural problems to refuse psychiatric treatment, cause social trouble or be aggressive. Here community care soon ran up against its limits. It had serious downsides for some patients: social isolation, abandonment, and lack of day activities. Also there was the public nuisance caused by people like the homeless mentally ill, alcoholics, and drug addicts. Pleas for more coercion in social-psychiatric care and for a vigorous public mental health policy put earlier ideals of emancipation and self-determination into question. Basically, these ideals lacked any relevance for those suffering from serious psychiatric disorders, who were incapable of taking care of themselves and who could not assert their needs. For them, social reintegration was no real option, and these patients were living proof that mental illness and full citizenship were hard to reconcile.

The optimism that had prevailed since the 1950s in the psycho-hygienic outpatient facilities, with their belief in stimulating individual self-development, had partly been facilitated precisely because there was a strong tendency to keep patients with serious psychiatric disorders out of the system. Emphasizing their identity as welfare services, the psychotherapeutic institutes, centres for family and marriage problems, and child guidance clinics, catered for a clientele with a variety of psycho-social and existential problems and they focused on the improvement of people's welfare, social participation, and assertiveness. A psychological perspective coupled with various 'talking cures' had increasingly set the tone in these facilities. Clients were expected to have some capacity for introspection, verbal skills, initiative, and willingness to change, and this automatically excluded those who were seriously mentally ill.

However, when in the 1990s social psychiatry was prioritized and outpatient facilities merged with psychiatric hospitals, the emphasis shifted to people with more serious and unmanageable mental disorders. These people did not respond to ideals of personal change and liberation, and mental health workers resorted to the more modest objectives of trying to limit or alleviate mental suffering so far as was possible and to control the symptoms. Notwithstanding the increasing use of pharmaceuticals for treating mental illness, various social, psychological, and behavioural therapies remained in use, but they were now not so much directed at personal development as at acquiring social and practical skills to cope with life, for better or worse.

The 1960s' and 1970s' ideology of individual liberation and emancipation was called into question in yet another way. Influenced by epidemiological research, which showed a high frequency and increase of psychic disorders among the population, reports and policy recommendations on mental health care in the 1990s pointed to an array of social developments that were likely to trigger psychiatric problems: individualization; the currently high demands on people's social skills and mental resilience; a diminishing sense of security and safety in society; and, above all, an assumed loss of shared norms and values, and of social responsibility and cohesion.[39] Evidently, in mental health care the optimistic view of society, in which emancipated and motivated citizens tried to solve problems together in mutual interaction, had been replaced with concern about the loss of community spirit and public morality.

In society at large, the citizenship ideal that had been promoted in mental health care since the 1960s was called into question too. Politicians and intellectuals took stock of the 1960s and 1970s, and concluded that the legacy of the period was largely negative. Anti-authoritarian celebration of individual freedom, they argued, had degenerated into a lack of self-restraint, erosion of social responsibility, and an exaggerated assertiveness exclusively based on rights rather than duties. Welfare state provision had resulted in calculating behaviour and improper use of benefits.

The resultant climate of opinion contained contradictory impulses. On the one hand, as a consequence of neo-liberal politics of deregulation and privatization, emphasis shifted from public care facilities to insistence on the self-reliance of citizens in the community and in the market. The new standard of good citizenships was an autonomous self-development of independent individuals, using their own talents and efforts, with a minimum of interference from government or welfare agencies. On the other hand, concerns about weakening social cohesion and the downgrading of the public domain brought a tendency to stress other civic virtues: social responsibility, moral sense, and a devotion to duty.[40] The taboo on coercion and enforced obligation began to recede, for instance, in efforts aimed at the reactivation of the unemployed and the integration of ethnic minorities and migrants. The government, and other voices in civil society, urged a restoration of norms and values, a repressive approach to previously tolerated (mis)behaviour, and a hardening of criminal law.

Notes

1. H. R. van Gunsteren (1994), 'Four Conceptions of Citizenship', in B. van Steenbergen (ed.), *The Condition of Citizenship* (London, Thousand Oaks, CA, and New Delhi: Sage Publications), pp. 36–48.
2. For the conceptualization of citizenship I rely on R. Beiner (ed.) (1995), *Theorizing Citizenship* (Albany, NY: State University of New York Press); B. S. Turner (1989), 'Outline of a Theory of Citizenship', *Sociology* 24, pp. 189–217.

3. These models of self-development are borrowed from J. W. Duyvendak (1999), *De planning van ontplooiing. Wetenschap, politiek en de maakbare samenleving* (Den Haag: Sdu Uitgevers).
4. H. Oosterhuis and M. Gijswijt-Hofstra (2008), *Verward van geest en ander ongerief. Psychiatrie en geestelijke gezondheidszorg (1870–2005)* (Houten: Nederlands Tijdschrift voor Geneeskunde and Bohn Stafleu Van Loghum), pp. 64–70, 207–21.
5. S. Stuurman (1991), 'Het einde van de productieve deugd', *Bijdragen en Mededelingen betreffende de geschiedenis der Nederlanden* 106, pp. 610–24.
6. H. te Velde (1993), 'How High did the Dutch Fly? Remarks on Stereotypes of Burger Mentality', in A. Galema, B. Henkes, and H. te Velde (eds), *Images of the Nation. Different Meanings of Dutchness 1870–1940* (Amsterdam and Atlanta: Rodopi), pp. 59–79.
7. H. Oosterhuis (2005), 'Insanity and Other Discomforts. A Century of Outpatient Psychiatry and Mental Health Care in the Netherlands 1900–2000', in M. Gijswijt-Hofstra, H. Oosterhuis, J. Vijselaar, and H. Freeman (eds), *Psychiatric Cultures Compared. Psychiatry and Mental Health Care in the Twentieth Century: Comparisons and Approaches* (Amsterdam: Amsterdam University Press), pp. 73–102; Oosterhuis and Gijswijt-Hofstra (2008), *Verward van geest en ander ongerief*, pp. 375–412, 664–731, 1449.
8. F. J. Soesman (1908), *Hygiène van den Geest. Tucht als middel tegen Zenuwzwakte* ('s-Gravenhage: W. P. Van Stockum en zoon); J. H. van der Hoop (1922), *Zelfverwerkelijking in Mensch en Gemeenschap* (Baarn: Hollandia).
9. In the 1970s the number of public institutes for psychotherapy would rapidly increase from three to 25. Oosterhuis and Gijswijt-Hofstra (2008), *Verward van geest en ander ongerief*, pp. 664–731, 965–94, 1449.
10. J. C. H. Blom (1981), '"Jaren van tucht en ascese". Enige beschouwingen over de stelling in Herrijzend Nederland 1945–1950', *Bijdragen en Mededelingen betreffende de Geschiedenis der Nederlanden* 96(2), pp. 300–33
11. Oosterhuis and Gijswijt-Hofstra (2008), *Verward van geest en ander ongerief*, pp. 621–54.
12. P. Luykx and P. Slot (eds) (1997), *Een stille revolutie? Cultuur en mentaliteit in de lange jaren vijftig* (Hilversum: Verloren).
13. T. de Vries (1996), *Complexe consensus. Amerikaanse en Nederlandse intellectuelen in debat over politiek en cultuur 1945–1960* (Hilversum: Verloren).
14. M. Gastelaars (1985), *Een geregeld leven. Sociologie en sociale politiek in Nederland, 1925–1968* (Amsterdam: Sua); E. Jonker (1988), *De sociologische verleiding. Sociologie, sociaal-democratie en de welvaartsstaat* (Groningen: Wolters-Noordhoff).
15. Oosterhuis and Gijswijt-Hofstra (2008), *Verward van geest en ander ongerief*, pp. 629–37, 645–65.
16. F. J. J. Buytendijk (1958), *De zin van de vrijheid in het menselijk bestaan* (Utrecht and Antwerpen: Uitgeverij Het Spectrum); F. J. J. Buytendijk (1950), *Gezondheid en vrijheid* (Utrecht and Antwerpen: Uitgeverij Het Spectrum).
17. See G. Brillenburg Wurth et al. (1959), *Geestelijke Volksgezondheid.* Nederlands Gesprekcentrum Publicatie No. 17 (Kampen: J. H. Kok; Utrecht and Antwerpen: Het Spectrum; Den Haag: W. P. van Stockum en Zoon).
18. J. Kennedy (1995), *Nieuw Babylon in aanbouw. Nederland in de jaren zestig* (Amsterdam and Meppel: Boom); H. Righart (1995), *De eindeloze jaren zestig. Geschiedenis van een generatieconflict* (Amsterdam and Antwerpen: De Arbeiderspers).
19. J. A. Weijel (1970), *De mensen hebben geen leven. Een psychosociale studie* (Haarlem: Bohn).
20. G. Blok (2004), *Baas in eigen brein. 'Antipsychiatrie' in Nederland, 1965–1985* (Amsterdam: Uitgeverij Nieuwezijds).
21. Oosterhuis and Gijswijt-Hofstra (2008), *Verward van geest en ander ongerief*, pp. 775–811, 941–94.
22. Ibid., pp. 971–94.
23. Weijel (1970), *De mensen hebben geen leven.*
24. Van den Bergh et al. (1970), pp. 10, 21, 26.

25. Ibid.; G. van Beusekom-Fretz (1973), *De demokratisering van het geluk* (Deventer: Van Loghem Slaterus).
26. Oosterhuis and Gijswijt-Hofstra (2008), *Verward van geest en ander ongerief*, pp. 953–65.
27. R. H. van den Hoofdakker (1971), *Het bolwerk van de beterweters. Over de medische ethiek en de status quo* (Amsterdam: Van Gennep), p. 50.
28. M. Thomson (2000), 'Constituting Citizenship. Mental Deficiency, Mental Health and Human Rights in Inter-war Britain', in C. Lawrence and A-K. Mayer (eds), *Regenerating England. Science, Medicine and Culture in Inter-war Britain* (Amsterdam and Atlanta: Rodopi), pp. 231–50.
29. H. Oosterhuis (1992), *Homoseksualiteit in katholiek Nederland. Een sociale geschiedenis 1900–1970* (Amsterdam: Sua); Oosterhuis and Gijswijt-Hofstra (2008), *Verward van geest en ander ongerief*, pp. 664–731.
30. H. Oosterhuis (2005), 'Outpatient Psychiatry and Mental Health Care in the Twentieth Century. International Perspectives', in Gijswijt-Hofstra, Oosterhuis, Vijselaar, and Freeman (eds) (2005), *Psychiatric Cultures Compared*, pp. 248–74.
31. L. Halman, F. Heunks, R. DeMoor, and H. Zanders (1987), *Traditie, secularisatie en individualisering. Een studie naar de waarden van Nederlanders in een Europese context* (Tilburg: Tilburg University Press).
32. See, for example, R. Castel (1976), *L'ordre psychiatrique. L'âge d'or de l'aliénisme* (Paris: Minuit); R. Castel, F. Castel, and A. Lovell (1982), *The Psychiatric Society* (New York: Columbia University Press).
33. M. Foucault (1975), *Surveiller et punir. Naissance de la prison* (Paris: Gallimard); idem (1976), *Histoire de la sexualité*, vol. I: *La volonté de savoir* (Paris: Gallimard).
34. E. Freidson (1970), *Profession of Medicine. A Study of the Sociology of Applied Knowledge* (New York: Harper and Row); I. Zola (1972), 'Medicine as an Institution of Social Control', *Sociological Review* 20, pp. 487–504.
35. M. Foucault (1979), 'Governmentality', *Ideology and Consciousness* 6, pp. 5–21; G. Burdell, C. Gordon, and P. Miller (eds) (1991), *The Foucault Effect. Studies in Governmentality* (London: Harvester/Wheatsheaf); P. Miller and N. Rose (2008), *Governing the Present. Administering Economic, Social and Personal Life* (Cambridge and Malden, MA: Polity Press).
36. N. Elias (1969 [1939]), *Über den Prozess der Zivilisation* (Bern: Francke); A. de Swaan (1988), *In Care of the State. Health Care, Education and Welfare in Europe and the USA in the Modern Era* (Cambridge: Polity Press); C. Wouters (2007), *Informalization. Manners and Emotions since 1890* (London: Sage).
37. A. de Swaan, R. van Gelderen, and V. Kense (1979), *Het spreekuur als opgave. Sociologie van de psychotherapie 2* (Utrecht and Antwerpen: Het Spectrum), pp. 28–34; A. Giddens (1990), *The Consequences of Modernity* (Cambridge: Polity Press).
38. Oosterhuis en Gijswijt-Hofstra (2008), *Verward van geest en ander ongerief*, pp. 1048–85.
39. See, for instance, Scenariocommissie Geestelijke Volksgezondheid en Geestelijke Gezondheidszorg (1990), *Zorgen voor geestelijke volksgezondheid in de toekomst. Toekomstscenario's geestelijke volksgezondheid en geestelijke gezondheidszorg 1990–2010* (Utrecht and Antwerpen: Bohn, Scheltema en Holkema);.
40. H. Wigbold (1995), *Bezwaren tegen de ondergang van Nederland* (Amsterdam and Antwerpen: De Arbeiderspers).

9

Human Sciences, Child Reform, and Politics in Spain, 1890–1936

Till Kössler

Hardly any other social group at the end of the nineteenth century was subjected to such intense scientific and public scrutiny as children. New branches of science devoted themselves to the study of childhood and children, and new groups of child experts sought to influence the social organization of childhood. In the recent past a growing number of scholars has begun to write the history of child studies, especially in the US and Britain.[1] However, while we already know quite a lot about the development of child psychology as a field of study, we are just starting to understand the political, social, and cultural implications of the scientification of childhood. This is particularly the case for the history of countries other than the US and UK, in which the strength of the national child study movements has resonated in historical scholarship since the 1960s.[2]

Some sub-areas of the history of childhood have, however, have made the role of the new children's sciences the subject of a more detailed examination. This applies, first, to works dealing with the appearance of the child protection movements in the nineteenth century. The emergence of a new understanding that children were in need of protection was strongly influenced by the new medical and psychological sciences.[3] Secondly, scholars have looked at how psychological knowledge, and especially methods of mental testing, have informed educational politics.[4] Beyond that, most points of departure for a history of the scientization of childhood arise from studies on parenting handbooks, an important medium in the twentieth century for the diffusion of expert knowledge into a willing readership of parents. In the case of Germany, the historian Miriam Gebhard has recently argued that a new scientific conception of early childhood education was reflected in parenting handbooks from the 1890s, though it was only from the late 1920s that it was sought after and received by bourgeois parents on a broad front.[5]

The present case study of the political dimensions and consequences of the establishment of the new children's sciences in Spain between the end of the nineteenth century and the 1930s pursues a different approach. It relates the triumph of the scientific knowledge about children not only to changing forms of pedagogy, but also interrogates the political and social consequences of this development. It regards the success of the new knowledge about children not as a largely neutral development, but rather as an eminently political phenomenon, one that profoundly transformed contemporary notions about the child, the societal organization of childhood, and thus invariably also the organization of society as such. These transformations did not take place without conflict, but led to vigorous political clashes. The diffusion of new knowledge about children was intimately linked with projects of social reform and political change.[6]

The present essay seeks to make a contribution to the growing field of research into the social and political repercussions of the development of modern knowledge-societies and the scientization of the social in the twentieth century.[7] An examination of the forms and political consequences of a scientization of childhood in Spain promises new insights into the societal reach of the modern sciences and the political forces they unleashed. The Spanish case provides a particularly good object for studying the scientization of the social since it shifts the view of scholarship beyond the classic core countries of Western modernity to developments in countries that were not among the centres of the new production of knowledge in the human sciences. We still know very little about the scientific, political, and social dynamic that was triggered by the reception of the new human sciences in these countries.

The history of Spain is generally seen as an example of the instability of political systems in the transition to modernity and as case study for the political and ideological polarization of society in the crisis years of classic modernity. After defeat at the hands of the United States in the Spanish–American war of 1898, Spain had become a second-tier European power, characterized in the decades that followed by unstable political structures and severe social conflicts. A coup by General Primo de Rivera in 1923 put an end to the moribund constitutional monarchy. But his military dictatorship, which had launched itself under the banner of modernizing the country, lasted only a few years: in the spring of 1931, following mass protests, it was replaced by the democratic system of the Second Republic, though this was likewise unable to achieve political stabilization of the country. In July 1936, another coup attempt by the military triggered the Spanish Civil War.[8] Childhood and education were especially contested issues in the political clashes over a renewal of society. After 1900, progressive, secular educators and Catholic child experts stood in implacably opposed camps and tried, in a long-contested struggle, to realize their respective ideas of a modern educational system.[9]

To date, historical scholarship has attributed the political battles and clashes over education policy largely to crises of structural modernization, and has interpreted the political conflicts as a struggle between the forces of modernization and those of tradition. By contrast, the discussion that follows will propose an expanded view of the conflicts, one that considers not only social and economic factors, but also changes in the systems of knowledge. It will be argued that by taking into account the role of the new human sciences, and specifically the scientization of childhood, one can uncover new dimensions of the contestation and offer alternative explanations for Spain's conflict-ridden history in the twentieth century.

Two questions structure this essay. First, there is the issue of the reach and political charge of the new children's sciences in Spain. Did all parties concerned with educational policy receive and adopt the new knowledge of children in the same way, or was it closely tied in with specific political/ideological worldviews and programmes? That is to say, did the dividing lines between a secular, progressive-republican camp and a Catholic-national one correspond to a confrontation between scientific modernity, on one side, and traditional beliefs, on the other? Secondly, there is the issue of the political dynamics of the new human sciences. Did they contribute to a radicalization of the political clashes, or did they tend to attenuate them? Did the new knowledge promote radical concepts of social reform and thus contribute to destabilizing the various political regimes? Or did it tend to promote the consensus between the political actors concerning the goals of social reform and contribute to the emergence of a new class of experts removed from party-political fighting?

At this point it seems appropriate to say a few words about terminology and concepts. For our purposes, the phrase 'scientization of childhood' refers to the successful acceptance of scientific, empirical research as the basis for the social organization of childhood and the interactions with children. One important aspect of this scientization is the enhanced importance of a new group of experts on children. In this sense, the scientization of childhood is always a societal phenomenon that extends beyond the scientific field itself. It generates new answers to the question about how the individual and the social order should be mediated, and it implies new claims on the state and society. This understanding of scientization displays structural similarities with the model of 'domestication' (*Verhäuslichung*) that Jürgen Zinnecker, an expert in the history and theory of pedagogy, has articulated as a secular trend of childhood in modernity.[10] The concept of 'domestication' also deals with new tendencies towards a societal shaping of childhood in the twentieth century through the growing integration of children into pedagogically controlled spaces. It thus provides an analytical matrix within which the consequences of scientization processes can be discussed. However, while Zinnecker imbues the concept with a strongly teleological perspective, he also suggests a largely anonymous structural process. By contrast, the more narrowly circumscribed concept of scientization adopted here makes it possible to enquire closely into political actors and political driving forces behind the change of childhood and society.

The rise of children's sciences and the triumph of a new conception of the child

The beginnings of a systematic empirical study of children lie in the eighteenth century. Two starting points for a scientization of childhood can be identified. First, doctors for the first time took a systematic interest in the child's body and its peculiarities. While the practical returns remained very modest for quite some time, what did take hold was an awareness that the child's body required different medical treatment from an adult's. Beginnings were also made in the field of child psychology. The first educators began to record the development of their own children in diaries in order to derive general principles from these observations. However, a coherent field of research was yet to emerge from these initial steps.[11]

It was only in the decades after 1880 that there was a breakthrough on an international level of child research as an independent, though still very heterogeneous, field of research. In the interplay of societal and scientific dynamics, the individual child and its psyche moved into the centre of political and scientific interest in a new way. Various factors played a role in this process. Only brief mention will be made of changes within the bourgeois family, in which the child, in the course of the nineteenth century, was understood in a new way as an emotional enrichment of the adult world. This development, however, heightened the interest of parents in the inner world of children and in their development.[12]

In addition, growing national rivalry and rapid urbanization in Europe directed attention in a new way to children and their healthy development. At a time of heightened international competition among nation states and empires, children drew new attention as a national resource that had to be cultivated and developed.[13] In order to protect the lives of children better and cultivate them into robust citizens, new medical knowledge about the peculiarities of the child's body and its development was necessary, the kind of knowledge that was being collected in the new discipline of paediatrics.[14] Finally, the new societal interest was grounded in a view of childhood – developed in Romanticism – as a more primal and authentic form of life. Childhood exerted a new fascination because of the naiveté, naturalness, and unaffectedness that were ascribed to it. It seemed to represent an access – lost to the adult human – to the sources of human existence, to the 'primitive and original in human nature', and it became a new place of yearning in times when the dark sides of the bourgeois design of society were being discussed in a new way.[15]

The interest in child medicine and the child's psyche was further reinforced by the implementation of universal compulsory schooling in the leading industrialized nations. The concentration of children in the schools made the state of children's health administratively and publicly visible. School hygiene began to address the condition of classrooms, school furniture, and school architecture. At the same time, the physical and mental differences between children of the same age became apparent as a pedagogical issue and boosted the demand for knowledge about children.[16]

After 1869, the new societal attention to childhood encountered an independent dynamic in the human sciences, which were newly discovering childhood as a topic of investigation. An important starting point for the new child sciences lay in Darwinian evolutionary biology. On the basis of the thesis about the correspondence between phylogenesis and ontogenesis, physiologists and human biologists in the wake of Darwin were hoping that a study of early childhood would provide insights into the early development and elementary constitution of humans. These investigations were followed with interest also by educators and psychologists, who in turn were hoping that the study of the 'child's soul' would offer clues for educational practice.

Against this varied backdrop, the international study of children gained momentum as an independent field of research in the 1880s. The number of publications and journals surged, and the Anglo-American world in particular witnessed the emergence of numerous organizations in which child researchers and interested laypeople collaborated to expand the knowledge of childhood. The designation of the new field of knowledge as 'pedology' or 'child studies' indicates that the various scientific disciplines that study children and childhood today were still little separated institutionally and conceptually. Child medicine, developmental psychology, empirical learning research, and the currents of a new, empirical pedagogy were still closely intertwined.[17]

Child study developed and popularized a new empirical-scientific view of childhood and children, and thus fundamentally transformed the debates about childhood both in expert circles and the media. Three new approaches must be highlighted. First, child psychology moved the individuality of the child into the foreground and thus departed from older, typologizing accounts of childhood, as for example in the popular doctrine of temperaments.[18]

Secondly, a significant intellectual fascination of the new child research lay in seeing childhood primarily as a psycho-biological developmental stage. Using developmental psychology, it refuted potent religious theses about an inborn innocence or sinfulness of the child by emphasizing the special pre-moral character of early childhood and highlighting anew the importance of inheritance as a factor in the development of the mind.[19] In this context, 'energy' became the new guiding concept of childhood. The influential American child researcher Stanley Hall went so far as to compare the educator with a hydraulic engineer who builds an irrigation system in which children's energies are put to sensible and beneficial use. The crucial issue in education was 'to direct the primitive energies of human nature'.[20]

Thirdly, the importance of physical and emotional factors gave a new focus to studies of the development of the child. The more comprehensive image of the child, which also took into account sensory impressions and emotions, proved far more suitable for interpreting and describing the behaviour of children than the dominant rationalistic concept of the child's soul.[21]

The reception of the international child sciences in Spain

Spain was not among the centres of the new child sciences. The Iberian Peninsula initially did not experience an institutionalization of the movement in research centres and organizations comparable to that in Germany, France, the UK, and above all the United States. Still, the fledgling sciences and a new understanding of the child did gain importance in Spain after 1900.[22]

The reception and appropriation of international child research in Spain cannot be understood without reference to the country's existential crisis at the turn of the twentieth century. Traumatic defeat in the Spanish–American war of 1898, with the loss of remaining overseas colonies, triggered in Spain an intense debate about the possibilities of national regeneration and a reform of society. With these exceedingly varied debates, Spain leapt into the twentieth century. Education and childhood quickly became important topics in these debates, since all political forces were hoping to initiate a fundamental reform of Spanish society precisely through a reshaping of child-rearing and education.[23]

The crisis of 1898 was the starting point for a comprehensive engagement with international child research. A new generation of educators strove to catch up with the research in other countries and to convey their findings to a larger Spanish public. That 1898 was a turning-point year is already evident from the basic concept of 'child psychology'. When child psychology was talked about in pedagogical journals in the 1890s, these were still entirely discussions of older concept focused on the child's spirit and morally charged concepts.[24] It was only after 1898 that the new, more comprehensive meaning of the term asserted itself in Spain.

After 1900, many pioneering works of the child sciences appeared in Spanish translations. William Preyer's 1882 study *Die Seele des Kindes* [The Soul of the Child], which was fundamental to the development of the discipline, was first published in Spanish by a Madrid publishing house in 1908; Gabriel Compayré's important 1893 study *L'évolution intellectuelle et morale de l'enfant* [The Intellectual and Moral Development of the Child] was released in Spanish in 1905 and appeared in at least three editions by 1920.[25] However, the translating activity was by no means uniform, let alone comprehensive. The works of probably the most famous international child scientist, the American Stanley Hall, for example, were not translated until the 1920s; the important studies of Edward Thorndike did not find their way into Spanish at all.[26]

At least as important as these translations, however, was the fact that Spanish writers discussed the findings of international research in influential pedagogical journals. As early as 1899, Pedro de Alcántara Garcia, the publisher of the important reform-oriented educational journal *La Escuela Moderna*, presented his readers with a lengthy overview of the international child science and called for its implantation in Spain. Even the journal of the Spanish teachers' association, *El Maestro Español*, not known within historical scholarship for its reform orientation, opened up its columns to the new child scholarship in 1900.[27] The driving forces behind the reception were for the most part individuals like the secondary school teacher Martin Navarro Flores, who translated Preyer's book and promoted the new science in books and essays from the turn of the century.[28] There were also the rudiments of an institutionalizing of child research. As early as 1911 in Barcelona, which rapidly established itself as the centre of child study in Spain, a Museo Pedagógico Experimental was established, which dedicated itself to developing and diffusing experimental pedagogy in the country.[29]

Alcántera García and Navarro Flores were liberals who were personally closely linked to the liberal educational reform institutions. But an openness to new knowledge about children was not tied to liberal political-ideological attitudes. Catholic intellectuals, who were sharply opposed to the liberal and progressive circles when it came to education policy, also occupied themselves with the new sciences. Rufino Blanco y Sánchez, probably the most famous and influential

Catholic educator in Spain before the 1930s, commented on the significance of psychology for pedagogy as early as 1888. After 1900, as well, he returned repeatedly to topics of developmental psychology, and his large-scale anthropometric studies of school children met with a broad media interest.[30] The two leading Catholic educational and cultural journals, the Jesuit *Razón y Fe* and *Revista Calasancia*, also published essays on the new children's studies before World War I. Father Rogelio Gutiérrez, a member of the influential Piarist Order, which ran many schools in Spain, was, by his own account, so thrilled by his first encounter with the new science that he devoted all his energy thereafter to promoting and spreading it.[31]

By the 1930s, the psycho-biological view of childhood had taken firm root among Catholic educators. The journal *Atenas*, founded in 1930 and soon seen as the official organ of Catholic educational circles, came out in favour of the comprehensive introduction of psychological child research. Authors pointed to the complexity of the inner life of children, which had supposedly been little considered so far, called for research into 'the individuality of the pupil, or, in other words, his individual psychology', and demanded that Catholic education 'be guided by input from child psychology'.[32] The intensive reception went hand in hand with the triumph of a new, psycho-biological view of childhood across the boundaries that separated the camps with respect to education policy. Primarily moral conceptions of childhood even receded into the background in the Catholic-conservative spectrum. The apodictic verdict of the liberal educator Gerardo Rodríguez García that the child was 'more a physiological than a moral being' met with approval among educational experts of all political camps.[33] A psychobiological understanding of childhood also carried the day in significant Catholic reform circles. The semantic field of energy replaced the previous moralistic vocabulary.[34]

In Spain, the new child sciences were thus not linked with a specific political project, but were compatible with very different positions on social and educational policy. Their reception became increasingly proof of a modern way of thinking about education, and thus also a distinguishing characteristic by which a new generation of educators sought to set themselves apart from the status quo of educational policy and to prove their modernity. Both liberal-progressive as well as Catholic-national circles grappled intensively with the new knowledge about children in order to provide a solid foundation for their efforts at social reform. Both sides of the divide over educational policy believed that only a modernization of the image of the child and education on a scientific basis would make it possible to influence the next generations successfully. In this sense the ideological quarrel favoured the establishment of a new understanding of childhood in Spain.[35]

The expansive dynamic of child psychology and childhood reform

The circumstances in which international child research was received and adopted by the two leading camps on educational policy did not mean, however, that the child sciences contributed to a depoliticization of child matters in Spain. On the contrary, their political effects were highly contradictory and also included an aggravation of political conflicts. To understand these effects it is necessary, first of all, to take a look at the relationship between child sciences and child reform debates in Spain, which were characterized by an expansive dynamic.

Child protection was the dominant topic of the journalistic childhood debates after 1900. Across many countries at the end of the nineteenth century, a perception of children as an endangered population group and simultaneously – especially in the form of street children – a danger to the social order established itself. The new journalistic interest certainly cannot be attributed solely to the emergence of the new child sciences. The concentration of children in the rapidly growing cities and the amplified problems of controlling the emerging generations this entailed also played an important role.[36] But the public activities of the new child scientists unquestionably reinforced the social interest and gave new concerns about children a voice and specific form. Child experts, whose numbers rose substantially in the new century through the establishment of new children's institutions, quickly gained influence in the Spanish public after 1898, and especially after 1914. It was they who defined problems in the media, set the goals of reform, and commented on steps of reform as these were taken. They translated a widespread concern over children into a public agenda and imparted a political urgency to it.[37]

Child experts were important protagonists in the child reform movements that arose first in the US and Great Britain. Spain had gone along with these new developments since the 1880s. Here, too, the starting point were impulses from individual reformers.[38] But it was only after the crisis of 1898 that the state institutionalized child protection in a series of statutory regulations. The most important turning-point was the passage of a child protection law (*Ley Protectora de la Infancia*) in January 1908. The law regulated child labour for the first time, subjected wet nurses to public inspection, declared the protection of mothers and infants to be public tasks, and, finally, asserted that public authorities had a duty to care for street children and juvenile criminals. To implement the various decrees, the government created on a national level a Child Protection Council (*Consejo Superior de Protección a la Infancia*), which was to act as a connecting link between the state institutions and private charities.[39] Parallel to the legal anchoring of child protection, the large cities and private reform groups founded a series of new institutions intended to improve the health and care of young children. For example, in 1914 the city of Madrid opened the *Institución Municipal de Puericultura* as a municipal institution for the protection of mothers and infants.[40]

The child protection debates were not static, however. Rather, they were characterized by an expansive dynamic – driven by advances in developmental psychology, in which new fields of child protection were continually discovered and placed in the hands of the public authorities. The findings of the expanding field of developmental psychology suggested in the 1920s an expansion of child protection by means of the already established fields of health care and child criminality. In addition to the basic right to life and care, the right of the child to develop its personality and the necessity to protect the child's psyche also moved into the centre of public debates. Shaping children into new, better humans demanded more than the satisfaction of basic physical needs. It required also the guarantee of a free development of the child's personality, and the protection of childhood as a special psychic-emotional phase of life. The experience of childhood, the new generation of child experts were convinced, must be happy and carefree, if growing children were to become psychically stable adults.[41]

The press directly embraced and reinforced the expansive logic of the international child reform debates. In the reporting of the 1920s, joy (*alegría*) became – alongside physical health – the most important indicator of a successful childhood. The media increasingly evaluated childhood reform efforts according to whether they also satisfied children's need for affection. For example, in January 1932 Manuel Bueno noted, in the conservative magazine *Blanco y Negro*, a new understanding of the contemporary Spanish child, one that had also changed politics towards children:

> It is only fair to state that society is showing itself more inclined by the day not only to watch over the child with protective regulations, but also displays the desire to make its life pleasant, to both protect (*amparar*) and amuse (*divertir*) the child.

These steps were evidence for the growing civilization of humanity, but they were not yet sufficient:

> All children shall have the right to the same affection (*caricia*) and praise (*halagos*). It is not enough to give the child only bread and a roof over its head; it is equally necessary to fill its soul with joy.

Joy was indispensable for children lest they despair of life.[42]

The expanded demands on childhood reform and child protection were reflected especially in the depiction of model institutions, which, from the end of the 1920s, were no longer described only as hygienic establishments promoting the health of the child, but also as happy places that took the psychic and emotional needs of growing children seriously. In 1927, for example, *Blanco y Negro* published a lengthy article about a children's home in Berlin-Lankwitz, which had the most modern sanitary installations and a large garden, but which also paid attention to the psyche of children, with the result that 'the joyfulness of good health and of freedom' characterized the home.[43] Successful childhood reform – and this was the almost unanimous opinion of the Spanish press at the beginning of the 1930s – had to ensure the protection and promotion of the child's body as well as of children's feelings and the child's personality.

Political radicalization from the spirit of childhood reform

It is important to emphasize once again that all camps of the political public embraced the expansion of what was subsumed under child protection and child rights. In the political clashes over childhood, the confrontation was not between a modern, scientific position and a traditional position sceptical about science. The intense political conflicts that were kindled around questions of childhood and education were not based on divergent attitudes towards the new knowledge about children. Rather, it was the scientization of childhood itself that drove an intensification of the political-cultural battles over children and social reform, and which also fuelled a critique of the Spanish state by sections of the journalistic public. However, the political consequences were not straightforward. In an ambivalent way the rise of the child sciences also attenuated the political confrontation and gradually removed questions over child reform from the clash of political opinions. In the 1930s, both developments, which will be examined successively in what follows, combined into a contradictory mixture.

The new child sciences popularized a picture of the child as an emotionally complex being in need of protection. Accordingly these sciences not only elevated the claims made on education and child-rearing, but also intensified the clashes over education policy. These clashes revolved no longer merely around the curriculum of what was taught at school, but in a new way around the entire personality of the child and the protection of the child's emotional world against harmful impressions. The new, comprehensive picture of the child's psyche had its counterpart in a heightened political interest in the shaping of childhood.

This is revealed by examination of the two primary camps when it came to social policy. Among liberal and progressive educators, the attitude towards Catholic child education became noticeably radicalized from the turn of the century and into the 1930s. The older criticism of Catholic education as tradition-bound and a brake on social progress increasingly expanded in the 1920s, under the influence of the new child sciences, into a critique of the formation of the religious personality. According to the new charges, the latter violated the principles of modern hygiene and deformed the bodies and souls of children.[44] More so even than in hygienic shortcomings, anti-clerical politicians perceived a scandal in the deformation of the child's psyche by a dogmatic education grounded in fear. As early as 1924, a progressive educator had argued that action against the confessional schools could not be justified if the question was merely one of technical instruction (*instrucción*). But since the schools also influenced the personality of their pupils in a dogmatic way, church-based education was an obstacle to the democratic project of creating free citizens, and therefore had to be ended. Development of a free personality was possibly only in opposition to the Church.[45] The attitude of the liberal party after the turn of the century, which had been concerned merely with limiting the influence of Catholicism, but not with a comprehensive expulsion of the Church from the educational system, contrasted markedly with a republican anti-clericalism of the 1930s, which sought to expunge confessional education as such.[46] It was precisely new notions of education as processes, which encompassed the 'whole child', that imparted urgency to anti-clerical education policy in the eyes of many republicans. Radical educational anti-clericalism was in this way an offspring of modern child psychology.

Catholic child experts discussed the threats to childhood in much the same way as their republican opponents, except that they wanted to protect the child's psyche against what they regarded as the corrupting political reach of the republicans. Especially after the founding of the Second Republic in 1931, Catholic circles were afraid that the child's psyche would be impaired and deformed by the direct confrontation with political demonstrations and the political propaganda of republican and leftist forces. As early as May 1931, for example, the Catholic journal *Pro Infancia* criticized the distribution of Communist brochures to children as a contamination of the child's imagination. During the years of the Republic, Catholic educators continuously looked askance at the political left, which was allegedly inciting children to 'hate and political passions'.[47]

However, the scientization of childhood not only provided new fuel for the political battles between the various camps, but also had important political consequences for the attitude of child experts and reform groups towards the Spanish state. While commentators had still welcomed the early steps of reform as hopeful signs of a modernization of the Spanish polity, the discrepancy between new visions of childhood reform and a status quo in Spain increasingly seen as untenable became the focus of reporting in the 1920s. Detailed descriptions of childhood reform and new model institutions in Berlin and Paris aroused in Spanish readers, and especially in reform circles, new expectations and intensified the claims that were made on the state and its institutions. The treatment of children became increasingly a criterion for Spain's level of civilization and progressiveness compared to the rest of the world. In the face of the steps towards reform in others countries as reported in the media, the Spanish state of affairs seemed particularly desolate and the situation of Spanish children particularly untenable. The accounts of model institutions, cast in an admiring tone, were constantly complemented by reports that highlighted the inadequacies of the Spanish situation, lamented the lack of a political will for reform, and culminated in pleas to step up the efforts to realize the new possibilities of child reform. The demand 'that we must try to elevate child protection here at home to the level of the most advanced countries' took on new urgency and political explosiveness. A gap opened up between the notions of what was possible and the material and administrative capacities of the Spanish state to realize these notions. The national and international debates about child reform significantly increased the reform pressure on the Spanish state.[48]

Against this backdrop, demands for fundamental political and social change gathered momentum on both sides of the political spectrum. On the republican left, following the establishment of the Republic in April 1931, there was initially overflowing optimism.[49] Reform initiatives by the new republican government were presented as the beginning of a new time of realizing the rights of children and as the foundation stone for a comprehensive renewal of Spanish society.[50] A renewal of childhood reform on the basis of scientific insights was elevated into the trademark of the new Republic. The enormous expectations that were invested in the Republic are revealed, for example, in an article in a popular magazine from the beginning of 1932: it argued that the right measures by the Republic could heal the psychological damage to children and transform even mentally handicapped juvenile offenders into normal children and good citizens. In the reporting, the new children's establishments represented laboratories of the future republican-democratic society.[51]

After 1933, however, the depiction of childhood in the left-liberal press became noticeably gloomier. The Republic, too, in spite of considerable efforts to fulfil the demands of child reform, proved unable to meet the boundless hopes of the childhood reformers. On the contrary, republican intellectuals soon noted with horror that child misery was assuming unimagined dimensions. Instead of the hoped-for new, healthy, and happy children, they saw in the streets of the big cities mostly ill-treated, abandoned, and psychologically deformed children.[52] This situation had immediate repercussions for their attitude towards the Republic. It was precisely progressive educators and left-republican oriented teachers' associations who, against this background, demanded an intensification of the pace of reform and more radical measures against the Catholic educational system, which was identified as the main brake on social modernization.[53] The radical reform impulse of these groups then manifested itself in the brief but exceedingly intensive phase of childhood reform in the republican cities in the first months of Civil War, when church institutions were seized, dozens of new child facilities were established, and in this way the presumed omissions of decades were to be rectified in the shortest possible time.[54]

Like their progressive opponents, Catholic child reformers also criticized republican politics, though they did so with a markedly different accent. Catholic accounts attacked the Republic especially because it was supposedly unable to effectively protect Spanish children and, beyond that, because it itself posed a threat to children since it confused the child's mind and unleashed harmful emotions. The depiction of children as the victims of the political upheavals of the Republic became a widespread topos in the Catholic press, and it came with new demands for a decisive intervention against the political indoctrination of children, 'the kindling of hate in the hearts of children, this despicable and scandalous corruption of the child's mind'.[55] A prophylaxis against revolution, in the view of a growing number of Catholic commentators, could be permanently successful only if the republic was prevented from getting its hands on children. In this regard, effective child protection invariably took on – similar to what happened on the political left – a fundamental political dimension. Genuine progress in the area of child reform could be expected only following a fundamental change in the political regime.

All in all, the change in the public debates over childhood reform was paradoxical. Precisely the triumph of new images of an apolitical, carefree childhood increased the political pressure for reform and promoted calls for an intensified intervention in childhood. Children, so it appeared to both progressive and Catholic commentators in the mid-1930s, needed more far-reaching political measures to be able to grow up as free and unburdened – also by politics – as the new models of childhood called for.

Countervailing trends: Scientization as a brake on politicization

The tendencies towards politicization must not be underestimated. Yet, one must not fail to note that a number of factors worked against an ideological polarization of child experts and a fundamental politicization of childhood reform. The political effects of the new child sciences were not as one-dimensional as this account may have suggested so far. The scientization of childhood not only promoted a fundamental politicization of childhood reform; at the same time, and paradoxically, it also worked in the direction of constraining the political dynamic. A brief account of a few facets of this development will conclude this essay.

To begin with, one must point to contradictions in the debates about state interventions in childhood. For a large group of moderate reformers argued that childhood had to be protected not only against specific ideological influences, but also against any and all political-ideological reaches. Not least in response to the day-to-day political battles of the early 1930s, they argued in favour of a categorical separation of childhood and politics, and with that for a clear differentiation of the domains of child experts and politicians. They demanded that schools and the education of children be kept out of the daily political battles as much as possible. For example, one essay in the *Revista de Pedagogía* on the eve of the Civil War called for

> protecting the free development of the spiritual life of children. ... We must safeguard the child against every possible impact of force on its development – be it dogmatic or political; we must prevent it from becoming an instrument of the contending confessions and parties.[56]

This position was initially found above all in the liberal, moderate-republican spectrum of opinion and was unacceptable to the Catholic Church at the time, since religious instruction, too, was seen as a disturbance of the free development of personality. Still, this attitude supported in political praxis a professionalization of child experts and teachers as politically independent experts beholden solely to pedagogical and psychological knowledge, and limited the extent to which educational policy could over-determine the lives of children.

Both teachers and the personnel of the new child institutions developed a self-understanding as independent child experts that stood in opposition to political attempts to mobilize them. Invoking one's own scientific qualification was not only the basis of demands for a higher social status, but it could also serve, especially in politically turbulent times, to ward off political claims and attempted interventions in one's own sphere of work, and to legitimate one's activity beyond political regime changes.[57] In particular, many of the medical and socio-pedagogical child experts working in the new child institutions cannot be assigned unequivocally to one political camp or another. For example, nationally known doctors such as Cesár Juarros (1879–1942) and Enrique Suñer Ordoñez (1878–1941), who had made a name for themselves since the turn of the century as experts in the field of child medicine, published in both the left-liberal and the militantly Catholic press.[58]

The growing demand for expert knowledge in political-journalistic camps across the spectrum of opinion, and the interest in shared problems served here as a bridge across political-ideological boundaries. Even though child experts were closely tied into social policy, they succeeded in being perceived first and foremost as experts removed from the clash of political opinions. This can be also seen from the fact that numerous republican child experts were able to continue their work without interruption in the early Franco period. Even child experts who had politically exposed themselves before 1936 as pro-republican were able to carry on with their work after 1939, if only after a more or less long phase of privations and uncertainty. The child psychologist José Mallart, for example, before 1936 an important contributor to progressive pedagogical journals and during the Civil War briefly the head of the Psychotechnical Institute in Madrid, was initially stripped of his post by the Franco regime, but in 1943 he was able to return to his old position as head of the Institute.[59]

Conclusion

The child sciences that established themselves in the nineteenth century as scientific disciplines transformed the societal conceptions of childhood on a broad front and across the boundaries of political and ideological camps. As a result, they influenced not only the educational system and pedagogical models, but also, as this essay has sought to show, the debates about childhood reform and children's rights. In the process the societal appropriation of new knowledge about children kindled a contradictory political dynamic. Initially it intensified the cultural battles over education and childhood. The societal importance of childhood as a phase of life rose, and the fundamental vulnerability of childhood suggested greater exertions to protect children against false influences and ways of life. The triumph of new knowledge about childhood was very closely tied to new, heightened demands made on the treatment of children and the organization of childhood.

At the same time, the defining of expanded standards of what constituted a successful childhood by the new child sciences, and the implementation of these standards as the guiding perspectives of child reform in the media, increased the demands on state and society to shape childhood in keeping with the new standards. The state and municipalities saw themselves confronted with lofty expectations of providing effective protection for children and promoting their development through the establishment of modern children's institutions. The increased importance of the child sciences in public debates and the broadening of the concept of child protection that went along with this became politically an explosive issue in Spain, since the country did not have the administrative infrastructure, a sufficient number of well-trained experts, and the material means to rapidly implement the demands of the international community of experts. In this way, the dynamic of the international child sciences and child debates contributed their part to destabilizing the political system in Spain. The new knowledge promoted on both sides of the political spectrum the quest for new forms of government that would be better able to foster and protect children – if need be also through palpable interventions in families and public life. The child sciences thus inspired the formulation of political utopias, whereby the precise form of the new system could be imagined in very different ways. At the same time, however, the reception of the child sciences also set clear boundaries to a politicization of childhood and society by implying a clear separation of politics and education, and creating a new class of experts who thought of themselves first and foremost as professional social reformers beyond the everyday political and ideological battles. The juxtaposition and intertwining of radicalization and attenuation represent the characteristic of the childhood and social reform debates of the early 1930s.

The subsequent history of the child sciences in Francoist and post-Francoist Spain has been barely studied. However, a few lines of development can be adumbrated in a preliminary look forward. Initially, the victory of the nationalist troops under Franco put a forceful end to the ideological quarrel between the two camps on education policy. The banning of a large number of pedagogical journals and the enforced conformity in many research institutions and public opinion led to a clear impoverishment of the debates on childhood. This reduced the necessity for the new rulers and the Catholic-Francoist educational experts to engage in an intensive discussion with the international sciences. The early phase of Francoist rule was characterized by attempts to develop a national-Spanish child psychology whose foundation was the thesis of a special national character of Spanish children and which also incorporated biologistic approaches.[60]

However, contrary to what some scholars of education have claimed with respect to pedagogy, there was by no means a complete self-isolation of the country against international developments. For example, already before the end of the Second World War one can detect a new interest in the pedagogical and psychological debates in the United States, with the Catholic educational circles in the US taking on the role of a 'translator'.[61] In this context, the attempts at a national child psychology seem to have lost influence. Rather, there is much to suggest that the regime in the course of the 1950s endeavoured, on the one hand, to assert its modernity vis-à-vis the Western democracies, and yet, on the other, to impart a clear Spanish accent to international developments. For example, the regime took the 'Declaration of the Rights of the Child' adopted by the UN General Assembly in 1959 as an occasion to introduce a 'Day of the Child' in May 1962, which was intended to portray Spain as a country that was particularly attentive to its duty to offer children an 'environment of affection and both moral and material security'. Through the international document, contemporary notions of childhood, which were based not least on ongoing developments in international developmental psychology, were diffused throughout Spain.

However, the regime continued in its efforts to retain interpretational supremacy over the childhood debates and to deploy the debates to secure its power. Many aspects of the contemporary debates in child psychology, such as John Bowlby's attachment theory, which posited a close mother–child bond as the precondition for the development of a stable personality, could in fact be adopted and propagated by the regime as confirmation of its Catholic family policy.[62] It is therefore also true for the period after 1945 that the international debates of child psychology were politically polyvalent and in no way suggested specific societal and political models. Finally, the answer to the question to what extent the transition to democracy in the 1970s constitutes a new phase in the relationship between child psychology, child experts, and society must be left to future research.

Notes

1. See, for example, N. Roberts (2004), 'Character in the Mind. Citizenship, Education and Psychology in Britain, 1880–1914', *History of Education* 33, pp. 177–97; A. Wooldridge (1994), *Measuring the Mind. Education and Psychology in England, 1860–1900* (Cambridge: Cambridge University Press). For German development, see A. Schulz (2003), 'Der "Gang der Natur" und die "Perfektibilität" des Menschen'. Wissensgrundlagen und Vorstellungen von Kindheit seit der Aufklärung', in A. Schulz and L. Gall (eds), *Wissenskommunikation im 19. Jahrhundert* (Stuttgart: Franz Steiner Verlag), pp. 15–39.
2. For the German and Spanish cases, see H-E. Tenorth (2008), *Geschichte der Erziehung. Einführung in die Grundzüge ihrer neuzeitlichen Entwicklung*, fourth expanded edition (Weinheim and Munich: Juventa); A. Escolano Benito (2002), *La educación en la España contemporánea. Políticas educativas, escolarización y culturas pedagógicas* (Madrid: Biblioteca Nueva); M. de Puelles Benítez (1999), *Educación e ideología en la España contemporánea*, fourth edition (Madrid: Ed. Tecnos).
3. P. S. Fass (1997), *Kidnapped. Child Abduction in America* (New York: Oxford University Press); J. M. Hawes (1991), *The Children's Rights Movement. A History of Advocacy and Protection* (Boston: Twayne).
4. K. Dombkowski, A. S. Nawrotski, and M. Vinovskis (2004), 'Social Science Research and Early Childhood Education. A Historical Analysis of Developments in Head Start, Kindergartens, and Day Care', in H. Cravens (ed.), *The Social Sciences Go To Washington. The Politics of Knowledge in the Postmodern Age* (New Brunswick, NJ: Rutgers University Press); G. Sutherland (1984), *Ability, Merit and Measurement. Mental Testing and English Education 1880–1940* (Oxford: Clarendon Press).
5. M. Gebhardt (2010), *Die Angst vor dem kindlichen Tyrannen. Eine Geschichte der Erziehung im 20. Jahrhundert* (Munich: DVA); P. N. Stearns (2003), *Anxious Parents. A History of Modern Childrearing in America* (New York: New York University Press); A. Hulbert (2003), *Raising America. Experts, Parents, and a Century of Advice About Children* (New York: Knopf).
6. On the political dimensions of the new human sciences, see also A. Reckwitz (2006), *Das hybride Subjekt. Eine Theorie der Subjektkulturen von der bürgerlichen Moderne zur Postmoderne* (Weilerswist: Velbrück Wissenschaft); N. Rose (1996), *Inventing Our Selves. Psychology, Power, and Personhood* (Cambridge: Cambridge University Press).
7. See the Introduction and the essay by L. Raphael in this volume. See also M. Szöllösi-Janze (2004), 'Wissensgesellschaft in Deutschland. Überlegungen zur Neubestimmung der deutschen Zeitgeschichte über Verwissenschaftlichungsprozesse', *Geschichte und Gesellschaft* 30, pp. 277–313.
8. W. L. Bernecker (2010), *Geschichte Spaniens im 20. Jahrhundert* (Munich: C. H. Beck); M. Vincent (2007), *Spain 1833–2002. People and State* (Oxford: Oxford University Press); J. Álvarez Junco and A. Shubert (eds) (2003), *Spanish History since 1808* (London: Arnold).
9. See J. de la Cueva and F. Montero (2007), *La Secularización conflictiva. España (1898–1931)* (Madrid: Biblioteca Nueva), and M. Suárez Cortina (ed.) (2001), *Secularización y laicismo en la España contemporánea* (Santander: Sociedad Menendez Pelayo).
10. J. Zinnecker (1990), 'Vom Straßenkind zum verhäuslichten Kind. Kindheitsgeschichte im Prozeß der Zivilisation', in I. Behnken (ed.), *Stadtgesellschaft und Kindheit im Prozeß der Zivilisation. Konfigurationen städtischer Lebensweise zu Beginn des 20. Jahrhundert* (Opladen: Leske + Budrich), pp. 142–62. A similar perspective is pursued also by

S. Mintz (2004) in his impressive account *Huck's Raft. A History of American Childhood* (Cambridge, MA: Belknap Press).

11. Schulz (2003), 'Gang der Natur', pp. 19–23; J. Schlumbohm (1998), 'Constructing Individuality. Childhood Memories in Late Eighteenth-century "Empirical Psychology" and Autobiography', *German History* 16, pp. 29–42; R. Spree (1992), 'Shaping the Child's Personality. Medical Advice on Child-Rearing from the Late Eighteenth to the Early Twentieth Century in Germany', *Social History of Medicine* 5, pp. 317–35.
12. From the extensive research literature, see especially V. Zelizer (1985), *Pricing the Priceless Child. The Changing Social Value of Children* (New York: Basic Books); G-F. Budde (1994), *Auf dem Weg ins Bürgerleben. Kindheit und Erziehung in deutschen und englischen Bürgerfamilien 1840–1914* (Göttingen: Vandenhoeck and Ruprecht); C. Heywood (2007), *Growing Up in France. From the Ancient Regime to the Third Republic* (Cambridge: Cambridge University Press).
13. Wooldridge (1994), *Measuring the Mind*, pp. 19–24; N. Roberts (2004), 'Character in the Mind. Citizenship, Education and Psychology in Britain, 1880–1914', *History of Education* 33, pp. 177–97, at 179f.; H. Cunningham (1991), *The Children of the Poor. Representations of Childhood since the Seventeenth Century* (Oxford: Blackwell).
14. S. A. Halpern (1988), *American Pediatrics. The Social Dynamics of Professionalism, 1880–1980* (Berkeley: University of California Press). On the history of the discipline, see also A. R. Colón (1999), *Nurturing Children. A History of Pediatrics* (Westport, CT: Greenwood Press). On intelligence research, see J. Carson (2006), *The Measure of Merit. Talents, Intelligence, and Inequality in the French and American Republics, 1750–1940* (Princeton: Princeton University Press).
15. C. Steedman (1993), *Strange Dislocations. Childhood and the Idea of Human Interiority, 1780–1930* (Cambridge, MA: Harvard University Press); M. S. Baader (1996), *Die romantische Idee des Kindes und der Kindheit. Auf der Suche nach der verlorenen Unschuld* (Neuwied: Luchterhand).
16. H. Cunningham (1998), 'Review Essay. Histories of Childhood', *American Historical Review* 103, pp. 1195–1208, at 1201; Wooldridge (1994), *Measuring the Mind*, esp. pp. 53–6, 73–110.
17. M. Depaepe (1993), *Zum Wohl des Kindes? Pädologie, Pädagogische Psychologie und experimentelle Pädagogik in Europa und den USA, 1890–1940* (Weinheim and Leuven: Leuven University Press).
18. W. Ament (1906), *Die Seele des Kindes. Eine vergleichende Lebensgeschichte* (Stuttgart: Kosmos), pp. 44, 113.
19. Schulz (2003), 'Gang der Natur', pp. 32f.; Wooldridge (1994), *Measuring the Mind*, p. 47; Carson (2006), *The Measure of Merit*, pp. 166f.
20. F. Buisson (February 1900), 'La Educación de la Voluntad (Conclusión)', *La Escuela Moderna* 107; S. Hall (1915), 'Nuevas ideas sobre la infancia', *Boletín de la Institución Libre de Enseñanza* 669, pp. 353–6.
21. T. Kössler (2008), 'Die Entdeckung der Kindheit im Gefühl. Kinderpsychologie und das Problem kindlicher Emotionen (1880–1930)', in U. Jensen and D. Morat (eds), *Rationalisierungen des Gefühls. Zum Verhältnis von Wissenschaft und Emotionalität 1880–1930* (Paderborn: Fink), pp. 189–210; P. N. Stearns and T. Haggerty (1991), 'The Role of Fear. Transitions in American Emotional Standards for Children, 1850–1950', *American Historical Review* 96, pp. 63–94.
22. A first overview of child research in Spain is provided by Depaepe, *Zum Wohl des Kindes?*,, pp. 116f.
23. A. Mayordomo Pérez (2007), 'Regeneracionismo y educación. La construcción pedagógica de la sociedad y la política', in V. L. Salavert Fabiani and M. Suárez Cortina (eds), *El Regeneracionismo en España* (Valencia: Universitat de València), pp. 165–206; T. García Regidor (1985), *La polémica sobre la secularización de la enseñanza en España, 1902–1914* (Madrid: Fundación Santa María).

24. Symptomatic of this is the uncritical reprint of a series of essays on the 'psychology of the child' by J. Sanz del Río from the 1860s in the liberal *Boletín de la Institución Libre de Enseñanza* of 1892: J. Sanz del Río (1892), 'Psicología del niño. El espiritu en el niño', *Boletín de la Institución Libre de Enseñanza* 16, pp. 81–3.
25. W. Preyer (1908 and later), *El alma del niño* (Madrid: Daniel Jorro); G. Compayré (1905 and later), *La evolución intelectual y moral del niño* (Madrid: Daniel Jorro).
26. The first detailed engagement with Hall dates from 1915: S. Hall (1915), 'Nuevas ideas sobre la infancia', *Boletín de la Institución Libre de Enseñanza* 669, pp. 353–6.
27. P. de Alcántara García (1899), 'De los estudios llamados de Pscicología infantil', *La Escuela Moderna* 104; A. Cervera y Royo (February 1900), 'Apuntes pedagógicos: De la sensibilidad', *La Escuela Moderna* 107; M. Chico y Suárez (9 June 1900), 'Cooperación del alumno en la obra educativa (IV.)', *El Magisterio Español.*
28. See, for example, M. Navarro (1904), 'La Paidología. Su historia y su estado actual', *Boletín de la Institución Libre de Enseñanza* 28, pp. 72–7, 100–5.
29. E. Collelldemont and Á. C. Moreu (2007), 'El Mueso Pedagógico Experimental de Barcelona. Enclave para una historia de los pequeños museos pedagógicos', *Historia de la Educación* 26, pp. 471–82.
30. R. Blanco y Sánchez (1888), *Nociones de psicogenesia aplicada a la pedagogía* (Madrid: Julian Palacios); idem (1911), *Paidología, paidotécnica y pedagogía científica* (Madrid: Imprenta de la Revista de Archivos).
31. See especially R. Gutierrez (1915), 'Paidología y Paidotécnica', *Revista Calasancia* 3, pp. 410–16; E. U. de Ercilla (September 1910), 'El problema psicofisiológico de la enseñanza (Parts I and II)', *Razón y Fe* 28.
32. As an example, see the discussion by J. Álvarez on *Fundamentos psicológicos de la escuela activa* at the national Catholic education week in 1933/4: J. Sotillo (January 1934), 'La Tercerca Semana de Estudios Pedagógicos', *Atenas* 36.
33. G. Rodríguez García (1921), *Lo que debe saber un padre para educar bien a su hijo* (Madrid: Libería de los Sucesores de Hernando), p. 21.
34. M. Asunción M. and O. de Urbina (February 1916), 'Reflexiones sobre dirección y educación del niño desde su más corta edad', *Revista de Educación Familiar* 2; Doctor Rosa (November 1917), 'La cólera en el niño (I.). Sus Causas', *Revista de Educación Familiar* 21.
35. On the consequences for educational praxis that must be left out here, see T. Kössler (2009), 'Towards a New Understanding of the Child. Catholic Mobilization and Modern Pedagogy in Spain (1900–36)', *Contemporary European History* 18, pp. 1–23.
36. On the quantitative increase of street children in Madrid from the end of the nineteenth century, see J. M. Borrás Llop (1996), *Historia de la infancia en la España contemporánea (1834–1936)* (Madrid: Ministerio de Trabajo y Asuntos Sociales), p. 43; P. Muñoz López (2001), *Sangre, amor e interés. La familia en la España de la Restauración* (Madrid: Marcial Pons Historia), pp. 296f.
37. See as an example Doctor Fausto, *ABC*, 22 June 1908.
38. Borrás Llop (1996), *Historia*, pp. 35–7; 'Al Empezar', *ABC*, 8 March 1908. On the context, see also Muñoz López (2001), *Sangre*, pp. 288f.
39. 'La Protección a la Infancia y el Trabajo de los Niños y de las Mujeres', *ABC*, 29 January 1908. On the early child protection movement, see also Muñoz López (2001), *Sangre*, pp. 297f., 348.
40. A. Prieto (23 October 1928), 'Una institución que ha criado de balde 2900 niños', *Estampa.*
41. A. Zozaya (22 December 1920), 'Del Ambiente y de la vida: ¿Incorregibles?', *El Mundo Gráfico.*
42. M. Bueno (3 January 1932), 'Paisajes Sociales: El Derecho a los juguetes', *Blanco y Negro.*
43. 'Por Amor al Niño, por Regina', *Blanco y Negro*, 9 January 1927.
44. On the descriptive tradition, see the comprehensive reports about the schools in Spain by L. Bello, for example, 'Visita de Escuelas: Seo de Urgel. Un maestro. 93 curas', *El Sol*, 28 January 1931.

45. B. Cabrera (1924), 'La reforma de la segunda enseñanza', *Revista de Pedagogía* 3, pp. 180–6.
46. On the change of political goals, see also M. Álvarez Tardío (2002), *Anticlericalismo y libertad de conciencia. Política y religión en la Segunda República Española (1931–1936)* (Madrid: Centro de Estudios Políticos y Constitucionales), pp. 15–17.
47. A. Iniesta (December 1934), 'Perfiles: Mirada angustiosa a la moral infantil', *Atenas* 46.
48. A. Cacho y Zabalza (24 January 1931), 'La Guardería infantil de Vallehermoso', *Estampa.* Here we can note only as an aside that similar dynamics can be observed also in other countries. See, for example, C. Ipsen (2006), *Italy in the Age of Pinocchio. Children and Danger in the Liberal Era* (New York: Palgrave Macmillan), pp. 162–95.
49. A good introduction to the history of the Second Republic is still S. G. Payne (1993), *Spain's First Democracy. The Second Republic, 1931–1936* (Madison: University of Wisconsin Press). For a more recent account, G. Ranzato (2006), *El ecplise de la democracia. La Guerra civil española y sus orígenes, 1931–1939* (Madrid: Siglo XXI).
50. See, for example, H. Peñaranda (4 July 1931), 'La Casa de los niños', *Mujer.*
51. 'Como y por que se hacen delincuentes los niños', *Ahora,* 31 January 1932.
52. A. de C. (31 March 1935), 'En favor de los niños abandonados o explotados', *Crónica.*
53. V. Z. (3 January 1933), 'La géstion ministerial. ¿Han triunfado las izquierdas o las derechas?' *El Magisterio Español.* This attitude is clear in the journals of the progressive teachers' associations, such as the socialist *Nuestro Escuela. Revista de los trabajadores de la enseñanza.*
54. On this, see T. Kössler (2008), 'Children in the Spanish Civil War', in M. Baumeister and S. Schüler-Springorum (eds), *'If You Tolerate This...' The Spanish Civil War in the Age of Total War* (Frankfurt and Chicago: Campus), pp. 101–32.
55. Editorial: 'La corrupción de la infancia', *Ellas,* 11 November 1934.
56. 'Notas del mes', *Revista de Pedagogía* 174, June 1936.
57. See T. Kössler (2011), 'Widerspenstige Kinder. Katholizismus und die Suche nach einer neuen gesellschaftlichen Ordnung in Spanien' (Habilitation manuscript, Ludwig-Maximilians University, Munich), Chapter 8 with numerous references.
58. 'Encuestas de Crónica. Cuáles son, a su juicio, los seis problemas más urgentes que debe resolver el Gobierno provisional de la República? Doctor Juarros', *Crónica,* 24 May 1931; KAY (25 October 1934), 'La Vida de nuestros hijos: Puericultura, Ciencia Moderna', *Esto*; E. Suñer (December 1933), 'Herencia y eduación (I)', *Atenas* 35; E. Suñer (6 July 1936), 'Los derechos del niño y los deberes de los padres', *La Voz.*
59. See F. Pérez Fernández (2003), 'José Mallart en la psicología española. Balance de una andadura intelectual', *Revista de Psicología General y Aplicada* 56, pp. 149–56, especially 152.
60. The most influential attempts were A. Vallejo Nágera (1941), *Niños y jóvenes anormales* (Madrid: Sociedad de Educación Atenas) and J. Álvarez de Cánovas (1941), *Psicología pedagógica. Estudio del niño español* (Madrid: Espasa-Calpe).
61. M. del Pozo Andrés and J. F. A. Braster (2006), 'The Reinvention of the New Education Movement in the Franco Dictatorship (Spain, 1939–1976)', *Paedagogica Historica* 42, pp. 109–26. Overall see also T. Kössler (2010), 'Erziehung als Mission. Katholische Sozialisation und Gesellschaft im frühen Franquismus (1936–1950)', in K. Tenfelde (ed.), *Religiöse Sozialisation im 20. Jahrhundert* (Essen: Klartext), pp. 193–217.
62. 'Los Derechos del niño', *ABC,* 22 July 1961; 'Ante del Día Universal del Niño', *ABC,* 6 May 1962.

10
Narcissism: Social Critique in Me-Decade America

Elizabeth Lunbeck

It is a commonplace of social criticism that America has become, over the past half-century or so, a nation of narcissists. Greedy, selfish, and self-absorbed, we narcissists are thriving, the critics tell us, in the culture of abundance that is modern, late-capitalist America. The disciplined, patriarchal Victorianism under which our stalwart forebears were raised has purportedly given way to a culture that makes no demands, while at the same time promising to satisfy our every desire. Self-indulgence has displaced self-control; plenitude reigns where privation was once the norm. The harsh Freudian super-ego, forged in Oedipal conflict with the powerful paterfamilias, has yielded to a permissive and undemanding simulacrum of the same.

The autonomous, inner-directed, and fully realized bourgeois male figure, exemplar of a lost golden age, has been superseded by the dependent, other-directed and essentially empty woman who, in popular renditions of this tale of cultural decline, finds her calling in shopping to her heart's content at the local mall. Her continuous primping and preening evidence of her essential vacuity, of her desire for nothing more than to be loved and admired, makes her the latter-day counterpart of Sigmund Freud's good-looking female narcissist – in his estimation 'the purest and truest' type of woman met with. Blissfully self-contented, loving themselves more intensely than any others, such women, Freud wrote in his landmark 1914 essay, *On Narcissism. An Introduction*, 'have the greatest fascination for men'.[1]

'Narcissism' has proven the pundits' favourite psychoanalytic diagnosis, a morally freighted term with appealing classical resonances, a highfalutin name for the old-fashioned complaint that modernity consists in a loosening of restraint and that modern satisfaction consists, as Philip Rieff put it nearly 50 years ago, in nothing so much as a 'plenitude of option'.[2] First used by Freud in 1910 to explain the male homosexual's puzzling self-love, his taking of himself instead of woman as sexual object,[3] the term 'narcissism' entered the popular lexicon in the mid-1970s as a shorthand for me-decade excess, and was quickly grafted onto perhaps the most compelling of the master narratives fashioned to account for the nation's purportedly sorry state. This narrative of declension, in its various registers, took us as a people from sturdy production to meaningless consumption, from small-town *Gemeinschaft* to anonymous *Gesellschaft*, from David Riesman's introspective bourgeois to his superficial glad-hander, from the self-denying hysteric of Freud's day to the pampered and indulged narcissist of our own.

This is a powerful story that has informed a major tradition of American social criticism, from Riesman's *The Lonely Crowd* (1950) and William H. Whyte, Jr's *The Organization Man* (1956) to Daniel Bell's *The Cultural Contradictions of Capitalism* (1976) and Christopher Lasch's *The Culture of Narcissism* (1978).[4] Collectively, the critics have painted a compelling portrait – well known, widely accepted, rarely interrogated – of a nation in cultural and characterological decline. That we, as moderns, are characterized by our lack of authenticity and genuineness, and that this lack is manifest in the predominance of false over 'real' selves, is an unexceptionable claim.[5] That this is traceable to the conditions of modern life – its instabilities and superficiality, its elevation of consumption over sturdy production – is a commonplace. That we're a nation of self-involved navel-gazers adrift in a culture of narcissism is a banal cliché that few would bother to contest. Any number of indictments of modernity – and specifically of its American variant – assert that an empty culture such as ours, which valorizes a mindless consumption and promotes the indulging, rather than the harnessing, of all sorts of impulses, not only tolerates but produces the figure of the narcissist, whose impoverished moral sensibility serves vividly to illustrate the depths to which we have fallen.

Through the 1950s and 1960s, a number of psychiatrists and psychoanalysts made much the same argument about the American self. Like the critics, and often in dialogue with them, they argued that the modal American social character had recently changed, adducing as evidence the appearance of a 'new type of patient' in the consulting room and hospital ward. As clinicians told it, this new patient suffered not, like Freud's patients, from repression and the sexual conflicts it occasioned – for, it was widely asserted, social prohibitions of all sorts had been weakened by the mid-century period – but from a range of vaguely defined complaints: feelings of emptiness, aimlessness, futility, and discontent. A series of now classic psychoanalytic papers registered the demise of the Freudian patient with her hysterical symptoms and the ascendancy of the chronically maladapted, character-disordered new patient.[6] The exemplary patient of modernity, it was widely agreed, was no longer the hysteric with her florid conversion symptoms, but the empty, grandiose narcissist. Among analysts and critics, then, it quickly became orthodoxy that the modal patient as well as the modal American character had decisively changed.[7]

'Narcissism', as both term and concept, was hardly new in the 1970s, but it was by any measure newly ubiquitous, both in psychoanalysis and in social criticism. Long a focus of psychoanalytic interest, narcissism – its development, its prevalence, its moral valence, its defining features – was at this point, for the first time, situated at the very centre of creative and contentious analytic debate. American social scientists eagerly incorporated the term into their already well-developed critiques of American affluence and abundance. And a public, possessed of what one British critic noted was a puzzling 'appetite for self-excoriating self-examination', made bestsellers of the very books that condemned them as empty selves and mindless consumers.[8]

'Narcissism' was a flexible enough term to encompass a spectrum of usages from the most narrowly professional, through the middle-brow social-scientific, to the most expansively popular. In the orthodox Freudian's idiom, for example, it could serve as a technical term to characterize the distribution of libido in the subject. To the social scientist, narcissism denoted a lamentable Tocquevillian excess of individualism at the expense of the imagined collective. Finally, in the popular press the term served as shorthand for an unseemly attention to the self – as in, for instance, solipsistically 'getting your head together', as a 1976 article in *Newsweek* was titled, in any of the 8000 ways to deepen one's consciousness the magazine claimed were currently on offer.[9]

In 1979, in a fitting capstone to the decade, President Jimmy Carter, after a crash course in sociology – ranging from selections on 'the problems of affluence' drawn from Alexis de Tocqueville's *Democracy in America*, first published in 1835, to Lasch's newly published jeremiad, widely reported to be the President's favourite – took to the airwaves, charging that the 'worship of self-indulgence and consumption' had displaced Americans' once-strong commitment to hard work, close-knit communities, and faith in God. 'Owning things and consuming things', he claimed, was inadequate to 'fill the emptiness of lives' devoid of meaning: America was in the throes of a full-bore spiritual and cultural crisis. Identity, he further charged, 'is no longer defined by what one does, but what one owns'. Part cultural critic, part preacher-in-chief, Carter in this speech – quickly dubbed the 'malaise' speech – channelled an old charge with a new twist, placing narcissism front and centre on the national agenda.[10]

Lutz Raphael, in his chapter in this volume, outlines four different configurations of the relationship between social science knowledge, on the one hand, and society, politics, and economics on the other, suggesting that in the 1970s a defining therapeutic ethos – centred on the individual and ways of conscripting his or her potential for the benefit of the greater whole – displaced an earlier postwar emphasis on social engineering in the service of planned modernization. The debut of the narcissistic critique of American culture is surely part of this story, a critique centred on individuals' characterological deficiencies and the collective consequences of these for the body politic. In its overwhelmingly negative estimation of individuals' potential for therapeutically prescribed uplift, the critique might best be considered the dark double to the optimistically tinged therapy-as-social policy that Raphael sees as ascendant in the 1970s, caustic and reproving where the latter was upbeat and sanguine.

Still, the fact that its cultural purchase was so quickly established – and has proven long-lasting, even to this day a staple of social-scientific and popular criticism – suggests it was exemplary of the new therapeutic configuration, heralding its consolidation and enabling its continued hegemony. Further, the protean quality of narcissism, a category that conceptually and rhetorically brought within a single beguiling compass sites as disparate as the analyst's couch and the president's bully pulpit, suggests it was one of those volatile and flexible ideas-as-flashpoints that, Raphael argues, generated 'interdiscourses' in which are knitted together the *esoterica* of the human science disciplines and the everyday language of political and moral debate. The term 'narcissism' proved a resource not only for psychoanalysts, who reoriented their discipline around it in the 1970s, but also for a range of nay-sayers, from the social scientists who saw in it the spectre of a therapeutic ethos gone amok to the proverbial man-in-the-street looking to narcissism to account for the baleful consequences of too much abundance and too little self-restraint.

In what follows, I examine narcissism's public fortunes in the 1970s, the decade to which it lent its name, looking at narcissism's sudden rise to prominence, its conscription into well-established narratives of national decline, and its all-purpose utility as social critique. I then examine the apparent seamlessness of the convergence between the psychoanalytic and the sociological in the figure of the narcissist that made a debut on the public stage, addressing the question of why it was that the exemplary narcissist was cast first and foremost as a mindless consumer and not, for instance, as the charismatic CEO who operates at the edge of destructiveness, a type that has recently been appearing with increasing frequency in popular print and web-based discussions of narcissism. The widely accepted assertion that America was a nation of narcissists, while drawing some of its language and concepts from psychoanalysis, was always more about the state of the American character than it was about psychiatric diagnoses and analytic understandings. Once delineated, narcissism in the public realm was, at times, freed from its moorings in psychoanalysis, a flashpoint for a moralizing and politicized commentary.

Narcissism between clinic and culture

Narcissism's public debut, facilitated by both critics and clinicians, may be located in the 1970s. It was 'the decade of the new Narcissism', proclaimed the *New York Times*. The term was suddenly everywhere, or so it seemed, offering critics a beguiling new language in which to voice venerable complaints.[11] Cassandras of cultural decline, from both left and right, mobilized in support of the contention that a 'new narcissism' was ascendant.[12] Too many Americans, they charged, were bristling at tradition, solipsistically seeking fulfilment at the feet of New Age gurus while spurning engagement with the common good. Among these Cassandras, no one was more important than Lasch, a gifted polemicist and formidable public intellectual who, throughout the decade, issued dire assessments of the nation's fall in tandem with narcissism's rise. In the hands of Lasch and other critics, narcissism usefully brought together under one snappy rubric the rampant individualism, the spiritual questing, the preoccupation with self, and the flight from commitment that, they argued, were newly prevalent.

Lasch was hardly the first to chart the precipitous decline of the American self, and neither were his readers the first to revel in accounts of their own deficiencies. In the 1830s, for example, Tocqueville had noted of the Americans that 'life would have no relish for them if they were delivered from the anxieties that harass them, and they show more attachment to their cares than aristocratic nations to their pleasures', an observation that elicited from David Riesman, writing in 1950, the characteristically dour comment that all that had changed in the interim was 'that the sphere of pleasures had become a sphere of cares'.[13] Americans like to think of themselves as particularly alert to the question of national character, and in their explorations they have tended more often to compare their fellow citizens in the present to forebears in the past than to inhabitants of other nations or cultures[14] – a hermeticism that may partly account for the dismissiveness, hostility, and what one commentator has described as a complete lack of comprehension characterizing British reactions to Lasch's book,[15] and may also have had the effect of underscoring the theme of decline running through the literature. In the American tradition of writing on national character, the manifest historicity of approach signalled by Lasch's dictum that 'every age develops its own particular forms of pathology, which express in exaggerated form its underlying character structure'[16] coexists comfortably with the patently ahistorical invocation of timeless truths accompanied by contemporary commentary, of which Riesman's invocation of Tocqueville is but a particularly pertinent instance. Such writing, then, can feel at once startlingly new and reassuringly familiar.

This was certainly the case with *The Culture of Narcissism*. Beneath the glitzy overlay was an argument adumbrated 20 years earlier, by the psychoanalyst Allen Wheelis in a book with mass-market appeal, *The Quest for Identity*. His vantage point 'behind the couch', Wheelis's subject was 'the malaise of our times' – foreshadowings of Jimmy Carter here – and its expression in modern Americans' deficient 'sense of self', their lack of meaning and purpose, their 'futility, emptiness and longing'.[17] The social character whose emergence Wheelis traced in 1958 looks much like the narcissist Lasch limned in 1978. Disinclined to fully commit himself to anything, governed by an ethos of flexibility and adjustment, Wheelis's modern American is nothing like this figure's grandfather – Wheelis locates his golden age 50 years in the past, when 'work, thrift, and will' were the individual's touchstones – but is the characterological twin of Lasch's bored and vaguely dissatisfied avatar of me-decade narcissism, plagued still by feelings of futility and emptiness.[18] Lasch's narcissist is as 'weary and skeptical' as Wheelis's silently despairing questor, and as haplessly without conviction in the face of life's 'enduring meaning and purpose'.[19] Both Wheelis and Lasch trade in oppositions between a stolid past and an unstable present, between a corresponding culturally regnant strength and plasticity of character, but what is for Lasch the solid ground – the lost golden age – of American individualism is in Wheelis's hands already shifting sands.[20] The simple point here is that appeals to a mythic past, common in the social character genre, are more persuasive as rhetoric than as history. More interesting is that the argumentative ground Lasch's book occupied was well tilled, worked over by a generation of social theorists less visible than he would become.

The critical success of *The Culture of Narcissism* was thus partly premised on the broad-brush familiarity of its social critique. The fact that, by the time of the book's publication, the narcissist had assumed shape as an identifiable figure within psychoanalysis was also critical to its singular success. In the psychoanalysis of the 1970s, narcissism was suddenly all the rage, considered by many among the 'most important discoveries' in the analytic realm since Freud's death.[21] Major works by Otto Kernberg and Heinz Kohut, psychoanalytic theorists of the new narcissism, drew both professional and popular attention. On the one hand, the analytic journal literature was newly awash in papers on narcissism;[22] on the other, *People Magazine* showcased Kernberg and Kohut on such topics as 'why some people can't love' and the diminishing quality of family life.[23] Lasch skilfully mined psychoanalytic thinking drawn from a wide range of sources to produce a manifestly convincing and theoretically persuasive account of precipitous national decline, featuring the narcissist as leading actor. Asserting his clinical bona fides in footnotes packed with references to the latest analytic journal literature, Lasch reproved his fellow critics for their tendency 'to dress up moralistic platitudes in psychiatric garb' and for interpreting narcissism in existential terms as 'the metaphor of the human condition'. He alone among them would eschew easy moralism and attend to 'clinical fact', invoking, in passing, such specialized analytic *arcana* as 'normal primitive narcissism' and 'grandiose object images'.

The portrait painted by Lasch is concise and devastating. His narcissists are Kernbergians in their grandiosity and manipulativeness, Kohutians in their heightened self-esteem and fantastic omnipotence. Their inner life impoverished, they deploy their considerable charm in parasitical relation to others whom they can only experience as sources of 'approval and admiration'. Consumed by rage and envy, they entertain fantasies of 'wealth, beauty, and omnipotence'. Lasch's narcissist was, on the one hand, a disturbed, if high-functioning, psychiatric patient and, on the other, a terrifying Everyman in whom we were invited to see our shallow, superficial, and exploitative modern selves.

Lasch brought cultural criticism and clinical insight within the same compass, with explosive results. Once the book was published, talk of his 'culture of narcissism' was suddenly everywhere. One can still find testimonials to his brilliance and prescience on the web today. 'He created a concept,' enthuses one recent self-styled critic of self-indulgence, reviewing the book on amazon.com. 'This is Freud that works!' adds another, even as others chime in that the book's pervasive 'Freudian jargon' is off-putting and confusing. Readers continue to be persuaded, pointing to the self-gratification, self-indulgence, neediness, and pseudo self-awareness Lasch lambasted as symptoms of the cultural disaster he prophesied more than 30 years ago, lauding him for his profound, penetrating, and 'definitive indictment of American society'.[24]

Few books have enjoyed this sort of staying power. The commentator who noted of Lasch, immediately following the publication of his book, that 'he has preached back to us precisely what we already believed'[25] was surely on to something. 'This is hardly original stuff,' sniffed *Newsweek* in 1979.[26] 'It is pop-sociology at its best,' adds an admiring recent reader, making a similar point in a celebratory register.[27] In the guise of public intellectual, Lasch imagined himself close kin to the Columbia historian and public intellectual Richard Hofstadter, explaining that the latter's work in summing up 'a way of looking at history that was already familiar' enjoyed 'some kind of mythic resonance'.[28] Lasch was, if anything, an even more adept mythographer than his mentor Hofstadter, fashioning himself as a gloomy, curmudgeonly Jeremiah delivering what one critic called 'a civilized hellfire sermon'[29] that defined a decade in a beguiling word – a remarkable achievement in itself – and that speaks to many still.

Critics complained of Lasch's 'penchant for sweeping generalizations',[30] of his aggrieved, relentless tone,[31] of the patent absurdities that punctuated the book. Psychoanalysts faulted his 'warmed-over Marxism and warmed-up Freudianism',[32] his nostalgic and sentimental view of the past,[33] and his treatment of narcissism as an 'explanatory portmanteau',[34] the source of every conceivable ill. More than a few observers pointed to the irony of Lasch conscripting for his own rhetorical purposes the psychoanalytic perspective he so roundly condemned in the book as evidence of the nation's decline.[35] Just as ironic and commented upon, Lasch, the dour critic of celebrity culture, was celebrated in the pages of *People*. None of these inconsistencies and paradoxes detracted from the book's impact. Its enduring appeal is to be found, as our critic suggested, in its very familiarity, in Lasch's adroit summoning up of well-established and culturally resonant antinomies in the service of his own myth-making.

Objects and things

Among the antinomies in which Lasch traded, none had deeper cultural resonance than that between abundance and plentitude, on the one hand, and asceticism and privation, on the other. *The Culture of Narcissism* appeared at a particularly vexed point in the history of intellectuals' engagement with modern consumer culture – a robust tradition inaugurated by Thornstein Veblen's *Theory of the Leisure Class* (published in 1899) and reaching its twentieth-century apogee in Bell's *Cultural Contradictions of Capitalism* (1976) from the right, and Richard Sennett's *The Fall of Public Man* (1977) from the left. The book appeared when the proposition that material abundance results in the impoverishment of the self had achieved near-axiomatic standing in popular and professional social science, and this was critical in shaping its reception.

Sociologists, historians, and cultural critics have spoken with one voice in decrying the evisceration of the self that they agree has been the price of the American nation's prosperity. It is worth pointing out that this linkage of outer affluence and inner poverty is not as self-evident as it has come to appear. One might imagine – indeed many have – the lives of individuals being enhanced and enriched by freedom from want and privation and by the capacity to possess goods that ease the burdens of day-to-day life, to which formerly only a fortunate few had access: everything from inexpensive books to pre-sewn clothing to reliable means of transport. Feminist observers, in particular, have pointed to the lesser burdens shouldered by the housewife, formerly engaged in all manner of home production, in the age of the machine. From the late nineteenth century to the present, however, a range of critics has cast a disapproving eye on the cornucopia of things capitalism has brought within reach of the masses and has taken aim at the promise our consumer society holds out: that the way to happiness is to be found in choosing from among the ever-expanding array of dazzling goods with which the market tempts us. The spectre of the self, reduced – as it was put presciently by the novelist Henry James in 1881 – to but 'a cluster of appurtenances', an assemblage of the *things* it possesses, has long haunted social commentary.[36]

Psychoanalysts have portrayed the narcissist's inner landscape in the bleakest of terms, as altogether devoid of the good sustaining objects – persons, among them parents, lovers, and intimate friends, as they are experienced internally – that, as they see it, animate the interiority of most individuals. The critics who, in the 1970s, began to characterize the narcissist, in Jamesian terms, as the exemplary consumer gorging on the fruits of capitalism in a vain attempt to fill the void within in effect conflated the analyst's internalized 'objects' with the vernacular's inanimate things. This move allowed them to highlight the narcissist's putative quest for material gratification, which they claimed to see in her frantic compulsion to consume, while completely overlooking the deep-seated asceticism of need, the renunciations of people and things in the name of self-sufficiency, that their analytic contemporaries were highlighting as symptomatic of the condition. The critics who cast the narcissist as leading actor in their already elaborated dramas of cultural declension thus construed narcissism as a realizable state of satiation and fullness, as a repudiation of the nation's sturdy Puritan past, while minimizing the dynamic of deprivation and renunciation – of needs, of attachments – that clinicians saw as constitutive of the disorder.

Narcissism is a concept that invites such plundering. Freud's fellow psychoanalysts were reportedly disturbed and bewildered by his 1914 essay on it, and its paradoxes and ambiguities continue to haunt their disciplinary descendants, who have complained volubly about the term's protean nature and lack of precision.[37] The capaciousness that was so vexing to clinicians offered an opportunity to critics. But the point here is not simply that they exploited a purely clinical category and, in so doing, degraded it.

For one, it should be noted that clinicians were among those who conscripted the narcissist into their own, clinically inflected condemnations of the 1970s culture of gratification. The point is rather that, from the start, analysts have conceptualized narcissism in a morally laden socioeconomic register, which is organized around concepts of privation, renunciation, and dependency on the one hand, and abundance, gratification, and independence on the other. These terms are strikingly similar to those in which critics couched their condemnations of the modal modern American.

Further, while from the time narcissism was first delineated, psychoanalysis has been divided in its theories and norms of practice – and, increasingly through the twentieth century, in its organizational structures – along the lines of this opposition between privation and gratification, for the most part only the gratification side of the divide has been visible to critics, who have largely missed the asceticism underlying Freud's patently liberationist vision of the modern self. David Riesman has given us an ascetic Freud whose economics of affect is governed by the laws of a scarcity outdated even in his own time,[38] and Philip Rieff likewise discerned a genuine affinity between Freud's psychology and that of the ancient Stoics.[39]

But theirs were singular, glancing dissents from a developing critical orthodoxy, peculiarly American, that focused primarily on the ways in which psychoanalysis loosened the restraints imposed on individuals by the Victorian social and sexual order, releasing them to find themselves and realize their inner natures, free of all forms of established authority. The commonality of concern – even, at times, of language – in analysts' and critics' parallel conversations about the modern self smoothed the way for easy acceptance of the latter's seamless linking of social and individual pathology in the person of the narcissist. But the commonality obscured as much as it enabled, in particular the fact that the emotional economy of the critic's modal American and the analyst's narcissist were not the same but, rather, mirror images of one another.

Analytic stoicism met cultural asceticism in Lasch's work, which brought the two parallel strands of clinical and cultural commentary into common conversation. Commentators writing before he published *The Culture of Narcissism* in 1978 charged that American culture was becoming increasingly narcissistic. In a 1973 essay prophesying that the 1970s would 'come to be known as the Me Decade', Tom Wolfe skewered the newly emergent penchant for unceasing 'analysis of the self', terming the impulse to do so narcissistic and linking it to the postwar prosperity that was endowing so many with the leisure to dwell on themselves.[40] Wolfe's indictment was followed two years later by Peter Marin's widely cited, plangent analysis of the beleaguered, solipsistic retreat into the self that, he argued, portended the rise of a 'new narcissism'.[41] Others chimed in, branding the United States a nation of narcissists and seeing its young as deformed by a media-fuelled collective narcissism that emptied the self while mandating reverence for 'the image'.[42] But it was Lasch's work that would define the decade, his name that beyond psychoanalysis would be long associated with narcissism.

Material girls or God-men?

Until quite recently, the narcissist of popular discourse has been cast as a feminized figure – as either homosexual or female. The homosexual narcissist, first limned by Freud in his essay of 1910 on Leonardo da Vinci, enjoys an extended old age, if not in professional publications, then on the Internet. The female narcissist, first glimpsed in Freud's 1914 essay, likewise lives on as an avatar of pathological emptiness, vanity, and self-absorption. Over the course of the last decade, however, another narcissist has come to the fore, the narcissist as visionary and charismatic leader. This narcissist's appeal – located both in the person of such exemplars of the type as the celebrity CEOs Steve Jobs and Larry Ellison and in the pleasure of contemplating his paradoxical hold on our collective imagination – has recently outstripped that of his feminized counterparts in the marketplace of ideas as, increasingly, the press focuses on the narcissism of the powerful.

Lasch briefly but acutely discusses the narcissist-as-manager (not yet promoted to the top spot), invoking his skill in 'the management of personal impressions', his need for admiration and his rule by fear, his ability to con, push around and otherwise manipulate subordinates who willingly submit to his authority.[43] But this figure had little traction at a time when the consumer society critique, with which Lasch's book was immediately identified, was ascendant. The shopaholic woman was a far more appealing target than was the powerful man for moralists of right and left alike.

The powerful-man-as-narcissist who now regularly appears in the news, so manifestly a product of his times, is not an altogether new figure. The first fully realized narcissist of the psychoanalytic literature was male, not female, and came, in 1913, from the pen not of Freud but of his London-based colleague and future biographer, Ernest Jones.[44] Twenty years later, Freud sketched the outlines of the male narcissist in a tantalizingly brief passage as a leader, a 'personality' aggressively poised to take on the world.[45] The portraits both Jones and Freud offered have a remarkably contemporary feel.

Jones's narcissist is a study in the masculine megalomania and exaggerated self-confidence that, yoked to a magnified desire to please, could propel a man to worldly success. He is superficially sociable but at bottom aloof and emotionally inaccessible. Manifestly indifferent to the estimation of others, he is in fact hungry for praise and admiration. Ostensibly self-effacing, he is in truth vain and exhibitionistic. He fancies himself omniscient, and, loath to acknowledge there may be things he does not know, devalues new information and rejects as worthless any advice tendered to him. Intolerant of authority, he acts to ensure that all are subject to his control. Jones maintained that this figure suffered from a 'God-complex', from a curious unconscious fantasy in which he was identified with God – calling to mind the Oracle employee's endlessly reiterated observation of the company CEO Larry Ellison that 'the difference between God and Larry is that God does not believe he is Larry'.[46]

Jones, too, elaborated on the distinction between God and his God-men, suggesting it was to be found in the 'colossal *narcissism*' of the latter, the 'excessive admiration for and confidence in one's own powers, knowledge, and qualities' that was their defining characteristic: in a word, their omnipotence. And he homed in on the puzzling paradoxes around which such men's characters were constructed – 'an exaggerated *desire to be loved*' coupled with a glorious independence of anyone else's opinion; dependent on the admiration of others, they yet preferred to 'reign in solitary isolation'. Jones claimed to have encountered this interesting constellation of traits in a dozen different analysands, successful businessmen and professional psychologists among them, and subsequent analyses only underwrote his conviction that he had outlined this character accurately.[47]

Freud's male figure is similarly characterized by his independence and self-sufficiency. He is not burdened with a super-ego or conscience and, indifferent to society's standards, he is not open to intimidation either. A visionary leader endowed with 'a capacity for vigorous action', he can mobilize others in the service of positive or negative ends. As Freud famously wrote,

> people belonging to this type impress others as being 'personalities'; they are especially suited to act as support for others, to take on the role of leaders and

> to give a fresh stimulus to cultural development or to damage the established state of affairs.[48]

Both Freud's and Jones's male narcissists are strong figures, possessed of worldly efficacy. Both all but disappeared, mentioned in the analytic literature from time to time but not highlighted. Jones's male figure was displaced almost immediately by Freud's fascinating female and developmentally stunted homosexual narcissists, and Freud's male narcissist was a cipher in the literature until the revival of interest in narcissism – especially among the powerful – in the 1970s and beyond. Freud never mentioned nor cited Jones's paper, and Jones, ever eager to please, took on Freud's conceptualization of narcissism as his own, trumpeting it in 1920 as among the most important recent advances in psychoanalysis and proclaiming as fact that 'analyses of the subject' demonstrated that narcissism and homosexuality were 'extraordinarily closely related'.[49] And, within psychoanalysis, so they were for the next 50 years. The weak, ineffectual, and feminized narcissist first glimpsed in Freud's *Leonardo* would long figure centrally in the analytic literature. That Jones so willingly abandoned his own precisely drawn characterology in favour of Freud's speculations is exemplary of the abjectness he assumed all along in his 30-year-long relationship with Freud. Jones, playing the courtier to Freud's sovereign, was fully complicit in the relegation of his own figure, so resonant with later conceptualizations, to the status of tantalizing historical footnote.[50]

But the relegation was premised not only on the vagaries of biography but also, importantly, on the particular ways in which psychoanalytic knowledge and social criticism were woven together in the 1970s, especially – and explosively – by Lasch. As I have suggested above, the fact that both psychoanalysts and critics traded in economic concepts paved the way for this interweaving, such that the consumerist interpretation of national and characterological decline gained immediate and lasting assent. From the 1970s through to the end of the century, the stock narcissist of the popular imagination suffered from vanity, entitlement, and a surfeit of self-regard, and her harm to the body politic consisted primarily in her flouting repeatedly asserted aspirational norms of asceticism in the marketplace and communalism in the public square.[51] Her – or, in the guise of the homosexual, his – predominance occluded the narcissism of the powerful man, to which psychoanalysts were increasingly turning their attention.

This figure of the vacuous consumer can still be found in popular books with titles like *The Narcissism Epidemic. Living in the Age of Entitlement* and *The Mirror Effect. How Celebrity Narcissism is Seducing America* that excoriate Americans for their inflated self-esteem, self-absorption, shallow values, and materialism, and for specific sins ranging from 'showing off' on social networking websites, to taking advantage of easy credit, to submitting to injections of Botox. Narcissism in this line of argument explains everything that is wrong with America's present condition and is used to blame all of us – from the teenager posting a flattering picture of herself on her Facebook page to the middle-aged purchaser of expensive cookware – for the sorry state in which our nation finds itself.[52]

It is notable that the narcissism of the powerful is in this line of argument largely exempted from the moralism and blame that is so freely distributed around. As the economic crisis has scrambled the terms in which the consumerist imperative has been framed – with many arguing that recovery depends on the willingness of individuals to spend, such that behaviour that just ten years ago threatened to undermine the body politic is now necessary to its survival – the narcissism of the relatively powerless has increasingly come to share the rhetorical stage with the narcissism of the fascinatingly but destructively powerful. This narcissist is overwhelmingly cast as male, with the explanation that it is men who, more easily than women, can wield the sort of influence that is the vehicle through which the condition is expressed. The man who skilfully deploys his personal magnetism to bind slavishly admiring subordinates to his person while at the same time devaluing, even psychologically brutalizing them, is now a staple on the Internet and the central player in the literature of the pathological workplace. Emotionally shallow and self-centred, dependent on a steady supply of adulation and tribute from others whom he devalues and depreciates, his charm covering a basic ruthlessness and desire to control, our contemporary narcissist is as likely as Jones's to enjoy success in business and academe.

The psychoanalyst Michael Maccoby, in his popular book *The Productive Narcissist*, published in 2003, champions this narcissist as visionary leader, as the personality type best suited to inspire and motivate followers in times of change and flux. Arguing against a tradition in the literature of management that highlights consensus and celebrates relational skills that are coded feminine, Maccoby fashions a militantly masculine programme for the aspiring leader – a mix of independence, charisma, and passion – while warning that if he is to succeed, such a leader would be well advised to enter psychoanalysis 'to overcome vital character flaws'.[53] Mobilizing the couch in the service of the corporation, Maccoby gives us a narcissism loosed from the consumerist critique whose captive it was for 20 years, a narcissism as puzzlingly paradoxical – in its contradictory fusion of the irresistibly charismatic on the one hand and the brutally exploitative on the other – as the character type Jones and Freud glimpsed nearly a century ago.

Narcissism as a form of social critique continues to fascinate. Shelves of books instruct us how to deal with the narcissists in our lives, the mothers, fathers, lovers, and bosses from whom we cannot escape even as they make our lives hell. *Freeing Yourself from the Narcissist in Your Life*; *Narcissistic Lovers. How to Cope, Recover and Move On*; *Children of the Self-Absorbed. A Grown-Up's Guide to Getting Over Narcissistic Parents*: the list is long, the operative concepts variants of how to manage, recover from, and otherwise deal with, or distance yourself from, what one title terms 'infuriating, mean, critical people'. And, a regular stream of books and articles by Michael Maccoby and his acolytes reminds us that while we may neither like nor trust the visionary executives – the big innovators like Bill Gates, Steve Jobs, and Jack Welch – who 'change the world', we can't do without them. 'We are in a time of great upheaval that needs visionary leaders and charismatic personalities,' writes Maccoby, summoning up Freud's portrayal of narcissistic types suited 'to take on the role of leaders'.[54]

We are a long way in both the self-help and the management literature from the one-dimensional, frivolous, and self-indulgent female shopaholic of the 1970s and 1980s. These contemporary narcissists whom we must endure, whether in the privacy of our homes or in the corner office at work, are dangerously charismatic creatures who entice us into glorious submission before viciously turning on us. They are every bit as paradoxical in their make-up and impossible to resist as were Jones's God-men and Freud's 'personalities'. The fact that anyone can now plumb their depths with a few clicks of the mouse provides us with an instance of the translation of scientific knowledge into everyday know-how that Lutz Raphael sees as exemplary of social science inquiry in the last 150 years.

Notes

1. S. Freud (1914), 'On Narcissism. An Introduction', in J. Strachey (ed.), in collaboration with A. Freud, assisted by A. Strachey and A. Tyson, *The Standard Edition of the Complete Psychological Works of Sigmund Freud* [hereafter *SE*], 24 vols (London: Hogarth Press and the Institute of Psycho-analysis, 1953–74), vol. 14, pp. 69–102, at 88–9.
2. P. Rieff (1987 [1966]), *The Triumph of the Therapeutic. Uses of Faith after Freud* (Chicago: University of Chicago Press), p. 26.
3. S. Freud (1910), *Leonardo da Vinci and a Memory of His Childhood, SE* 11, pp. 59–137.
4. D. Riesman, in collaboration with R. Denney and N. Glazer (1950), *The Lonely Crowd. A Study of the Changing American Character* (New Haven: Yale University Press); W. H. Whyte, Jr (1956), *The Organization Man* (New York: Simon and Schuster); D. Bell (1976), *The Cultural Contradictions of Capitalism* (New York: Basic Books); C. Lasch (1978), *The Culture of Narcissism. American Life in an Age of Diminishing Expectations* (New York: W. W. Norton).
5. On the 'real self', see E. Lunbeck (2000), 'Identity and the Real Self in Postwar American Psychiatry', *Harvard Review of Psychiatry* 8, pp. 318–22.
6. Among the classic articles are L. Stone (1954), 'The Widening Scope of Psychoanalysis', *Journal of the American Psychoanalytic Association* 2, pp. 567–94; B. R. Easser and S. R. Lesser (1965), 'Hysterical Personality: A Re-evaluation', *Psychoanalytic Quarterly* 34, pp. 390–405; and H. W. Loewald (1979), 'The Waning of the Oedipus Complex', *Journal of the American Psychoanalytic Association* 27, pp. 751–75.
7. For fuller treatment of this argument, see E. Lunbeck (2006), 'Borderline Histories. Psychoanalysis Inside and Out', *Science in Context* 19, pp. 151–73.
8. P. Conrad, writing in *The Observer* in 1980, quoted by B. Richards (1985), 'The Politics of the Self', *Free Associations* 1 D, pp. 43–64, at 46.
9. K. L. Woodward (6 September 1976), 'Getting Your Head Together', *Newsweek*, pp. 56ff.
10. On the speech, see D. Horowitz (2005), *Jimmy Carter and the Energy Crisis of the 1970. The 'Crisis of Confidence' Speech of July 15, 1979* (Boston, MA: Bedford and St Martin's). See also K. Mattson (2009), *What the Heck Are You Up To, Mr. President? Jimmy Carter, America's 'Malaise', and the Speech that Should Have Changed the Country* (New York: Bloomsbury).
11. W. K. Stevens (30 November 1977), 'Narcissism In the "Me Decade"', *New York Times*.
12. On the appearance of 'new narcissism', see J. F. Battan (1983), 'The "New Narcissism" in 20th-Century America. The Shadow and Substance of Social Change', *Journal of Social History* 17, pp. 199–220.
13. Riesman (1950), *Lonely Crowd*, p. 148.
14. R. Wilkinson (1988), *The Pursuit of American Character* (New York: Harper and Row).
15. Richards (1985), 'Politics of the Self', especially pp. 46–7.
16. Lasch (1978), *Culture of Narcissism*, p. 41.
17. A. Wheelis (1958), *The Quest for Identity* (New York: W. W. Norton), pp. 9, 18, 87.

18. Ibid., pp. 85ff., 19; Lasch (1978), *Culture of Narcissism*, pp. 36ff.
19. Wheelis (1958), *Quest for Identity*, pp. 87–8.
20. Lasch opens *The Culture of Narcissism* with an invocation of the dream of world power expressed in Henry Luce's 1941 'American century' contrasted to the current malaise: p. xiii.
21. C. K. Hofling and R. W. Meyers (1972), 'Recent Discoveries in Psychoanalysis. A Study of Opinion', *Archives of General Psychiatry* 26, pp. 518–23.
22. Before 1958, for example, 34 articles with 'narcissism' or 'narcissistic' appeared in the journal literature; by 1978, that number had increased to nearly 200.
23. An interview with Kernberg appeared in the 9 July 1979 issue of *People Magazine*, and with Kohut in the 20 February 1979 issue.
24. Quotations from reader reviews of *Culture of Narcissism* on www.amazon.com.
25. H. Allen (24 January 1979), 'Doomsayer of the Me Decade. Christopher Lasch on America as a Nation of Narcissists', *Washington Post*, B1.
26. Cited in ibid.
27. D. H. Wrong (Summer 1978), 'Bourgeois Values, No Bourgeoisie? The Cultural Criticism of Christopher Lasch', *Dissent*, pp. 308–14, at 310, writes of his impression that he'd 'been listening to Lasch's bill of indictment for most of my life', adding 'and I wasn't born, alas, yesterday'.
28. C. Blake and C. Phelps (1994), 'History as Social Criticism. Conversations with Christopher Lasch', *Journal of American History* 80, pp. 1310–32, at 1317.
29. F. Kermode (14 January 1979), 'The Way We Live Now' (review of Lasch 1978, *Culture of Narcissism*), *New York Times* 1, pp. 26–7.
30. M. Kammen (1979), 'A Whiplash of Contradictory Expectations', *Reviews in American History* 7, pp. 452–58, at 456.
31. L. Menand (11 April 1991), 'Man of the People', *New York Review of Books*, 38(7), pp. 39–44.
32. M. R. Green (1981) (review of Lasch, 1978, *Culture of Narcissism*), *Journal of the Academy of Psychoanalysis and Dynamic Psychiatry* 9, pp. 330–1, at 330.
33. Richards (1985), 'Politics of the Self'.
34. E. M. Weinshel (1981) (review of Leo Rangell (1980), *The Mind of Watergate. An Exploration of the Compromise of Integrity*), *International Journal of Psychoanalysis* 8, pp. 121–4, at 122.
35. Allen (24 January 1979), 'Doomsayer of the Me Decade'; Richards (1985), 'Politics of the Self'; Menand (11 April 1991), 'Man of the People'.
36. The passage – cited by both J.-C. Agnew (1983), 'The Consuming Vision of Henry James', in R. W. Fox and T. J. J. Lears (eds), *The Culture of Consumption. Critical Essays in American History, 1880–1980* (Pantheon: New York), pp. 65–100, at 85, and M. Orvell (1989), *The Real Thing. Imitation and Authenticity in American Culture* (Chapel Hill: University of North Carolina Press), p. 65 – is from James's 1891 novel, *The Portrait of a Lady*. The words are Madame Merle's, her character in Agnew's estimation 'perhaps one of the most celebrated instances of an achieved marketplace identity': 'What shall we call our "self"? Where does it begin? Where does it end? It overflows into everything that belongs to us – and then it flows back again. I know a large part of myself is in the clothes I choose to wear. I've a great respect for *things*! One's self – for other people – is one's expression of one's self; and one's house, one's furniture, one's garments, the books one reads, the company one keeps – these things are all expressive!' James contrasts this nascent consumerist construction of the self with his heroine Isabel Archer's self, free of material encumbrances: 'I know that nothing else expresses me. Nothing that belongs to me is any measure of me.' *The Culture of Consumption* is foundational to the critical assessment of consumption in United States history that took shape in the 1980s. See also L. B. Glickman (ed.) (1999), *Consumer Society in American History. A Reader* (Ithaca, NY: Cornell University Press) for a way into the massive historical literature on the American culture of consumption. In this volume, see especially Agnew,

'Coming Up for Air. Consumer Culture in Historical Perspective', for a brilliant overview and critique (as well as, notably, a muted *mea culpa*) of changing fashions in historians' assessments: pp. 393–7.

37. See E. Jones (1955), *The Life and Work of Sigmund Freud*, vol. 2 (New York: Basic Books), p. 302, on the reactions of Freud's contemporaries.
38. D. Riesman (1950), 'The Themes of Work and Play in the Structure of Freud's Thought', *Psychiatry* 13, pp. 1–16.
39. Rieff (1987 [1966]), *Triumph of the Therapeutic.*
40. T. Wolfe (23 August 1976), 'The Me Decade and the Third Great Awakening', *New York Magazine*, pp. 26–40.
41. P. Marin (October 1975), 'The New Narcissism', *Harper's*, pp. 45–56.
42. J. Hougan (1975), *Decadence. Radical Nostalgia, Narcissism, and Decline in the Seventies* (New York: William Morrow), esp. pp. 151–5. Lasch (30 September 1976), 'The Narcissist Society', *New York Review of Books*, pp. 23, 15, stakes out his particular take on narcissism as 'the key to the consciousness movement and to the moral climate of contemporary society'.
43. Lasch (1978), *Culture of Narcissism*, pp. 44–5, 186.
44. E. Jones (1913), 'The God Complex. The Belief That One Is God, and the Resulting Character Traits', in Jones (1951), *Essays in Applied Psychoanalysis*, vol. 2 (London: Hogarth Press), pp. 244–65.
45. Freud (1931), 'Libidinal Types', *SE* 21, pp. 217–20, at 218.
46. Cited in M. Maccoby (2000), 'Narcissistic Leaders. The Incredible Pros, The Inevitable Cons', *Harvard Business Review* 78(1), pp. 68–77. The quip is all over the Internet, and provides the title for M. Wilson (2002), *The Difference between God and Larry Ellison. Inside Oracle Corporation* (New York: HarperCollins).
47. Jones (1913), 'God Complex'.
48. Freud (1930), *Civilization and Its Discontents, SE* 21, pp. 83–4; idem (1931), 'Libidinal Types'.
49. Jones (1920), 'Recent Advances in Psycho-Analysis', *International Journal of Psychoanalysis* 1, p. 168.
50. A. Phillips (1994), 'Freud and Jones', in *On Flirtation* (Cambridge, MA: Harvard University Press), pp. 109–21, brilliantly captures the dynamics of the two analysts' 30-year-long relationship.
51. For one example, see J. M. Twenge and W. K. Campbell (2009), *The Narcissism Epidemic. Living in an Age of Entitlement* (New York: Free Press), p. 63, citing R. Putnam (2000), *Bowling Alone. The Collapse and Revival of American Community* (New York: Simon and Schuster), p. 63, in support of her contention that ramped-up self-esteem went hand in hand with declining community involvement.
52. These figures can be found in Twenge and Campbell (2009), *Narcissism Epidemic.*
53. M. Maccoby (2003), *The Productive Narcissist. The Promise and Peril of Visionary Leadership* (New York: Broadway Books).
54. M. Maccoby (3 March 2003), 'The Narcissist-Visionary. How to Stop Worrying and Learn to Love Your Difficult Boss', *Forbes*, 36, http://www.maccoby.com/Articles/onmymind.shtml [accessed 23 September 2010].

Part III
Polling, Marketing, and Organizations

11

Hearing the Masses: The Modern Science of Opinion in the United States

Sarah E. Igo

Gleaning the opinions of ordinary individuals became the object of an astonishing number of enterprises during the twentieth century: journalists hoping both to pinpoint and expand their readership, corporations to tap into and stoke consumers' desires, state agencies and political candidates to read and influence the public mood. Focused on the rise of modern public opinion polling, this chapter tracks the creation by self-professed experts in the United States – and, somewhat later, in other Western 'mass' societies – of a science of opinion. It also explores the justifications for this development, and particularly the argument that empirically determining the sentiments, tastes, and will of 'the people' would lead to a more responsive, transparent or efficient society and body politic.

Such arguments, as we shall see, did not always persuade either ordinary citizens or political elites. Nevertheless, by the mid-twentieth century, Americans – and increasingly, national populations around the globe – would take part in, and depend upon, opinion research as part of the common-sense functioning of their societies. The striking mix of distrust and dependence that greeted modern polling is a distinctive feature of its history, illuminating the ways that scientific instruments can enter and transform public life even without gaining full acceptance from their users.

This is true even today, in the United States and other industrial nations, where scientifically collected opinion data inform understandings of consumer behaviour, voter preferences, race and ethnic relations, social attitudes, and patterns of work, religious and family life. Practitioners have called the sample survey 'arguably the single most important methodological development and contribution of twentieth-century social science', and survey-based social science has been brought to bear on a bewildering array of problems, from financial expectations to pedagogical effectiveness, and from sexual habits to military morale.[1] Opinion polls are an omnipresent tool of governments as well as demographers, political campaign strategists, public policy institutes, economists, activists, and advertisers. It is difficult to think of an area of modern life untouched by the science of opinion. And yet, serious doubts persist from many quarters about the polls' scientific accuracy as well as their civic value.

This essay seeks to dislodge so-called scientific polling from its position of familiarity, sketching why this kind of surveying came about when it did, how it intersected with larger political and economic developments, and what some of its implications were for modern publics. It will then turn to the early years of 'scientific polling' in the United States in order to trace the ways that a new technology helped to remake not just politics and public life, but the 'public' itself, definitively if unevenly scientizing the social.

The sample survey arrives

What makes the modern opinion poll 'modern' is the technique of statistical sampling. Determining the precise scientific relationship between a part and the whole, a selected sample and the entire universe of possible subjects, is what distinguished twentieth-century survey research from social surveys of the nineteenth century, which took local communities or defined groups (London's poor, French working-class families, black Philadelphians) as their object.[2] In general, surveys in this earlier period that attempted to extrapolate from small numbers of responses were met with distrust.[3] A series of mathematical and conceptual developments across the late nineteenth and early twentieth centuries would, however, enable sample surveys to generalize about extremely large groups by collecting data from relatively few.[4] As Alain Desrosières describes it, this 'moved statistics from the ground of the study of wholes summarized by a single average (holism) to analysis of the *distribution* of individual values to be *compared*'.[5]

The impulses behind the sample survey were not, to be sure, simply mathematical. Measurable opinions became a valuable currency for a diverse set of parties in the US in the early decades of the twentieth century, an object of keen interest within universities, industry, and the federal government. Sociologists and political scientists developed new methods to measure subjective attitudes; corporations hired industrial psychologists to monitor workers' mental states in an attempt to improve productivity and stave off workplace conflicts; the Department of Agriculture surveyed rural opinion to inform its policy-making; and the military employed new statistical tools to assess fighting men's morale.[6] Commercial researchers were at the forefront of new techniques to quantify opinion in the hope that a systematic science of subjective desires would stimulate consumption and lead to untold profits.

James Beniger has described such developments as the demand for 'new technologies of mass feedback'. Such feedback would become increasingly vital to advertisers, state agencies, and politicians who wanted more and better data about the public in order 'to attract and hold its attention', to 'stimulate and control its consumption behavior', and to 'influence its opinions and its vote'. Beniger emphasized that these techniques were instigated not by people who wanted to 'speak their minds' but by corporations and governments seeking to sell to or otherwise profit from information about the mass public.[7] Market and opinion research in this sense were mechanisms that allowed elites to maintain at least an illusion of 'direct and continuing contact' with citizens, voters, and consumers, and to act upon (and benefit from) such knowledge.[8] Beniger was surely correct that new techniques for listening to the public 'speak' were inaugurated by powerful entities that had something to gain thereby. Even if opinion surveyors claim to be motivated by the public good, their primary patrons and beneficiaries have always been newspapers, marketers, politicians, and corporations.

One intriguing feature of most major early survey operations, even if not state-run, was that they imagined their scope to be *national*, rather than sub-national or international. This is no doubt a legacy of the long history of statistics as a 'science of state' and the fact that the earliest systematic surveys were conducted in order to raise militaries and levy taxes.[9] In other words, the tight link between nation states and statistics was not coincidental. Sampling techniques served various sectors of industrial society, revealing economic and demographic trends as well as otherwise imperceptible shifts in national attitudes, habits, and behaviours – all without requiring expensive and logistically complex tabulations of the entire population, as did national censuses.

Sample surveys were not the only means of assessing national sentiment in the early decades of the twentieth century. At the same moment that scientific polling was introduced there were competing efforts to document national citizenries. The left-leaning Mass-Observation project launched in Britain in 1937 to create an 'anthropology of ourselves' and some of the Works Progress Administration reportage in the US during the Great Depression are two examples. Although such ventures probed attitudes and opinions, attempting thereby to get a grip on national trends, they did not (primarily) take the form of the modern opinion survey. Instead they tilted in a more qualitative, even anecdotal direction, hoping to capture the complexity of local communities, social class relations, and individual personalities.[10] This tradition would persist in both journalism and social ethnography, but the sample survey – seemingly more objective and reliable – would eventually overtake these other means of locating the national mood.

Mass feedback technologies, to use Beniger's term, would help give rise to a peculiarly modern vision of the public as composed of anonymous, atomized individuals holding discrete views. Talal Asad has linked the rise of a new statistical concept of representativeness in the mid-twentieth century to the welfare state and the consumer market as well as the national election poll. All of these entities, he writes, 'constitute social wholes that do not depend logically either on the intimate experience of a given region or on the assumption of typical social actors'. Such means of knowing populations, he underlines, have been a central project of modern, liberal societies that have, in conjunction with such techniques, conceived of themselves as 'an aggregate of individual agents choosing freely and yet – in aggregate – predictably'. Asad adds that 'ways of statistical calculation, representation, and intervention have become so pervasive that capitalist social economies and liberal democratic politics are inconceivable without them'.[11]

Polling as social knowledge

By the middle of the twentieth century, the public opinion survey had become one of the most familiar, and protean, routes for producing empirical knowledge about the body politic. And yet, the social and epistemological transformations wrought by the arrival of scientific opinion polling have received surprisingly little attention from science studies scholars. One reason for this is polling's deceptively transparent, straightforward quality. On the face of it, public opinion polls dutifully record a population's composite responses to selected questions. In the words of Pierre Bourdieu, polling 'gives to the questions which "everyone asks themselves" ... rapid, simple and quantitative answers, in appearance easy to understand and to comment on'.[12] A technique this pedestrian has not attracted many historians or sociologists of knowledge.

A perhaps more fundamental reason is that opinion research is not in and of itself a well-demarcated field, and thus does not lend itself to conventional disciplinary history. Gathering and processing questionnaire data is part of the toolkit of many of the social sciences, notably political science, economics, and sociology; indeed, survey instruments have helped to shape each of those disciplines. But opinion research, at least in standard understandings, is not in and of itself a discipline. Nor are its practitioners a defined group and easy to track. Opinion researchers ply their craft for established survey and polling organizations but also for commercial, media, and political concerns; some are credentialed members of public opinion associations and academic departments, whereas others are marketers, audience researchers or contractors for hire. Although private commercial research and public opinion research have been tightly linked in methods as well as personnel, few scholars have attempted to analyze the field as a whole.[13] This has made it difficult to grasp the broad ramifications of survey instruments for the societies they survey.

A paradox of the kind of social knowledge derived from opinion polls, then, is its simultaneous ubiquity and invisibility. One might argue that it is precisely this paradox that gives polls their potency: the more that survey technologies are regarded as unremarkable and matter-of-fact, the more difficult it is to discern their capacity to define entire domains of social life. This is true even though, as suggested above, many sectors of public life and private enterprise depend on survey technologies, and even though these technologies are perhaps unparalleled in their ability to project 'objective' portraits of the citizenry.

Certainly, work in science studies has dealt with aspects of the history of sample surveys: the rise of probabilistic and statistical thinking,[14] the privileging of quantification as a 'technology of trust',[15] the role of commensuration and classification in making social life tractable,[16] the 'looping effects' of social scientific categories,[17] and even the genealogy of 'the modern fact'.[18] Key studies, usually proceeding from Foucauldian insights, have addressed the ways populations are rendered governable through mapping, measuring, and other descriptive techniques.[19]

Scholarship that probes how the sciences and social sciences create their objects of investigation has also led to important insights about public opinion polls. Some argue that such surveys do not simply report public views but instead generate entirely new entities. Philip Converse maintains that large numbers of the polled lack 'meaningful beliefs' on the topics at hand but offer answers to their questioners anyway, leading to 'nonattitudes'.[20] Pierre Bourdieu famously contended that polls manufacture public concerns that do not actually exist by asking questions that the polled would not themselves pose; in this sense, polls deliberately further elite rather than 'public' interests.[21] Others have suggested that surveys create public opinion as well as 'opinionated people'. As Thomas Osborne and Nikolas Rose write:

> public opinion is something that is demanded by the very activity of asking questions in surveys. That is, the existence of questionnaires and surveys *them selves* promote the idea that there is a public opinion 'out there' to be had and measured.[22]

Still others have detailed the question-wording effects and 'information effects' of public opinion polls.[23] And, of course, politicians and commentators from the early days of scientific polling onwards, and in every nation where polls have made headway, have warned that citizens themselves can be swayed, or silenced, by knowledge of others' opinions in what is known as the 'bandwagon effect'.[24] Polls, that is, are deemed responsible for the very shifts in opinion they merely claim to register. In this light, public opinion polls, rather than being mere recording instruments, appear extraordinarily productive of social attitudes, and indeed, social life.

What has been investigated in far less detail are the precise ways in which opinion surveys entered social life and the ways they worked in and on the public – the ostensible object of pollsters' calculations. How did a new tool, one that claimed both scientific and democratic credentials, alter that same public, and citizens' understanding of their place in it? Looking at the early years of polling, years when its very novelty and visibility provoked vigorous public debates, permits us to view this transformation in progress. It also allows us to perceive nuances not captured by broad statements about the engineering of society, raising the question: how, and how thoroughly, was the social made scientific? In the case of scientific polling, a complex brew of resistance and acceptance from citizens produced a partial, yet still profound, victory over earlier ways of knowing the public.

Using the first few decades of public opinion polling in the US as my example, I trace how one national population responded to the new science of politics and public opinion that pollsters proffered. Although polls would quickly enough become seamlessly interwoven with public life, the introduction of so-called scientific polling in the 1930s was deeply contentious. For citizens, the very prominence of polling triggered new arguments about, but also accommodations to, the statistics that claimed to represent them. Scientific polling, I argue, altered not only how citizens came to know the public mood but also how they placed themselves within the nation. At the same time, critiques of the new polls show how threatening they could be to older ways of participating in public discussion and assessing majority opinion. That such resistance persisted long after the apparent triumph of scientific polling reveals that scientization need not be thoroughly persuasive to do its work.

Making opinions count

Launched in the United States by George Gallup, Elmo Roper, and Archibald Crossley, 'scientific' opinion polls were a key episode – in fact, by some lights, *the* key episode – in the popularization of the sample survey.[25] These three pollsters, and opinion polling itself, made a dramatic entrance on to the national stage in 1936. Publicly challenging conventional wisdom and the famous straw poll conducted by the *Literary Digest*, each predicted Franklin Roosevelt as the victor over the favoured Alfred Landon in the presidential contest of that year. Their bet paid off, making the notion of a science of opinion polling credible among newspaper publishers and the general public.

Gallup's American Institute for Public Opinion (AIPO) and Roper's *Fortune* Survey quickly moved beyond the electoral field, discovering a ready market for syndicated opinion polls on a broad range of topics: labour unions and wartime mobilization, working women, and venereal disease. In this, they peered more deeply into the population than had earlier surveyors, not simply tallying external characteristics but instead individual beliefs and values. The questions put to scattered, anonymous citizens offered, in the words of George Gallup, a 'week-by-week picture of what Americans are thinking'.[26] In the years after 1936, many other individuals would enter the polling arena as would government agencies and major survey research organizations such as the National Opinion Research Center (1941), Columbia University's Bureau of Applied Social Research (1944), and the Survey Research Center at the University of Michigan (1946).

Although I will focus on the US case, it is important to note that scientific polling would not remain an exclusively American operation for long. In short order, Gallup himself would help establish polling affiliates abroad. By 1946, nine of these national opinion institutes were in full swing: in England, Canada, Australia, France, Norway, Sweden, Denmark, Finland, and Brazil. By that same year, Elmo Roper's International Public Opinion Research had gained a toehold in nine Latin American countries: Cuba, Mexico, Colombia, Venezuela, Peru, Uruguay, Brazil, Chile, and Argentina.[27]

Yet, despite this seemingly inevitable spread of national polls the world over, other polities were both more hesitant and slower to embrace them than was the US. In Britain, opinion polls did not become well ensconced until 1945, with the pollsters' correct prediction of a Labour victory.[28] In France it was the much later election of 1965 – in which De Gaulle won much less support from the electorate than expected, but almost exactly the amount that several polls suggested: 43 per cent – that finally and thoroughly ushered polls into French public life.[29] In West Germany, as Anja Kruke and Benjamin Ziemann detail in this volume, public opinion polls were introduced by the Western military Allies following the Second World War; but even after the establishment of German institutions, and into the 1950s, US technical experts remained critical to the enterprise and polling itself was regarded as a sort of American import.[30] In Australia, the US example was also significant: the Gallup Poll, even though in local hands, was 'the only survey of public opinion conducted on a regular basis for the press and the only regular survey conducted by any Australian body nationwide' from 1941 to 1971.[31]

This brief sketch cannot do justice to the considerable national variations in the adoption and shape of opinion polling – for example, the differences between official and private, state-run and media outfits. But, given that every industrial nation eventually came to rely upon systematic surveys to gauge its internal political dynamics, the early introduction and rapid spread of polling in the United States offers a compelling case of the public reception and effects of a new science of social life.

Vital to the ascension of polling in the US was the mingling of scientific and civic languages, and in particular the presentation of the public opinion poll as a democratic instrument indispensable to modern society. Pollsters were careful to describe sample surveys not as a radically new way of constituting the public but rather as a technical instrument for improving age-old democratic processes. This public relations turn was essential for the fledging polling profession, made up of experts who intervened in novel fashion in public affairs and dramatically reshaped the nature of social knowledge. Against charges to the contrary, surveyors claimed that their purpose was not the manipulation of opinion but objective knowledge 'about ourselves'. Polling data were billed as representing the public and clarifying its desires – and making government officials or politicians responsive to them. Moreover, opinion polls were framed as democratic in their treatment of the attitudes of all respondents (and, by extension, all citizens) as equivalent. Unlike older notions of public opinion as the considered views of the upper and middle classes, modern summations of the 'mass public' bestowed each citizen's preferences – at least, in the abstract – with identical weight.

Pollsters asserted a special kind of authority, subservient to 'the people' and pledged to precise and neutral measurement of their will. In this light, the expert collection of information about the public could be construed as a proxy for political inclusion and democracy itself. This kind of claim was made again and again as national polls were introduced elsewhere. The 'British Gallup', the British Institute of Public Opinion, established in 1937, was for example publicized as a way for 'forty-eight million people in the British Isles' to finally have some 'account' in the deliberations of leaders like 'Chamberlain, Hitler, Benes, Roosevelt'.[32] This easy equivalence between the polls and the people would be extremely useful in combating detractors. The French polling industry attacked critiques of opinion surveys (and Bourdieu's in particular) by claiming that 'in criticizing opinion polling, Bourdieu and others were in effect opposing the whole idea of consulting the citizenry'. Similarly, opponents of an election-season ban on polling, proposed in the 1970s by French political leaders, charged that the ban's supporters 'curiously resemble those of notables in the last century regarding the introduction of universal suffrage: "The voice of the ignorant will count for as much as mine? What a scandal!"'[33]

In the American case, opinion polls were from the beginning sold – literally, to corporate sponsors and rhetorically, to the public – as a scientific but also a democratic technology.[34] Purporting to be able to read the mind of the 'man on the street', pollsters made extravagant promises on behalf of their profession. In Elmo Roper's words, polling techniques were the 'greatest contribution to democracy since the introduction of the secret ballot'.[35] In George Gallup's, the new polls applied 'scientific methods to the old problem of finding out what the people of this free-thinking, free-speaking democracy wish to do with their society'.[36] Both Gallup and Roper contended that polls were *more* representative, *more* democratic even than elections since they ascertained the views of those who never made it to the voting booth. As such, the representative scientific survey could perfect, if not quite replace, representative government.

Precisely because they promoted themselves as the voice of the people, the best-known US pollsters offer a unique glimpse into the early public reception of scientific polling. As it turned out, responses to standardized questionnaires constituted only a small fraction of the sorts of opinions pollsters would be party to over the years. Inviting a certain intimacy and familiarity despite their status as national experts, Elmo Roper and George Gallup both corresponded regularly with that segment of the public who hoped to modify, amplify, join or otherwise correct their quantitative representations of the American public. The dialogue between these new experts and the public exposes the anxieties and ambivalence prompted by the arrival of a surveyed society.

Polls and their publics

For citizens of the early polling era, surveyors' technical methods and statistical findings were novel enough to comment upon. Through Gallup's AIPO surveys, proclaimed one observer in the *Forum and Century* in 1941, 'America has had a firsthand opportunity to become acquainted with its own mind'.[37] Critics and fans alike understood the polls as a form of reportage 'which has given the reading public data which it did not previously possess'.[38]

One powerful sign of the polls' growing authority was the fact that many hoped to make their voice heard in the new surveys. Some who wrote to the pollsters looked to extend a conversation that had already begun on their doorstep. Elizabeth Stein, who identified herself as a retired New York City teacher, began a letter to Roper by mentioning the poll she had just taken part in the day before, the topic of which had been radio and television programming. Stein's letter indicates that she attempted to engage her questioner on a much broader range of topics than the latter had intended: 'when I criticized the average music she was not sure if you were interested tho [*sic*] she agreed that in the main it was "noise"'. Stein went on to elaborate upon this opinion, and many others, for the pollster, including 'youth's taste for music', American cultural influence abroad, the violence of children's toys, and the quality of commercial advertisements. Such correspondents believed their opinions were valued by the experts. Like Stein, they assumed that since the pollsters had been curious about their views on one aspect of a social issue, they might well be interested in their views about others.[39]

Many more were moved to tell the pollsters how they *would* have spoken on the issues of the day had they only been asked – in one woman's words, to add her 'penny's worth to the discussion'.[40] Such writers acknowledged with little hesitation their desire to register their views with the experts. They also seemed convinced that they were in fact taking part in a 'discussion' that might influence the course of public opinion, rather than simply responding to preformulated questions. They wrote in unsolicited, as did 'A Contented Ford Employee' in 1947, to say to Roper: 'I did not happen to be one of those selected to answer questions on your recent [poll], but I would like to express my views.'[41]

Others responded not to specific surveys but instead – again, unbidden – offered their own singular views about which way public opinion was going to swing next. 'I wonder if the [R]epublicans [*sic*] really know what a beating they are going to take in the next election,' wondered one writer. Revealing the multiple other channels of opinion open to the polls' consumers, this woman told Gallup that her conversations with like-minded shoppers in the grocery store evinced great discontent with prices, the national budget, and taxes.[42] Not quite ready to cede judgment to the pollsters' new technology, she, like many others, evaluated the public's pulse herself and shared her insights with the experts.

This ocean of opinions that the polls triggered expressed something of their civic potential. In exchanges with the pollsters, as Gallup and Roper claimed in oft-repeated arguments about making the unheard 'articulate', some Americans may indeed have been provoked to engage more directly in political debates. E. Dayton, a reader who had encountered a Gallup Poll on juvenile delinquency in 1954, was prompted to write to its creator: 'I have never before written to any paper or persons expressing my views but I feel this time I must get it off my chest.' Once made aware of others' views, however impersonally, citizens may have gained a new sense of belonging to a broader public dialogue.[43]

But there was a critical difference from earlier workings of public opinion. Interactions between the public and the pollsters were not actually in public, nor did they aspire to alter fellow citizens' views. Individual opinions, whether asked for or not, were directed to national experts, and to them alone. A Gallup interviewer, writing in 1940, observed that respondents frequently had queries of their own while being questioned, wondering who would be elected, how polls worked, or simply 'what other men and women are thinking'. On the latter point, she observed,

> eight out of ten, after answering a question, will either ask directly what most people said about it or will remark indirectly, 'I suppose nobody else said that.' They are delighted if told that everybody said it. It makes them feel that they were right.[44]

Informed of others' views via national statistics, consumers of the polls would more and more frequently go through intermediaries in order to take part in the public.

Statistical scepticism

The desire to participate in the polls was a reaction to opinion surveys' growing presence and sway in daily life – their claim to speak for the national mood. Another, equally strong, reaction however was a deep scepticism about the capacity of opinion polls to register public sentiment faithfully.

To cite the most prominent example, polling's key methodology, scientific sampling, itself aroused public fear and distrust. The logic of sampling – the idea that the nation's spectrum of beliefs could be divined from interviews with as few as a thousand people – was never intuitive to many Americans, who saw it as fundamentally unrepresentative and undemocratic. As one woman wrote to Gallup,

> I notice you always say 'The Public' in the headline, but further down in very small print you say the question was put to a carefully selected sample of 1,536 persons. Now, how in the world, by any stretch of the imagination can 1,536 people be termed 'the public'?[45]

Another writer borrowed the language of the US Constitution to make the case: 'obviously, you haven't asked the majority of "We The People"'.[46]

Already in 1936 Gallup would write with some exasperation,

> Tell [the average man] that it is possible to select a few hundred or a few thousand persons in his state who represent the divergencies in point of view of the entire voting population of that state, and he will laugh, if he does not swear.[47]

Believing such attitudes threatened their entire enterprise, both he and Roper made strenuous efforts to educate the public about the validity of statistical sampling. But the frequency of the complaint makes clear that many citizens did not buy the conceit that a handful of respondents could stand in for their views, much less 'public opinion'.

A related reaction came from those who had never been questioned by Gallup or Roper, and believed this fact alone betrayed the polls' claims to representativeness. The pollsters had, through their own doing, invited such critique. Given the pollsters' many statements extolling the democratic nature of their surveys, many Americans found it surprising or even unjust that they had never been polled. One woman enquired of Gallup:

> I am 48 years old – Have lived in Calif., N.Y., Colo., and Massachusetts and *never* in my long life have I ever known anyone who was quizzed by you or even approached! *Who* do you quizz [*sic*] and where and when??[48]

A man wrote to Elmo Roper after hearing one of his radio broadcasts, to ask: 'Where do you get your information? I am 44 yrs old and have never been interviewed by a poll taker.' He went on to challenge Roper's figures, reasoning that the pollster hadn't queried anyone like him.[49]

In such responses, common-sense understandings of political representation clashed with the pollsters' scientific one. The letters that poured into Roper's and Gallup's offices testify to individuals' anger at, and disbelief in, the methods that produced majority and minority percentages. One such critic was a Connecticut woman who complained in 1951 that Elmo Roper was '98 per cent wrong' regarding American support for universal military training. He might speak for 'a certain group' (she suspected New Yorkers) but certainly not 'the American people'.[50] That same year, a man from Southern California heard and promptly dismissed Roper's statistics on majority support for the Truman administration versus General MacArthur during the Korean War. 'In Los Angeles county,' he asserted, 'it is 10 to 1 against the administration', and he did not require any sophisticated polling techniques to figure this out. 'You will learn that the public is still for him [MacArthur]. ... Your statement tonight about the People being of majority opinion on Adm[inistration] policy is your opinion and you should say so,' he argued.'[51]

These protests throw into sharp relief the challenge that statistical sampling faced in gaining public acceptance. They also reveal the enormous investment citizens deposited in their own readings of the political milieu. Ina Hazen from Youngstown, Ohio, for example, condemned one of Gallup's surveys showing support for Alaskan statehood. A vigilant poll-follower, she went on to say, 'I have yet to see a poll result published that agreed with my opinion or that of anyone I knew.'[52] The power of impersonal numbers claiming to represent the thinking of the American public was always most evident to those who disagreed with them.

Like Hazen, many who took up their pens to write to Gallup and Roper had an unwavering conviction in their grasp of national attitudes, and used their own statistics to buttress challenges to the polls. Weighing in on national debates about the infamous Communist-baiter Joseph McCarthy in the mid-1950s, one of Gallup's readers in the *Houston Post* charged, 'You are certainly in error as far as the Great State of Texas is concerned as to only 6% of the voters being for Senator Jo [*sic*] McCarthy.'[53] Another put the figures in the positive, but otherwise agreed, writing, 'if you took a true poll of the American people ... you would find them over 95% for Senator McCarthy'.[54] J. M. Mottey, this time responding to a poll in Florida's *St Petersburg Times* that ranked Eleanor Roosevelt as America's most admired woman, fumed,

> In regard to Mrs. Roosevelt you are crazy as a bed bug. She is made fun of and despised by 95% of the people. Your Polls are the biggest fakes I know of.[55]

Such critics adopted the pollsters' quantitative argot. But their exaggerated numbers – 95 or 98 percent of the population were nearly always purported to agree with them – and their belief that 'true' polls would bear them out expressed a stubborn confidence in their own judgments about where the majority stood.

What prompted so many to write to the pollsters to tell them they were wrong? And what explains their hostility towards polls they disagreed with? Certainly some simply could not see themselves in the numbers, or believed that pollsters' general categories could not capture the nuances of their individual views. But citizens' sense of the nation, it also seems clear, could be profoundly shaken by Gallup's and Roper's statistics. This may account for the passion with which they protested findings that threatened to undermine their own personal certainties. When presented with scientifically determined majority opinions that contradicted their own sense of the social and political mainstream, some rejected such data out of hand. For others it led to an uncomfortable recognition that they belonged in the minority.

These writers often stressed their local credentials and on-the-ground perspective to demote the polls. Offering proof of their broad social contacts or special insights, they accused the experts of being too far from grocery store gossip or the factory floor to get a true reading of public opinion. Evidently – given the volume of Gallup's and Roper's correspondence – many also believed that the old-fashioned technique of writing letters remained an effective channel for expressing public opinion. But the depth of many writers' antagonism towards the polls revealed misgivings that their ability to challenge scientific surveys might not last for long, and that old ways of assessing opinion were giving way to a new top-down regime of expert measurement.

Indeed, the polls' consumers, even the combative ones, furthered this process. In disputing results that did not match up with their sense of social reality, few rejected the survey enterprise per se. Instead, they insisted that those polls were biased, unrepresentative, or poorly conducted. Responses to Gallup's and Roper's facts thus hint at the increasing hold of a new social scientific technology, if not always the specific conclusions that flowed from it. But there was something else at work in the poll-watchers' responses. Many seemed to have internalized the statistical notion of the people's will, and wanted to make sure, somehow, that they belonged to it. The very rhetoric of opinion surveys meant that, whereas it was relatively easy for 'participants' to challenge specific poll findings, it was more difficult for them to launch effective critiques of a purportedly inclusive, not to mention scientific, technique for 'conjuring' the public's preferences. How to argue with a scientific technology that (theoretically) permitted each citizen a say, and allowed majority views to become visible?

As these examples show, scientific polls sparked new dilemmas for citizens. Many worried about the displacement of local beliefs by amassed national ones, the possibility of incorrectly registering collective trends, and the startling discovery that one might be attached to a minority cause. Apart from relatively infrequent elections, there had been no easy or straightforward way to measure one's place in the mass before 1936.

The power of aggregates

Gallup's and Roper's mailboxes reveal that pollsters did not fully manage to convince the public of the validity of their statistics. The same letters suggest, however, that whether sceptical or credulous about the surveyors' facts, citizens recognized the ability of polling data to persuade. Some were fearful that the frequent broadcasting of aggregate opinion wielded a potent power, creating the allure of universality and the desire to join the majority. Douglas Ward of California, for example, charged Gallup with 'a slick and nasty method of campaigning by using your figures to prove your man is a winner'. This he judged especially pernicious since 'everybody wants to be on the Winning Side'.[56] Another angry writer sent a letter to Gallup care of the 'American Institute for Measuring or Formulating of Public Opinion'.[57] To these critics, measuring *meant* formulating.

Striking in this regard was the fact that some correspondents hoped to influence the numbers, even if they didn't fully trust or understand the way that they had been produced. Charles Moore mused in a letter to Gallup in May 1956, 'I do not know where you get your figures.' Nevertheless, he wanted the pollster to 'add all four votes from my family' against the proposition of a visit by Soviet premier Nikita Khrushchev to the United States.[58] This equation between votes and poll responses was adopted by others as well, suggesting that responding to a poll was taking its place alongside voting as an imagined means of gaining political influence. A man by the name of R. F. Holmes who had seen a Gallup Poll in the *Los Angeles Times* in the mid-1950s fired off a missive to the pollster the same day, explaining, 'I want to get my vote in against the U.N. and all foreign entanglements.'[59] Still another reader, this one anonymous, sent a curt note to Gallup regarding a survey on daylight saving time: 'Sir, you may add 3 more no' to the tally.[60]

Such writers clearly desired to make their mark in whatever way they could on the pollsters' authoritative rendering of public opinion. In conflating votes and poll responses, they seemingly absorbed the pollsters' peculiar mode of representation, one based on the statistical sampling of a mass public. However, their urgent attempts to influence the numbers, whether part of a scientific sample or not, registered their ambivalence about – and perhaps their discounting of – those same techniques. As such, polling stood in tension with the notion of counting every citizen at the (physical) polls, as well as older understandings of democracy and public opinion. This dispute between the people and the pollsters, supposedly solved by the pollsters' scientific triumph in 1936, would never fully be overcome.

Despite the best efforts of pollsters to sell their craft, then, opinion polls always stood in uneasy relation to the American public. Distrust and dependence could go hand in hand. The best evidence for this came in the aftermath of the Dewey–Truman election of 1948, in which Harry Truman beat Thomas Dewey, the candidate predicted to win by all of the polls. Critics of all stripes were able to invoke 1948 to cast doubt on the pollsters' methods and findings, whether related to other political forecasts or social issue polls. To many, this single well-publicized mistake had the power to call into question the entire polling industry. 'You were wrong in the election and you are wrong now to tell the Congress we want any part of war,' wrote a woman protesting against Roper's assessment that Americans supported intervening in Korea in 1950.[61] One letter arrived at Gallup's AIPO office in 1952, rebuking his pre-election polls of that year. It was addressed simply to 'Wrong Again', Princeton, New Jersey.[62] Another writer requested that Roper never use the word 'accuracy' in connection with any of his findings. As he put it, 'your reputation and past record is too greatly fresh in our minds'.[63] For *this* segment of the public anyway, the pollsters' infamous error in 1948 was an invaluable plank in the discrediting of the polls more generally.

Nor surprisingly, pollsters would employ the method most familiar to them – public opinion surveys – to sound out the damage done by Truman's unfortunate defeat of Dewey. In their polls of the polls, researchers found that clients of survey agencies were divided over whether to continue their patronage. Members of Congress had become more hostile to the profession. Some newspaper editors claimed they could not foresee running polls in their pages for years to come, believing the failed forecasts had made not merely opinion surveys but the newspapers that printed them the targets of derision.[64]

And yet, in almost no time, the crisis had passed. 'Within six months,' noted one observer, 'public opinion polls were functioning at their 1948 levels.'[65] One explanation was the polls' blend of scientific and democratic rhetoric, promoted tirelessly since 1936. The second, and related, factor was the growing dependence of various sectors of US society on Gallup's techniques. Critiques of the polls from elites as well as ordinary individuals did not cease in 1948. But the idea that this particular mode of gathering and displaying social data could be banished from American life was becoming increasingly implausible. In 1953, one scholar found that even four to five years after the election disaster of 1948, 'the question of public opinion polling still strikes an intensely sensitive nerve among news executives of United States daily newspapers'. Significantly, though, he noted that 'whether a particular newspaper currently subscribes to or does not subscribe to a public opinion poll makes little difference in responses to the subject' – suggesting that use of the polls and faith in them was not at all the same thing.[66]

The same was true of the ways ordinary citizens approached the polls. For them, there was never a definitive moment at which the opinion surveyors' techniques gained legitimacy, when a general consensus affirmed the polls' validity and worth. Both before and after 1948, many citizens would find polling techniques suspect and their implications for public discourse distressing. To Robert Baker of San Francisco, polls were a sign of the downfall of modern life, catering to 'our almost universal demand for predigested thought and ready-made opinions'. But such critics did not call for an end to opinion surveys. The same man wrote to Elmo Roper in 1952, 'I do not believe that you should – or that you can – withdraw from the field of published political research. The demand would still exist, and the gap you would leave would be filled by someone else.' Having helped create the demand for a certain kind of knowledge, opinion surveys had become the only means to satisfy it.[67]

Here, George Gallup may have been correct when he speculated that even the volume of criticism levelled at the polls was 'a testimony to their importance'.[68] Americans' robust disagreements with particular findings and the techniques used to reach them revealed their investment in the polls – or at least, as *Time* magazine put it in 1948, the recognition that they were 'here to stay'.[69] Culturally persuasive or not, polls had effectively displaced other, earlier ways of gathering political and social information, becoming the only available technology for telling 'the public' what it collectively believed.

The surveyed society

After 1948 the pace and scope of public opinion polling in the US would only intensify. Already in the 1930s polling had penetrated the corridors of the White House, starting with approval from Franklin D. Roosevelt, an early and eager convert to survey data. In this, Roosevelt was hardly alone: from his tenure onwards, every president save Harry Truman relied on confidential surveys of opinion, whether to evaluate campaign strategies or to claim an independent base of support for his views.[70]

Opinion polling, along with the allied field of market research, expanded dramatically in the 1960s and beyond, as interest groups, political consultants, and television and news organizations entered the polling business. Not only presidential administrations but members of Congress, governors, and mayors regularly consulted or commissioned polls. Opinion polling was on its way to becoming a permanent aspect of US governance, colouring campaign strategies and political advertisements, public statements, and policy-making. By the opening years of the twenty-first century, several hundred survey organizations existed in the US. As already noted, scientific polling had also spread throughout the world. By the mid-1960s, the Gallup Poll alone had 32 affiliates and conducted polls in nearly 50 countries.[71] Questions remained, however, about the adequacy of polls to measure and inform public opinion. Political scientist James Fishkin, urging more deliberative and nuanced ways of researching and communicating opinion, argued that the 'people's voice' had been lost, not despite – but because of – the welter of public opinion polls.[72]

Despite their ubiquity, scepticism about opinion polls endures. Over half of Americans surveyed in 1985 claimed not to believe in the accuracy of polling techniques, and sampling in particular. Recent debates over employing statistical sampling in the US census have likewise exposed widespread doubts about the opinion pollsters' foundational technique.[73] Nevertheless, Susan Herbst writes, 'few candidates run for office, express their views on current issues, or change campaign strategy without consulting opinion polls'.[74] Recent, well-aired flaws in pre-election and exit polls notwithstanding, quantitative opinion surveys are a mainstay of public discourse. What remains striking about the reach of opinion surveys into daily life is how very uneven it is. Compared with other attempts at social engineering detailed in this volume, national polling is resilient and ever-expanding. Yet deep distrust continues to shape citizens' relationship with the data that claim to represent them.

Early opinion pollsters worked both in and on 'the public'. Although claiming to solve some of the problems of individuals' participation in increasingly complex, bureaucratic societies – to clarify and respond to what citizens wanted in an era of big government – opinion surveys have instead exposed rifts between democratic credos and technical expertise, and between the power of 'the public' and those who speak in its name. Uses of, and challenges to, statistical opinion data in the mid-twentieth century only begin to suggest the complicated relationship between polls and their publics, and between opinion surveys and modern modes of citizenship. These dilemmas, first raised in the 1930s by the likes of George Gallup and Elmo Roper, remain part and parcel of surveyed societies today.

Notes

1. J. S. House, F. T. Juster, R. L. Kahn, H. Schuman, and E. Singer, (eds) (2004), *A Telescope on Society. Survey Research and Social Science at the University of Michigan and Beyond* (Ann Arbor: University of Michigan Press), p. 453.
2. See, for example, Frederic LePlay's *Les Ourvriers Européens* (1855), Charles Booth's monumental 17-volume *Life and Labour of the People in London* (1889–1903), and W. E. B. Du Bois's *Philadelphia Negro* (1899).
3. Y. P. Seng (1951), 'Historical Survey of the Development of Sampling Theories and Practice', *Journal of the Royal Statistical Society*, series A, 114(2), p. 217. S. Herbst (2003), 'Polling in Politics and Industry', in T. M. Porter and D. Ross (eds), *The Cambridge History of Science*, vol. 7: *The Modern Social Sciences* (New York: Cambridge University Press), p. 577.
4. One step in this process, crucial for the pollsters, was the discovery of ways to estimate standard errors based on sample size in 1935. See F. F. Stephan (1948), 'History of the Uses of Modern Sampling Procedures', *Journal of the American Statistical Association* 43(241), pp. 12–39.
5. A. Desrosières (1991), 'The Part in Relation to the Whole. How to Generalize? The Prehistory of Representative Sampling', in M. Bulmer, K. Bales, and K. Kish Sklar (eds), *The Social Survey in Historical Perspective* (New York: Cambridge University Press), p. 235.
6. University of Chicago sociologists W. I. Thomas and Florian Znaniecki were among the first to quantify attitudes in *The Polish Peasant in Europe and America. Monograph of an Immigrant Group*, 5 vols (1 and 2, Chicago: University of Chicago Press, 1918; 3, 4, and 5, Boston, MA: Badger, 1919–20). On industrial psychology, see R. Gillespie (1991), *Manufacturing Knowledge. A History of the Hawthorne Experiments* (New York: Cambridge University Press). See also the useful summary of academic attitude research in the 1920s and 1930s in J. Converse (1987), *Survey Research in the United States. Roots and Emergence, 1880–1940* (Berkeley: University of California Press) pp. 54–86, and J. Platt (1996), *A History of Sociological Research Methods in America, 1920–1960* (Cambridge: Cambridge University Press).
7. J. R. Beniger (1986), *The Control Revolution. Technological and Economic Origins of the Information Society* (Cambridge, MA: Harvard University Press), p. 376.
8. For a specific discussion of feedback mechanisms and marketing, see K. S. Buzzard (1990), *Chains of Gold, Marketing the Ratings and Rating the Markets* (Metuchen, NJ: Scarecrow Press), pp. 3–4. See also P. Napoli (2003), *Audience Economics. Media Institutions and the Audience Marketplace* (New York: Columbia University Press).
9. M. Poovey (1998), *A History of the Modern Fact. Problems of Knowledge in the Sciences of Wealth and Society* (Chicago: University of Chicago Press). For an overview of state information-gathering projects, see P. Starr (1987), 'The Sociology of Official Statistics', in W. Alonso and P. Starr (eds), *The Politics of Numbers* (New York: Russell Sage Foundation), pp. 7–57.
10. 'Straw' polls, unsystematic tallies of voter preferences undertaken by American newspapers and private citizens beginning in the first quarter of the nineteenth century, predated such ventures and were commonplace up until the 1930s. See C. E. Robinson

(1932), *Straw Votes. A Study of Political Prediction* (New York: Columbia University Press), and S. Herbst (1993), *Numbered Voices. How Opinion Polling Has Shaped American Politics* (Chicago: University of Chicago Press), pp. 69–87. See also T. Harrisson and C. Madge (1939), *Britain by Mass-Observation* (Harmondsworth: Penguin); W. Stott (1973), *Documentary Expression and Thirties America* (New York: Oxford University Press). Mass-Observation did employ random sampling, but also direct observation, interviews, diary-keeping, open-ended questionnaires and so forth.

11. T. Asad (1994), 'Ethnographic Representation, Statistics, and Modern Power', *Social Research* 61(1), pp. 55–88; quotations on 74 and 77.
12. P. Bourdieu (1985), 'Opinion Polls. A "Science" Without a Scientist', reprinted in P. Bourdieu (1990), *In Other Words. Essays Toward a Reflexive Sociology*, trans. Matthew Adamson (Cambridge: Polity Press), p. 168.
13. A notable exception is Converse's excellent *Survey Research in the United States* (1967). See also D. Robinson (1999), *The Measure of Democracy. Polling, Market Research, and Public Life, 1930–1945* (Toronto: University of Toronto Press).
14. G. Gigerenzer, Z. G. Swijtink, T. M. Porter, L. Daston, J. Beatty, and L. Kruger (1989), *The Empire of Chance. How Probability Changed Science and Everyday Life* (New York: Cambridge University Press); I. Hacking (1990), *The Taming of Chance* (Cambridge: Cambridge University Press); T. M. Porter (1988), *The Rise of Statistical Thinking, 1820–1900* (Princeton: Princeton University Press); A. Desrosières (1998), *The Politics of Large Numbers. A History of Statistical Reasoning* (Cambridge, MA: Harvard University Press).
15. T. M. Porter (1995), *Trust in Numbers. The Pursuit of Objectivity in Science and Public Life* (Princeton: Princeton University Press). See also P. C. Cohen (1982), *A Calculating People. The Spread of Numeracy in Early America* (Chicago: University of Chicago Press).
16. G. C. Bowker and S. L. Star (1999), *Sorting Things Out. Classification and Its Consequences* (Cambridge: MIT Press); W. N. Espeland and M. L. Stevens (1998), 'Commensuration as a Social Process', *Annual Review of Sociology* 24, pp. 313–43.
17. I. Hacking (1995), 'The Looping Effects of Human Kinds', in D. Sperber, D. Premack, and A. J. Premack (eds), *Causal Cognition. A Multi-disciplinary Debate* (New York: Oxford University Press); idem (1986), 'Making Up People', in T. C. Heller, M. Sosna, and D. E. Wellbery (eds), *Reconstructing Individualism. Autonomy, Individuality, and the Self in Western Thought* (Stanford, CA: Stanford University Press). See also M. Foucault (1988), *Technologies of the Self. A Seminar with Michel Foucault*, ed. L. H. Martin, H. Gutman, and P. H. Hutton (Amherst: University of Massachusetts Press), and N. Rose (1990) *Governing the Soul. The Shaping of the Private Self* (New York: Routledge), especially pp. 213–58.
18. Poovey, *A History of the Modern Fact.*
19. The role of censuses in creating the 'imagined community' of the nation is of course spelled out in B. Anderson (1991), *Imagined Communities. Reflections on the Origin and Spread of Nationalism*, revised second edition (New York: Verso), pp. 163–85. See also J. C. Scott (1998), *Seeing Like a State. How Certain Schemes to Improve the Human Condition Have Failed* (New Haven: Yale University Press).
20. P. Converse (1964), 'The Nature of Belief Systems in Mass Publics', in D. Apter (ed.), *Ideology and Discontent* (New York: Free Press), pp. 206–61.
21. Bourdieu, 'Opinion Polls. A "Science" Without a Scientist'. See also idem (1979), 'Public Opinion Does Not Exist', in A. Mattelart and S. Siegelaub (eds), *Communication and Class Struggle* (New York: International General); J. Habermas (1970), 'The Scientization of Politics and Public Opinion', in *Toward a Rational Society* (Boston, MA: Beacon Press); and B. Ginsberg (1986), *The Captive Public. How Mass Opinion Promotes State Power* (New York: Basic Books).
22. T. Osborne and N. Rose (1999), 'Do Social Sciences Create Phenomena? The Example of Public Opinion Research', *British Journal of Sociology* 50(3), pp. 367–96; quotation on 387.
23. See J. R. Zaller (1992), *The Nature and Origins of Mass Opinion* (New York: Cambridge University Press), pp. 28–39; S. L. Althaus (2003), *Collective Preferences in Democratic Politics. Opinion Surveys and the Will of the People* (New York: Cambridge University Press),

pp. 277–82; and G. F. Bishop (2005), *The Illusion of Public Opinion. Fact and Artifact in American Public Opinion Polls* (New York: Rowman and Littlefield).

24. Scholars have also made this argument, including Elisabeth Noelle-Neumann, who was the founder and director of the Public Opinion Research Center in Allensbach, Germany. See Noelle-Neumann (1993), *The Spiral of Silence. Public Opinion – Our Social Skin*, second edition (Chicago: University of Chicago Press).
25. M. R. Frankel and L. R. Frankel (1987), 'Fifty Years of Survey Sampling in the United States', *Public Opinion Quarterly* 51(2): Supplement: 50th Anniversary Issue, S128. Some of the examples and analysis that inform the remainder of this chapter are drawn from S. E. Igo (2007), *The Averaged American. Surveys, Citizens and the Making of a Mass Public* (Cambridge, MA: Harvard University Press).
26. G. Gallup (1940), 'Polling Public Opinion', *Current History and Forum* 41 (February), p. 23.
27. S. C. Dodd (1946–7), 'Toward World Surveying', *Public Opinion Quarterly* 10(4), pp. 470–83.
28. L. D. Beers (2006), 'Whose Opinion?: Changing Attitudes towards Opinion Polling in British Politics, 1937–1964', *Twentieth Century British History* 17(2), pp. 177–205.
29. J. Cowans (2002), 'Fear and Loathing in Paris. The Reception of Opinion Polling in France, 1938–1977', *Social Science History* 26(1), pp. 71–104, at 86.
30. See the chapter by A. Kruke and B. Ziemann in this volume.
31. M. Goot (2010), '"A Worse Importation than Chewing Gum". American Influences on the Australian Press and their Limits – The Australian Gallup Poll, 1941–1973', *Historical Journal of Film, Radio and Television* 30(3), pp. 269–302; quotation on 273. Goot argues that it was not just the name of the Australian poll that was American, but that the confluence of 'interviewing, sampling theory, and news' in the 1930s had 'deep American roots'.
32. Beers, 'Whose Opinion?', p. 183.
33. Cowans, 'Fear and Loathing', pp. 94, 96.
34. See M. J. Hogan (1997), 'George Gallup and the Rhetoric of Scientific Democracy', *Communication Monographs* 64(2), pp. 161–79.
35. E. Roper (1944), 'What People are Thinking', *New York Herald-Tribune* (30 November).
36. Gallup, 'Polling Public Opinion', p. 23.
37. R. G. Hubler (1941), 'George Horace Gallup. Oracle in Tweed', *Forum and Century* 103(2) (February), p. 95.
38. L. Rogers (1941), 'Do the Gallup Polls Measure Opinion?', *Harper's* (November), p. 623.
39. Elizabeth A. Stein (New York City, NY) to Roper, 1 August (year unknown), Roper Papers.
40. Alice W. Baker (Redlands, CA) to Roper, 3 June 1951, Roper Papers.
41. Contented Ford Employee (Dearborn, MI) to Roper, 19 September 1947, Roper Papers.
42. Unsigned to Gallup, undated, George H. Gallup Papers, University of Iowa Libraries, Iowa City [hereafter, Gallup Papers].
43. E. Dayton to Gallup, 21 November 1954, Gallup Papers.
44. P. London (1940), 'Ringing Doorbells with a Gallup Reporter', *New York Times Magazine* (1 September), p. 9.
45. Elinor Nevins (Nashville, TN) to George Gallup, undated, Gallup Papers 2, her emphasis.
46. Anonymous to Gallup (1956?), Gallup Papers 2.
47. G. Gallup (1936), 'Putting Public Opinion to Work', *Scribner's Magazine* (November), p. 39.
48. Mrs Robert (W.?) to Gallup, 11 November 1955, Gallup Papers, her emphasis.
49. D. J. Stoner (Pittsburgh, PA) to Elmo Roper, 12 September 1952, Roper Papers.
50. Eleanor Holmes (Ansonia, CT) to Roper, 6 June 1951, Roper Papers.
51. W. C. Hollingworth (Long Beach, CA) to Roper, 3 June 1951, Roper Papers, his emphasis.
52. Ina F. Hazen (Youngstown, OH) to Gallup, 17 January 1955, Gallup Papers, her emphasis.
53. Joseph E. Glenn (Houston, TX) to Gallup, 23 November 1954, Gallup Papers.
54. 'Not-Your-Dirty-Biased Lies' to Gallup, undated, Gallup Papers.

55. J. M. Mottey to Gallup, January (1955?), Gallup Papers.
56. Douglas Denton Ward (Fontana, CA) to Gallup, 13 April 1956, Gallup Papers.
57. James A. Webb (Athens, GA) to Gallup, 23 June 1961, Gallup Papers 2.
58. Charles E. Moore to Gallup, 7 May 1956, Gallup Papers.
59. R. F. Holmes to Gallup, 24 October (1955 or 1956), Gallup Papers.
60. Anonymous to Gallup, undated, Gallup Papers, his emphasis.
61. 'An American Mother' to Roper, undated, 1950, Roper Papers.
62. Anonymous (Santa Clara, CA) to Gallup, 20 October 1952, Gallup Papers 2.
63. John Libko (New York City, NY) to Roper, 29 June 1952, Roper Papers.
64. See, for example, P. B. Sheatsley (1948–9), 'The Public Relations of the Polls', *International Journal of Opinion and Attitude Research* 2, pp. 453–68. For an earlier assessment of public opinion on polls, see E. F. Goldman (1944), 'Polls on the Polls', *Public Opinion Quarterly* 8, pp. 461–7.
65. M. Parten (1950), *Surveys, Polls, and Samples. Practical Procedures* (New York: Harper and Brothers), p. 3.
66. W. C. Price (1953), 'What Daily News Executives Think of Public Opinion Polls', *Journalism Quarterly* 30, p. 287; Merton and Hatt, 'Election Polling Forecasts', p. 214.
67. Robert A. Baker (San Francisco, CA) to Roper, 15 December 1952, Roper Papers.
68. G. Gallup (1949), 'The Case for Public Opinion Polls', *New York Times Magazine* (27 February), p. 11.
69. 'The Black & White Beans', *Time* (3 May 1948).
70. R. M. Eisinger (2003), *The Evolution of Presidential Polling* (New York: Cambridge University Press); R. W. Steele (1974), 'The Pulse of the People. Franklin D. Roosevelt and the Gauging of American Public Opinion', *Journal of Contemporary History* 9, pp. 195–216.
71. W. P. Davison (1968), 'Public Opinion', in *International Encyclopedia of the Social Sciences* 13, p. 189.
72. J. S. Fishkin (1995), *The Voice of the People. Public Opinion and Democracy* (New Haven: Yale University Press), p. 80.
73. See M. J. Anderson and S. E. Fienberg (1999), *Who Counts? The Politics of Census-Taking in Contemporary America* (New York: Russell Sage).
74. S. Herbst (1993), *Numbered Voices. How Opinion Polling Has Shaped American Politics* (Chicago: University of Chicago Press), pp. ix, 124, 127.

12

Observing the Sovereign: Opinion Polls and the Restructuring of the Body Politic in West Germany, 1945–1990

Anja Kruke and Benjamin Ziemann

It is hard to overestimate the impact of opinion polling on the political system in West Germany since 1945. Widespread measuring of political attitudes and perceptions with the statistical techniques of the random or quota-sample, pioneered by George Gallup in the USA since the 1930s, fundamentally changed the way in which political decisions were taken, how party allegiances were shaped and organized, and how different policies were designed and implemented. In addition, the use of polling techniques had a massive bearing on what is called public opinion or the public sphere, that is, a field that is constituted by observations of the political process, and by observations of these observations. When Gerhard Schmidtchen, an employee of the *Institut für Demoskopie* in Allensbach, published one of the first book-length reflections 'on the impact of polling on politics' in the Federal Republic in 1959, he stressed precisely this point. The political system had 'to face the fact', Schmidtchen argued, 'that an archaic age would come to an end, during which people believed that social insights ... would emerge automatically'. Polling, already 'an informal institution in the large democratic states', would replace human observation with a sophisticated 'technical apparatus' and thus create a crucial 'macro effect' on any future political history.[1]

The implementation of political opinion polling in West Germany after 1945 is thus a classic example of the 'scientizing of the social', in this case reflecting the increasing presence of opinion-polling experts in political parties, government institutions, and the mass media, and the concomitant relevance of the techniques, concepts, and strategies the pollsters had on offer.[2] It is, also in comparative perspective, rather straightforward to posit the tremendous impact of polling on the political system in the Federal Republic.[3] There is less certainty, however, with regard to the precise nature of this impact. This is partly related to divergent theoretical conceptualizations, as discourse analysis, anthropological approaches, and sociological systems theory stress different aspects of the transformation of the public sphere through polling.[4] Disagreement about the possibly beneficial or detrimental effects of opinion surveys on the political process is also a consequence of normative presuppositions in many contemporary accounts. Jürgen Habermas and his work on the 'Structural Transformation of the Public Sphere', first published in 1962, is only one, although seminal, example for a critique of polling that is based on normative assumptions. Attempts to gauge public opinion empirically through polling were for Habermas only the last step in the long demise of a well-educated and informed public that had simultaneously pursued and constituted reason. Ultimately, Habermas insisted, polling would lead to the 'social-psychological liquidation of the concept of public opinion as such'.[5]

In this chapter in a first step, we will eschew any normative stance with regard to the definition of the public sphere. Rather, we hope to offer sufficient contextual information to explain the specific trajectory for the introduction of polling in the Federal Republic. In a second step, we will accentuate the changing methodological and conceptual premises of the pollsters, a point that is often neglected in historical accounts of public opinion research, which present this technique as a more or less stable approach, based on the use of random sampling.[6] In a third step, we identify 'responsiveness' as a key issue in the critical reception of polling in the Federal Republic, and outline the significance of the mass media not only for the performative presence of polls, but also as bodies that increasingly commissioned their own surveys and thus added a new element to the triangular relation between parties/government, pollsters, and the public sphere. Some comparative perspectives, which try to relate this case study to other Western democracies, conclude the chapter.

Implementing the American science

The introduction of sample-based opinion polls in Germany is firmly situated in the post-1945 reconstruction of a democratic system. As the three Western Allies commissioned polls of their own and licensed German polling institutes, their main interest was to monitor the possible persistence of anti-democratic attitudes in a post-fascist society, and to ensure that Germans learned to appreciate the openness of a free society as quickly as possible. In the British occupation zone, a 'Survey Branch' investigated political attitudes as well as responses to the economic situation. From the beginning, most of the practical work was conducted on behalf of the British military by the private Emnid institute for public opinion research, founded in 1945. A similar pattern emerged in the American zone, where the military government OMGUS established an 'Opinion Survey Section' in October 1945. Until the founding of the Federal Republic in 1949, it conducted a large variety of polls with regard to the political attitudes of the German population, but also began to use German interviewers.[7] Soon it became clear that polls commissioned by the US military yielded highly distorted results, as resentment against the US and its role in the de-Nazification process was widespread. As the successor to OMGUS, HICOG, the US High Commission in the Federal Republic, thus decided in 1951 to outsource the survey staff to an institute led by Germans, called DIVO (Deutsches Institut für Volksumfragen).[8]

Even after German pollsters had established their own institutes, the 'Reaction Analysis Staff' of HICOG and its director, Leo Crespi, continued to offer informal and institutional support, in particular through contacts with American social scientists. In institutional terms, but also with regard to their professional expertise in the sampling and interview techniques, German pollsters were heavily dependent on US support. Even though polling was also supported in the British and French occupation zone, it is hence not surprising that, from 1945 well into the 1950s, opinion polling in the Federal Republic was, in the view of politicians, journalists, and a wider public, perceived as something American.[9] This perception, however, was not only based on the fact that sample-based polling had received crucial support from the Western Allies, and the US in particular. It also chimed in with the notion that polling in the US, based on a 'clever' sense of business as the pollsters would earn 'horrendous sums', more generally reflected the 'American mentality to cram even the most private aspects of life into a statistic or a table'.[10] On some occasions, outright political resentment fed into these perceptions. When the *New York Times* commented on results of a HICOG poll in January 1953, highlighting that National-Socialist ideological patterns had resurfaced among the Germans, a storm of furious reactions in the media dubbed polling with condescension the 'American method'.[11]

Such a critique of the Americanization of the public sphere by means of statistical calculation reveals some of the ambivalences in the reception of polling in 1950s Germany. Published in 1953, it was part of the first cover story in the widely read *Spiegel* magazine that featured a photo of an affable young woman, Elisabeth Noelle-Neumann. While the headline described her, in a metaphor that stressed the invasion of privacy as the main element of polling, as the 'peephole of the old man' (i.e. Chancellor Konrad Adenauer), the article introduced Noelle-Neumann as *the* most crucial element of Adenauer's successful Federal election campaign. The *Spiegel* quoted Fritz Erler from the defeated Social Democrats as admitting that the 'craftsman-style propaganda methods' of the SPD had not been able to compete with the 'scientific methods of modern polling'. It appears that, while polling was an 'American' science, the public face of polling was unmistakably German. The next *Spiegel* cover story on Noelle-Neumann, published in 1957 in the context of another Federal election, was entitled 'Mistress of Public Opinion', thus indicating the extent to which polling already shaped German politics.[12]

Elisabeth Noelle (1916–2010) had analysed Gallup polls as an exchange student in journalism studies at the University of Missouri in 1937–8. Back in Berlin, she gained her Ph.D. with a dissertation on the use of polls in US politics. Until liberation in 1945, Noelle was a loyal, conformist follower of the Nazi regime. Jointly with the journalist Erich Peter Neumann, who she married in 1946, Elisabeth Noelle-Neumann founded in 1947 the Institut für Demoskopie (IfD) in Allensbach. Licensed by the French military, the IfD quickly turned the sleepy village at the Lake Constance into a household name for opinion polling in the Federal Republic.[13] The long-term commercial success and political impact of the IfD rested to a large extent on a specific network of personal contacts. In 1951, Adenauer appointed the journalist Otto Lenz as an Undersecretary of State in the Chancellery, with a responsibility for public relations. In the same year, Lenz had met Neumann and his wife, and a close friendship between the two men developed. As a result, the IfD was the main pollster to conduct surveys for the 'Press- and Information Bureau' of the Federal government, with its overall expenditure almost doubling from 1958 to 1965.[14]

From 1951 to 1963, when the 'old man' Adenauer finally stood down as Chancellor, the close cooperation between the IfD and the government in the gathering and assessment of statistical evidence on public opinion yielded benefits for both sides. While the IfD established itself as one of the leading public opinion research institutes in the Federal Republic, the government could rely on extensive information on a whole raft of policy issues. The results of the polls and the accompanying reports were usually classified, hence accessible only to a small group of high-ranking government officials. Using polls as arcane information, the Adenauer government had a head start in the explicitly instrumental use of social science knowledge, employed to maximize power and political success.[15] The 1953 Federal election, mentioned above, is a crucial case in point. Until July 1953, approval rates for the CDU/CSU had stagnated at around 37 per cent. Based on opinion polls, worries about price inflation for consumer goods were identified as a crucial popular concern. Taking immediate action, in June and August the CDU parliamentary party pushed through two laws that reduced taxes on tobacco and coffee. These consumer goods had again been identified through polling as the most effective levers for such a move. In August, approval rates for the CDU shot up, preparing for a clear victory with 45.2 per cent of the vote.[16]

While the smooth cooperation between Chancellery and its pollsters (mainly the IfD, but also Emnid, which delivered data on purely economic issues) proved to be successful for both sides, Noelle-Neumann failed to secure recognition and a reputation for opinion polling in the academic field. Former émigré sociologists such as Theodor W. Adorno and René König were powerful players in the emerging field of university-led empirical social research, and in the main professional body, the Deutsche Gesellschaft für Soziologie (DGS), in particular.[17] While Adorno criticized the positivistic, theory-free approach of the pollsters, König, the pre-eminent academic promoter of empirical social research in the Federal Republic, was deeply suspicious about Noelle-Neumann's Nazi past, and dubbed her 'nicely interspersed with brown' remnants. Academic departments only began to host large-scale public opinion research from 1974, when the Forschungsgruppe Wahlen started its work at the University of Mannheim, securing substantial grants from the public television broadcaster ZDF for election coverage.[18]

Against the backdrop of the financial firepower of the Chancellery and its well-established cooperation with the IfD, the SPD struggled to utilize the analytical potential of the new social science properly. This was not only a matter of a comparative lack of financial resources. Well into the 1950s, leading SPD functionaries were still bound to Marxist notions of politics as a struggle between two solid blocs, a reactionary mass on the one side, and the working class as the epitome of progressive politics on the other. Communication with the voters was understood as propaganda, which had to inform, but not to manipulate. Beginning with the disastrous defeat in 1953, however, proponents of a party modernization came to the fore and initiated a number of scattered opinion polls. Even then Fritz Heine, in the party executive responsible for 'Advertisement and Propaganda', promoted, for the time being, a purely instrumental use of the new technique. Heine only started to change his mind when three young students of social and political sciences approached the party executive in 1955. Wolfgang Hartenstein, Klaus Liepelt, and Günter Schubert presented plans for the tailor-made use of polls, which could explore the reluctance of working-class people to vote for the SPD. Defeat in the Federal elections in 1957, when the CDU scored an absolute majority, led to the systematic use of polling data by the SPD.[19]

In January 1959, Hartenstein, Liepelt, and Schubert founded Infas (the acronym stands for 'Institute for Applied Social Science') as a commercial institute for opinion research. Well into the 1970s, Infas was closely linked to the SPD as its main pollster, continuously supplying a wealth of data to the party executive and to regional branches. In a memorandum written in 1959, the three summed up how their approach to using polling data differed from that taken by the government. In their view, the CDU had overstepped the boundaries of a 'legitimate' use of polling, 'manipulating the masses of voters' and 'discrediting alternative programmes' in a 'cynically chiselled plan'. In the hands of the SPD, polling had to be used not as 'hidden persuasion' (an allusion to Vance Packard's diatribe against the advertisement industry in *The Hidden Persuaders*, published in 1957), but according to a 'strategic plan', which would 'neutralize' the impact of the 'consumer advertisement' employed by the IfD. These ideas were explicitly based on the notion that the SPD, as a socialist party, had to make a rational, emancipatory use of social science. In order to achieve these aims, the memo sketched out a 'political-sociological overall balance sheet' of West German society. In addition, it also recommended the analysis of motive structures and symbolic attachments among the electorate.[20]

The observations and juxtapositions in this memo reflect the ambiguities of applied polling in the SPD during a crucial period of transition for the party. All three Infas-founders, to be sure, had studied in the US, and Hartenstein and Liepelt had even been together in Ann Arbor, Michigan, where they encountered the most advanced American opinion research. However, it would be wrong to describe their optimistic promotion of an advanced social science approach as a simple transfer of American ideas or as a straightforward 'Americanization'.[21] Clearly, with their focus on 'opinion leaders' they advocated ideas derived from Lazarsfeld's model of a two-step flow of communication. Promoting to target 'indifferent soft voters' with no clear-cut party allegiance, the three pollsters carefully translated American ideas into an ongoing process of party reform.[22] Still, the introduction of polling as a superior approach to attract new potential voters actively contributed to ongoing change in the SPD, from a party that advocated class struggle into a more pragmatic attitude, symbolized by the Godesberg party programme in 1959, which accepted the rapid social change in Germany since 1945.[23] But at least in their memo from 1959, the Infas-founders did cling on to the rationalistic notion that applied social science had to be only a means to a progressive end, not an end in itself.

Disaggregating the body politic

At the beginning of the 1960s, opinion polls were a firmly established element of the political system in the Federal Republic. A handful of commercial polling institutes, all of which handled contracts in market research for companies aside from their engagement with politics catered for the needs of government and the main political parties.[24] IfD and Infas were clearly associated with CDU and SPD respectively. Other pollsters such as Emnid, DIVO or Infratest, founded in Munich in 1947 with an initial focus on audience research for broadcasters, worked for a variety of clients. By then, opinion polls were an integral part of all aspects of the political process. At a basic level, the repetition of the *Sonntagsfrage* ('How would you vote if a Federal election were to be held next Sunday?') continuously charted party preferences. Specialist polls analyzed the public image of key politicians, and investigated attitudes with regard to specific bills or certain policy fields. With respect to the 1950s, for instance, it can be shown how the Christian Democrats learned to conceptualize German Angst as a crucial emotion, an insight that Adenauer converted into the deliberate use of 'security' as an overarching frame for both domestic and foreign policies.[25]

In the following, however, we prefer not to analyze the uses of polls with regard to certain policies, although such an analysis can yield crucial results.[26] Rather, we are interested in the changing methodological programmes that pollsters used to analyze the basic results of random or quota sampling. These changes do not only reflect attempts to refine the conceptual premises and to stay in touch with recent academic developments in social psychology and communication studies. Changing methodologies brought the constructivist implications of polling to the fore. Not only with regard to its political impact, but also through its conceptualization of the political, polling does more than merely feel the 'pulse of democracy', as the famous metaphor coined by George Gallup put it.[27] With the elaborate use of distinctions between different types of voters, and of different reasons for political attitudes, the pollsters actively constructed core elements of the political process and changed the ways in which politics were understood. Ultimately, they not only transformed the parties, but disaggregated and fundamentally reshaped the body politic in democratic societies.

When the three founders of the Infas-institute commenced their work for the SPD in 1957, their use of polling data was still tied to a consideration of statistical data on the social stratification of the populace. The local SPD branch had employed this approach with great success for the 1957 elections in the city state of Bremen. Relying on a breakdown of official data on employment, social strata, and affluence, Infas had prepared a '*Sozialstrukturanalyse*', a detailed analysis of the social structure in the individual boroughs in Bremen. These statistical background data were then used for targeted polls, in order to offer specific leverage for campaigning in particularly promising areas. A similar approach was later adopted by the party executive for the 1961 Federal elections. When Willy Brandt, the SPD candidate for Chancellor, toured the country during this campaign, his route was based on a typology of different constituencies provided by a 'structural atlas', which had been compiled by Infas.[28] Thus, in the early 1960s the pollsters still took their cues for the targeting of specific audiences from an understanding of society as organized into structurally defined social groups.

At the same time, however, they had already begun to undermine the notion that there was a clear-cut correspondence between social stratum and voting preference, a case that was particularly difficult to make in the SPD, where traditional ideas about the proletariat prevailed. In 1956, Hartenstein, Schubert, and Liepelt explained the significance of the 'soft voter' in a memorandum for the SPD party executive. They described him as 'pragmatic', and as someone who would 'not feel any loyalty to a particular party'. But as a 'changing element' in public opinion did in fact exist, the party could appeal to other strata than its core working-class constituency, as they explained in another memorandum. One recommendation was to target those parts of the Protestant middle strata who were close to Social Democratic ideas. Upon perusal of this memo, Fritz Heine opined that the party should, nonetheless, not cease to 'cling on to the estate (*Stand*)' of the workers.[29] This use of terminology is revealing. While polling suggested an increasingly flexible shifting of preferences between parties, Heine still conceptualized the working class as a stable entity with a fixed set of political interests.

But Heine's objection and misunderstanding was, in the end, that of a traditional socialist whose influence in the party was to fade away. Beginning in the mid-1950s, both Social Democrats and Christian Democrats began to conceptualize the electorate in terms of consumers in an open marketplace. While some of them had a clear preference for one particular brand of politics, others would be open for different offers and eventually change their minds. The terminology for the description of the *Wechselwähler* or floating voter varied, both between the parties and over time, although most labels tended to disregard the 'unpolitical' nature of this attitude. 'Soft' voter, the term used by Infas, was in this context still a rather neutral category. As a matter of routine, polls conducted for the CDU described them as 'Meinungslose', voters without opinion.[30] It was only in the early 1970s, years after the Christian Democrats had lost a considerable number of their core voters, and hence ceded power to the SPD in 1969, when a different approach was taken. Werner Kaltefleiter, a political scientist who had coordinated social scientific planning for the CDU in a separate research institute since 1970, suggested a more positive understanding. Contrary to earlier assumptions, *Wechselwähler* should not be seen as 'absolutely disinterested voters', but rather as 'critical voters' who would make up their mind through attention to newspapers and television. For those without political interests, Kaltefleiter coined the term *Grenzwähler* or 'borderline voters'.[31]

As the pollsters suggested the existence of floating voters, the parties were, at a basic level, ready to take a much more pragmatic approach with regard to key principles of parliamentary democracy. Henceforth, they were ready to accept that voting decisions were based on individualized premises, and not simply determined by social realities. And as they conceptualized politics as an open marketplace, they acknowledged the core reality of what increasingly turned into a two-party system, with 90 per cent of the votes cast for either CDU or SPD in 1972: that power could change hands when the voter decided. Thus, it was not by chance that both parties refined their understanding of the *Wechselwähler* when they faced devastating defeats – the SPD in the late 1950s, the CDU after 1969.

But the notion of the floating voter did not only stress the basic contingency of democratic politics. It also informed aggressive attempts to regain control of the public climate that informed these shifting allegiances. This was precisely the strategic stake for the 'Spiral of Silence'. When Elisabeth Noelle-Neumann first presented her theory of public opinion in 1974, she aimed to design a successful campaign strategy for the CDU in the upcoming Federal election of 1976. She analyzed the micro-climate of opinions shaped in face-to-face encounters. According to this theory, how the 'silent majority' of floating voters observed local opinion leaders and followed their lead would be decisive.[32]

The theory of the 'Spiral of Silence' was later revised, but meanwhile, from the late 1960s, another set of approaches to the study of data from random sampling had come to the fore. Since about 1960, Infas had started not only to acknowledge and define the existence of floating voters, but to go one step further in identifying specific target groups. In the early 1960s, these social groups were still rather conventionally defined, as more or less homogeneous social strata, which could be addressed with tailor-made propaganda. Towards the end of the decade, more complex approaches were employed by all the polling institutes. Based on a factor analysis, which identified a large number of hard and soft variables for voting decisions, and using insights from social psychology, Infas and Infratest began to present an ever more complicated patchwork picture of shifting voter allegiances and target groups. Using focus groups and group discussions in addition to the traditional sampling methods, Infratest pioneered an approach that identified value preferences and also included information on media usage.

This general shift towards the incorporation of social psychology into political polling culminated in an extensive special programme on 'political psychology', which the Chancellery under SPD leader Willy Brandt started in 1973 together with Infratest. These psychological approaches were already widely used in marketing and advertising. As a consequence, political polling became more akin to a form of market research. Conceptually, this shift led to a further disaggregation of the specified target groups. They were no longer fixed social groups like workers or women, but rather, for instance, 'pessimists' or 'authoritarians', categories used in 1974 to describe how and why some voters were disappointed by the Schmidt government. Thus, rather than class or social stratum, a certain set of expectations characterized these target groups.[33]

Another crucial shift in the respecification of sampling data occurred in the early 1970s. So far, the notion of structurally defined groups within the electorate had been refined and differentiated, leading to a rather complex pattern of groups that were identified by a multitude of variables. In a separate development, already beginning in the late 1950s, the pollsters had contemplated the significance of themes or issues. These surveys were thought to help define issues that could be used to demonstrate leadership. One possible outcome was 'followership by mistake', as Chancellor Ludwig Erhard discovered in 1965, when he coined his vision of an 'aligned society' (*formierte Gesellschaft*) that was characterized by a low level of democratic conflicts. Only after he had gone public did pollsters scrutinize popular attitudes, and strongly advised against using this concept.[34]

The linguistic aspects of political communication gained even more attention from the late 1960s, when value-laden issues such as concern about the environment or the terrorism of the Red Army Faction (RAF) came to the fore. These developments triggered polls that charted attitudes with regard to 'inner security' as a new field of domestic politics. In a situation when political strife intensified and the SPD was able, in 1969, to gain a majority of the vote, the Christian Democrats pioneered a new approach for the analysis of political semantics. In 1968 WIKAS, the scientific institute of the Adenauer Foundation, started to record how changes in social values were expressed in political terminology. This was basically a defensive reaction, based on the perception that both Social Democrats and the students of '68' were proponents of a semantic shift that had devalued traditional political concepts such as 'peace' or 'freedom'.

Testing the associations connected to certain terms, the pollsters of WIKAS developed an understanding of how concepts express and influence value systems. Based on these insights, Kurt Biedenkopf, general secretary of the CDU, urged the 1973 party conference to adopt his strategy of '*Begriffe besetzen*', that is, to take possession of concepts and thus wrestle them out of the hands of political opponents.[35] The Federal election in 1976 provided an opportunity to put these ideas into practice. The pollsters of the Allensbach Institute were sure to trace a significant current of conservative values in the electorate. Elisabeth Noelle-Neumann used this information to coin 'freedom or socialism' as the key slogan for the CDU, thus trying to tap into anti-communist sentiment.[36] Even though the CDU lost the election, the highly controversial slogan gained widespread attention, and still is one of the most effective in the – by now – six decades of electioneering in the Federal Republic.

Polling based on the random sample, to sum up, was not a static methodology that followed the same approach over the decades. Rather, we see constant efforts to redescribe both the electorate and the political process, based on a range of different theories and empirical methods that accompanied the core approach of random sampling. In the process, polling actively disaggregated the body politic. Expanding beyond their core expertise of observing voter preferences, the pollsters developed sophisticated techniques for an ever more differentiated analysis of shifting allegiances among the electorate. Abandoning the notion of fixed classes and contemplating fluid preferences, polling contributed towards an increasing volatility of the political process.[37]

'The rule of the 2000': Responsiveness and media realities

During the late 1940s and early 1950s, public debate was by and large positive about the claims of the pollsters to foster democracy. Powerful media outlets such as the *Spiegel*, as we have seen earlier, tapped into the by then already conventional narrative of the 'American' science, which would illuminate, using a representative sample, the preferences of ordinary citizens. This largely positive reception turned into widespread criticism during the Federal election campaign in 1957, when the media noticed that both major parties used polls in a highly instrumental fashion as a tool in order to maximize turnout. Tone and direction of this critique were set by the journalist Paul Sethe, who published an opinion piece on 'the rule of the 2000' on the front page of *Die Welt*, the conservative flagship newspaper of the Axel Springer media empire. The title formulation referred to the 2000 individuals who make up a random sample. While Sethe acknowledged the technical sophistication behind sampling, polls were an 'uncanny' procedure to him. Rather than putting trust into the elected leaders, the principles of parliamentary democracy would erode in an accelerating cycle of short-term responses to polling results. Sethe was concerned about 'endangered leadership', which was coming under pressure from 'fanatical disciples of Rousseau'. The original German term for leadership – *Führertum* – illuminates what was at stake. Twelve years after the collapse of the Third Reich, conservatives still longed for a strong government, while politicians, under the impact of the polls, seemed increasingly willing to give in to populist tendencies.[38]

Sethe was only one prominent voice in a broad stream of media coverage from 1957 to the early 1960s that was critical of the 'galluping consumption' of democracy through polling.[39] While the focus of these conservative criticisms was on the erosion of proper leadership through polling, academic contributions to the debate added another argument. In his 'critique of political polls', first published in 1957, political scientist Wilhelm Hennis also attacked the illusions of a purely quantitative approach that abandoned the 'humanist purpose' of all social science. From the normative viewpoint of an Aristotelian notion of politics as the pursuit of the good life, Hennis offered a trenchant criticism of the 'positivistic character' of polling. Public opinion was more than information, it was reasoning based on principles, and hence in essence qualitative, not quantitative.[40]

The pollsters, and the IfD in particular, which received the brunt of the criticism as it was the most visible representative of polling, were slow to react. Eventually, Elisabeth Noelle-Neumann responded to Hennis, pointing out that even while the masses were uninformed, they were still capable of learning, and that polling thus made a vital contribution to a broader debate about democratic politics.[41] Gerhard Schmidtchen, one of her employees, pursued a similar line of argument in his 1959 book on opinion polls. He was particularly adamant to reject the notion that politicians would eschew unpopular decisions when polls suggested a clear majority opinion. Two main examples served to drive home this point. First, the Basic Law (the West German constitution) had abolished the death penalty, even though the members of the Parliamentary Council in 1948 had commissioned a poll on this matter, according to which 74 per cent of all West German citizens favoured its retention. Even more striking was the example of Adenauer's plans for a nuclear armament of the *Bundeswehr*, one of the most controversial policy issues in the late 1950s. Despite widespread protests against these plans, backed up by polls that suggested a whopping 80 per cent majority against them, Adenauer stayed the course. Even more, he was able to defeat the Social Democrats, who led the protests of the 'Fight against Atomic Death' campaign, decisively at the ballot box.[42]

This conservative wholesale critique of opinion research, based on the charge that it would endanger democratic leadership, slowly faded away in the 1960s. Meanwhile, observers of the political system started to agree in the middle of the decade that opinion polling was there to stay, and that it had to be accepted as one crucial element of representative democracy. At the same time, in another crucial strand of events, the mass media displayed a more systematic interest in the results of polling, and in their potential value as news. The three Federal elections in 1965, 1969, and 1972 were crucial milestones in this development, which fundamentally changed the place of polls in the political process.

In 1965, the widely sold *Stern* magazine commissioned a series of polls on its own. Using its award-winning artwork department, the magazine presented Allensbach-data on the *Sonntagsfrage* in a series of articles with bar charts and other elements of visualization, which were then not yet familiar to the German reader. Election night in 1965 also brought for the first time televised computer projections, prepared for the ARD, the first public broadcasting channel, by the Infas-institute. The second channel, ZDF, presented projections drawn up by DIVO. In a separate and highly publicized event, a ZDF-presenter revealed forecasts by Emnid and Allensbach, which had been made 48 hours in advance. In a provocative U-turn, Noelle-Neumann had abandoned earlier forecasts of a 'horse race' and – rightly – predicted a comfortable majority for the CDU. In an embarrassing moment, Emnid-boss Karl Georg von Stackelberg had to admit defeat in this competition between the pollsters. One year later, Emnid was sold off to a commercial market research institute.[43]

The election night scandal in 1965 further increased the interest of both print and electronic media in using and representing polling data. The ZDF-Magazin, presented by veteran conservative broadcaster Gerhard Löwenthal, continued to commission its own polls for the Federal election campaign in 1969. The campaign in 1972, when the SPD government under Willy Brandt was re-elected, saw a hectic flurry of polls commissioned by both weekly news magazines, tabloid newspapers such as the *Bild-Zeitung* and, again, the two main public television channels.[44]

Since the 1972 elections, opinion polls have been a firmly established part of the media coverage during the election campaign. As the reach of electronic media among the population neared saturation point, with about 16 million television sets in German households, careful consideration of televised campaign events became a crucial task for the campaign planners in party headquarters, and thus for their pollsters. Conducting polls via telephone allowed a quick response to swings in the mood of the population and to media events. As the television took centre stage, timing was crucial. 'Timing' – in English! – was in fact one of the headlines in an assessment of the first two televised debates between Willy Brandt and Rainer Barzel, the CDU contender, during the 1972 campaign, prepared by Infratest for the Social Democrats. As only viewers with a strong interest in politics would watch a third debate beyond 9.30pm, Infratest recommended that Brandt should deliver all positively charged 'issues' – again in English! – before that time.[45]

From the 1950s onwards, control over polling data had switched from the Allies to the Federal government, and subsequently also to all major parties. Beginning in the mid-1960s, and with a first peak in 1972, the mass media wrestled control over polling from the parties. While government and parties continued to commission their own polls, the mass media became the key players in the public representation of polling data. This strategic advantage allowed them to claim a position as crucial intermediaries between the sovereign, that is, the people, and their preferences on the one hand, and the institutions of the *Parteienstaat* in the Federal Republic on the other, where the parties controlled the rules of the political game both through constitutional leverage and according to the system of party consensus that had emerged since the 1950s. As a result, a structural coupling between the mass media and the political system evolved from the 1970s, for which opinion polls provided the crucial link. The subsequent co-evolution of politics and mass media relied on both the constative and the performative qualities of opinion polls. Not only did the polls provide an accurate picture of public opinion, they could also be moulded and represented in highly dynamic visual forms.

The iconic moment for this trend was the introduction of the *Politbarometer* by ZDF television in 1977. This encapsulates the final and still ongoing phase of the relation between polling and politics. Based on data provided by an academic pollster, the Forschungsgruppe Wahlen, and embedded in the news coverage of a main public broadcasting channel, this monthly programme aimed and still aims to provide information on polls like the weather report, to which its title metaphorically refers: in an objective, cool, and almost clinical fashion, underscored by the design of the studio with chrome and clear horizontal lines, and through the even by German standards detached and rather dull presentation by the anchors, both men and women.[46] But the *Politbarometer* is not only the embodiment of a trading of polls as news that connects populace, parties, and government. Apart from turning polling data into a pervasive and apparently non-partisan resource of political discourse, the television broadcast also reinforced a general trend of political campaigning, which had come to the fore since the 1960s. Election campaigns in the Federal Republic were, in a stark and often noted contrast to the Weimar Republic, characterized by a style of representation that favoured sobriety and matter-of-factness (*Sachlichkeit*) over polarization and dramatization.[47] The *Politbarometer* also provided an antidote to a trend that has been described as 'politainment', that is, the proliferation of popular media formats in which a symbiotic entanglement between media and politics turns the latter into a superficial form of entertainment.[48] Ultimately, the increased presence of the media also forced conservatives to reconsider their position, rather than bemoan the volatility of public opinion. When Elisabeth Noelle-Neumann prepared the 'Spiral of Silence' as a book publication in 1980, she significantly changed her argument. While she had initially identified peer pressure in face-to-face interactions as the key force for the creation of a silent majority, she now turned her attention to the media and its manipulation of a climate of opinion.[49]

Comparisons and conclusions

The political history of the Federal Republic cannot be written without taking the impact of opinion polling into account. Within 15 years after their introduction, polls pervaded all aspects and institutions of the political system. They fundamentally changed the ways in which decisions were taken (politics dimension), how fields of political intervention were defined and approached (policy dimension), and how the core parameters of the political were discussed (polity dimension). In all these respects, the quantitative results of opinion polling provided the political process with 'new strategies of communication' and a 'new vernacular language', to quote the felicitous phrase by Theodore M. Porter.[50]

But before we try to tease out some more general conclusions with regard to the dynamics of scientization and theoretical premises for its study, we would like to highlight some peculiarities of the West German case in comparison with other countries. It has been argued that differences in the constitutional framework of politics bear out the national differences in the trajectory of opinion polling.[51] In our view, this seems to be only a minor factor, even though the technical intricacies of 'first-pass-the-post' go a long way in explaining the somewhat muted interest in polling in the UK, and it is clear that the Swiss system of popular referenda has stalled systematic interest in polling in that country for many decades.[52]

Other factors, though, played a much more important role in West Germany, in particular the specific constellation in which the polling institutes were set up, and how they were linked with the mass media. In France, by contrast, the private Institut français d'opinion publique (IFOP), founded by Jean Stoetzel in 1938, was until the early 1970s the single most important provider of polls, and had an exclusive publication contract with the newspaper *France-Soir*. Hence, there was widespread suspicion that IFOP would work on behalf of the government. Taking these perceptions into account, it is no surprise that, when the Parisian students attacked the political establishment of the Fifth Republic in May 1968, they also targeted IFOP. One of the famous woodcuts produced by the students of the École des Beaux Arts shows, under the heading 'refusez l'intoxication', an open skull in which liquid from a bottle with the label 'IFOP' is poured.[53] Accordingly, the most vociferous attacks against polling in France came from the political left, when Pierre Bourdieu criticized the illegitimate and unsystematic nature of the knowledge gathered by the pollsters, whom he derided as 'doxosophes'.[54] In West Germany, quite to the contrary, the critique of polling remained the domain of liberals and conservatives.

A comparison with Australia highlights the importance of the mass media. Polls were introduced there in 1941, when the Australian Public Opinion Poll (APOP) was registered as a private company. Founded by Keith Murdoch (1885–1952), APOP had permission to use the brand name 'Gallup poll' exclusively in Australia, and went on to have a monopoly on political opinion polls in the country until 1971. Murdoch and Roy Morgan, who operated the company on his behalf, publicized the results of the polls not only through the *Herald & Weekly Times*, the biggest newspaper in the country, but to some extent also through his television channels, until the cross-ownership of print and electronic media was outlawed in 1987. From the beginning until 1971, opinion polls in Australia were thus the monopoly of a conservative media conglomerate, which led to a certain bias against the Labor Party. The exploitation of polls by one media company for vested political interests continued when in 1985 Rupert Murdoch, son of Keith, introduced Newspoll, which quickly became the 'most important poll published by the Australian press'.[55]

Against this backdrop, a constant pluralism of pollsters, methodological approaches, and media outlets appears to be the most prominent feature of polling in the Federal Republic. The specific trajectory of polling in West Germany was partly a result of the initial Allied intervention, which tried to ensure that a variety of different institutional players embarked on the challenging task of gauging political attitudes in a post-fascist polity. Taming and normalizing the potentially eruptive consequences of German nationalism was clearly one of the intended side-effects of polling. And indeed, it was not any longer the pageantry of nationalist mass rallies that represented the unity of the German people, but rather the pie charts with polling results printed in newspapers. When Elisabeth Noelle-Neumann and her husband presented their key findings on 'the Germans' to an English readership in 1967, the cover image displayed the German colours – black, red and gold – but only in rather subdued fashion, hidden behind the clear rectangular lines of some graph paper.[56] And when, since the 1970s, the news media finally took control of the public representation of polls, the dominance of public broadcasting, and of the *Politbarometer* in particular, continued to detach the polls from excessive partisan usage.

Turning to the dynamics of scientization, it should first be noted that the notion of a 'discourse coalition' or 'reform coalition' of social scientists and progressive politicians who collaborate towards social progress is not really necessary and helpful to understand the rise of polling in the Federal Republic.[57] In a similar way to the German industrialists who had already displayed an early interest in market research in the 1930s, postwar politicians quickly developed their own, intrinsic interest in the new technology of sampling once it had been introduced by the Allies.[58] As their counterparts in the economic system, government officials and party leaders were keen to exploit the crucial strategic advantage polls provided: the chance to anticipate potential reactions of a volatile and unknown audience in a mass democracy, to which they did not have any direct access. The chronology for the implementation of polling thus also differs from the dynamics of scientization in the welfare states, where reform and planned modernization culminated in the late 1960s. Progressive politicians in the SPD and their pollsters shared the hopes for a wide-ranging planning of the democratic process, which peaked in 1970. But their interest in the new technology had been triggered much earlier, mainly for the mundane reasons of political survival.

Polling also differed from other practical applications of social science knowledge in terms of the politicization of experts. While polling was from the start a politically contested field, these conflicts did not have major repercussions in the academic realm. Most academic sociologists resented the 'positivism' of polling, and the pollsters were thus not able to gain an academic reputation. The polling institutes, to be sure, founded an association in 1955, called the Arbeitskreis deutscher Markt- und Meinungsforschungsinstitute. But the primary purpose of this endeavour was to maintain professional standards and to defend the trade against criticisms in the public. Only from the 1970s, once the Forschungsgruppe Wahlen was founded, were professors such as Erwin K. Scheuch and Rudolf Wildenmann able to represent empirical research on electoral sociology more forcefully in academic circles.[59]

In theoretical terms, a conceptual pluralism seems the best way to understand the significance of opinion polling in West Germany after the Second World War. From the perspective of Niklas Luhmann's sociological systems theory, polling provided, in the long term, a vital hinge for a structural coupling between the political system and the mass media. Moving beyond the main task of polls – to reduce the uncertainty of politicians about reactions of the public to their decisions – statistical data were increasingly relevant as newsworthy items, which the media could turn into a never-ending narrative about the shifting topics and allegiances in democratic politics.[60] If we analyze the period from 1945 to 1990 as a whole, this long-term trend can also be described, in a Foucauldian perspective, as the major contribution of polling to the governmentality of modern society.[61] As the increasingly complex theoretical models of the pollsters contributed towards a de-corporation of the body politic, polling allowed for a more flexible exercise of power in postwar democracy. Presented as results of opinion research in bar charts, even extreme political and moral opposites became mere endpoints on a sliding scale of gradual differences.[62] Opinion polls were thus part and parcel of a discourse of 'flexible normalism', which abandoned the notion of inevitable collisions between substantial political, religious or moral norms. Instead, polls functioned as purely descriptive signifiers of tendencies in collective behaviour.[63]

Notes

1. G. Schmidtchen (1959), *Die befragte Nation. Über den Einfluß der Meinungsforschung auf die Politik* (Freiburg im Breisgau: Rombach), p. 213.
2. See the chapter by Lutz Raphael in this volume.
3. Cf. A. Kruke (2007a), *Demoskopie in der Bundesrepublik Deutschland. Meinungsforschung, Parteien und Medien 1949–1990* (Düsseldorf: Droste); for comparison, see L. Blondiaux (1998), *La fabrique de l'opinion. Une histoire sociale des sondages* (Paris: Seuil); S. Herbst (1993), *Numbered Voices. How Opinion Polling has Shaped American Politics* (Chicago: University of Chicago Press); M. Goot (2010), '"A worse importation than chewing gum". American Influences on the Australian Press and their Limits. The Australian Gallup Poll, 1941–1973', *Historical Journal of Film, Radio and Television* 30, pp. 269–302; see also the chapter by Sarah Igo in this volume.
4. Cf. F. Keller (2001), *Archäologie der Meinungsforschung. Mathematik und die Erzählbarkeit des Politischen* (Konstanz: UVK); L. Lipari (1999), 'Polling as Ritual', *Journal of Communication* 49, pp. 83–102; B. Ziemann (2006a), 'Opinion Polls and the Dynamics of the Public Sphere. The Catholic Church in the Federal Republic after 1968', *German History* 24, pp. 562–86.
5. J. Habermas (1989 [1962]), *The Structural Transformation of the Public Sphere* (Cambridge: Polity), p. 240. For a systematic critique of Habermas's reification of the public sphere,

see H. Mah (2000), 'Phantasies of the Public Sphere. Rethinking the Habermas of Historians', *Journal of Modern History* 72, pp. 153–82.

6. S. Herbst (2003), 'Polling in Politics and Industry', in T. M. Porter and D. Ross (eds), *The Cambridge History of Science*, vol. 7: *Modern Social Sciences* (Cambridge: Cambridge University Press), pp. 577–90.
7. See A. J. Merritt and R. L. Merritt (1970), *Public Opinion in Occupied Germany. The OMGUS-Surveys, 1945–1949* (Urbana: University of Illinois Press).
8. Kruke (2007a), *Demoskopie*, pp. 40–3.
9. Ibid., pp. 43–9.
10. 'Was halten sie von Adenauer? Guckloch des alten Herrn: Meinungsforscherin Elisabeth Noelle-Neumann', *Der Spiegel*, 28 October 1953, pp. 11–16, at 12.
11. G. Baumert (1958), 'Bemerkungen zur Entwicklung und gegenwärtigen Stellung der sogenannten Meinungsforschung in Deutschland', *Kölner Zeitschrift für Soziologie und Sozialpsychologie* 10, pp. 383f.
12. *Der Spiegel*, 28 October 1953, p. 11; cf. the cover 'Herrin der Öffentlichen Meinung', *Der Spiegel*, 21 August 1957.
13. N. Kramer (2009), 'Elisabeth Noelle-Neumann. Die Demoskopin in der "Schweigespirale"', in T. Bauer, E. Kraus, C. Kuller and W. Süß (eds), *Gesichter der Zeitgeschichte. Deutsche Lebensläufe im 20. Jahrhundert* (Munich: R. Oldenbourg), pp. 133–49; on Neumann, cf. N. Grube (2004), 'Deutschlandkonzepte des Journalisten und demoskopischen Politikberaters Erich Peter Neumann 1938–1949', in G. Nickel (ed), *Literarische und politische Deutschlandkonzepte 1938–1949* (Göttingen: Wallstein), pp. 309–47.
14. On the Federal press office, see M. Weiss (2006), 'Öffentlichkeit als Therapie. Die Medien- und Informationspolitik der Regierung Adenauer zwischen Propaganda und kritischer Aufklärung', in F. Bösch and N. Frei (eds), *Medialisierung und Demokratie im 20. Jahrhundert* (Göttingen: Wallstein), pp. 73–120.
15. Kruke (2007a), *Demoskopie*, pp. 61–75.
16. Schmidtchen (1959), *Die befragte Nation*, pp. 156–61.
17. On empirical social research, see C. Weischer (2004), *Das Unternehmen 'Empirische Sozialforschung'. Strukturen, Praktiken und Leitbilder der Sozialforschung in der Bundesrepublik Deutschland* (Munich: R. Oldenbourg).
18. Kruke (2007a), *Demoskopie*, pp. 49–57, quote on 51.
19. Ibid., pp. 168–98.
20. W. Hartenstein, K. Liepelt, and G. Schubert (n. d. [1959]), 'Sozialdemokratie und Sozialwissenschaft. Bericht über unsere Erfahrungen und Vorschläge für die künftige Arbeit': Archiv der sozialen Demokratie Bonn (AdsD), PV, Büro Herbert Wehner, 2570; cf. Kruke (2007a), *Demoskopie*, pp. 198–209.
21. Cf. J. Angster (2003), 'Der neue Stil. Die Amerikanisierung des Wahlkampfs und der Wandel im Politikverständnis bei CDU und SPD in den 1960er Jahren', in M. Frese, J. Paulus, and K. Teppe (eds), *Demokratisierung und gesellschaftlicher Aufbruch. Die sechziger Jahre als Wendezeit der Bundesrepublik* (Paderborn: Schöningh), pp. 181–204.
22. Quotes: Hartenstein, Liepelt, and Schubert, 'Sozialdemokratie und Sozialwissenschaft', n.d. [1959]: AdsD, PV, Büro Herbert Wehner, 2570.
23. Kruke (2007a), *Demoskopie*, pp. 221–7.
24. On polling as market research, cf. A. Kruke (2007b), '"Atomwaffe im Propagandakampf"? Markt- und Meinungsforschung in Politik und Wirtschaft der frühen Bundesrepublik', in H. Berghoff (ed.), *Marketinggeschichte. Die Genese einer modernen Sozialtechnik* (Frankfurt am Main and New York: Campus), pp. 346–71.
25. A. Kruke (2007c), 'Western Integration vs. Reunification? Analyzing the Polls of the 1950s', *German Politics and Society* 25(2), pp. 43–67.
26. For polls on European integration, see ibid.; for polls on German rearmament: M. Geyer (2001), 'Cold War Angst. The Case of West German Opposition to Rearmament and Nuclear Weapons', in H. Schissler (ed.), *The Miracle Years. A Cultural History of West Germany, 1949–1968* (Princeton: Princeton University Press), pp. 376–408.

27. G. Gallup and S. F. Rae (1940), *The Pulse of Democracy. The Public-Opinion Poll and how it Works* (New York: Simon and Schuster).
28. Kruke (2007a), *Demoskopie*, pp. 209–13, quotes on 211, 213.
29. Ibid., pp. 338, 345.
30. Quote ibid., p. 335. For the discovery of the floating voter in the British Labour Party, see D. Wring (2005), *The Politics of Marketing in the Labour Party* (Basingstoke: Palgrave), pp. 49–53.
31. Kruke (2007a), *Demoskopie*, pp. 331–6, quote on 336.
32. E. Noelle-Neumann (1974), 'Die Schweigespirale. Über die Entstehung der öffentlichen Meinung', in E. Forsthoff and R. Hörstel (eds), *Standorte im Zeitstrom. Festschrift für Arnold Gehlen* (Frankfurt am Main: Athenäum), pp. 299–330, here 302, 308; see Kruke (2007a), *Demoskopie*, pp. 340–2.
33. Kruke (2007a), *Demoskopie*, pp. 268–71, 349–74.
34. Ibid., pp. 374–82.
35. Ibid., p. 157.
36. Ibid., pp. 155–60, 385–91.
37. Ibid., pp. 429–37.
38. P. Sethe, 'Die Herrschaft der 2000', *Die Welt*, 19 January 1957.
39. Kruke (2007a), *Demoskopie*, pp. 450–7, quote on 453.
40. W. Hennis (1957), *Meinungsforschung und repräsentative Demokratie. Zur Kritik politischer Umfragen* (Tübingen: Mohr), pp. 62–4.
41. Kruke (2007a), *Demoskopie*, pp. 458–9.
42. Schmidtchen (1959), *Die befragte Nation*, pp. 141, 157, 218–24.
43. Kruke (2007a), *Demoskopie*, pp. 467–76.
44. Ibid., pp. 480–9.
45. Infratest Sozialforschung, 'Einige Überlegungen zur Fernsehdiskussion der Parteivorsitzenden am 15.11.1972', 14.11.1972: AdsD, Depositum Albrecht Müller, 1/AMAD 000 222.
46. Kruke (2007a), *Demoskopie*, pp. 494–506. For the presentation, see also http://politbarometer.zdf.de [accessed 24 September 2010].
47. T. Mergel (2003), 'Der mediale Stil der "Sachlichkeit". Die gebremste Amerikanisierung des Wahlkampfes in der alten Bundesrepublik', in B. Weisbrod (ed.), *Die Politik der Öffentlichkeit und die Öffentlichkeit der Politik. Politische Medialisierung in der Geschichte der Bundesrepublik* (Göttingen: Wallstein), pp. 29–54.
48. A. Dörner (2001), *Politainment. Politik in der medialen Erlebnisgesellschaft* (Frankfurt: Suhrkamp).
49. E. Noelle-Neumann (1993), *The Spiral of Silence. Opinion – Our Social Skin* (Chicago: University of Chicago Press) [first German edition in 1980]. For a critical discussion, cf. C. Simpson (1996). 'Elisabeth Noelle-Neumann's "Spiral of Silence" and the Historical Context of Communication Theory', *Journal of Communication* 46(3), pp. 149–73.
50. T. M. Porter (1995), *Trust in Numbers. The Pursuit of Objectivity in Science and Public Life* (Princeton: Princeton University Press), pp. viii–ix.
51. See Blondiaux (1998), *La fabrique*.
52. A. Taylor (2003), '"The record of the 1950s is irrelevant". The Conservative Party, Electoral Strategy and Opinion Research, 1945–1964', *Contemporary British History* 17(1), pp. 81–110; L.D. Beers (2006), 'Whose Opinion? Changing Attitudes toward Opinion Polling in British Politics, 1937–1964', *Twentieth Century British History* 17, pp. 177–205; N. Stettler (1997), 'Demoskopie und Demokratie in der Nachkriegsschweiz. Die "Volksumfrage 1946" der Neuen Helvetischen Gesellschaft als demokratische Herausforderung', *Schweizerische Zeitschrift für Geschichte* 47, pp. 730–58.
53. Cf. J. Cowans (2002), 'Fear and Loathing in Paris. The Reception of Opinion Polling in France, 1938–1977', *Social Science History* 26(1), pp. 71–104, at 86–8, 92f. For the woodcut, see L. F. Peters (1968), *Kunst und Revolte. Das Politische Plakat und der Aufstand der französischen Studenten* (Cologne: DuMont Schauberg), n. p., image no. 190.

54. P. Bourdieu (1972), 'Les doxosophes', *Minuit* 1, pp. 26–45; see also P. Champagne (1988), 'Le cercle politique. Usages sociaux des sondages et nouvel espace politique', *Actes de la recherche en sciences sociales* 71/2, pp. 71–97.
55. See Goot (2010), 'American Influences', pp. 273–6, quote at 288.
56. See the jacket cover of E. Noelle and E.P. Neumann (1967) (eds), *The Germans. Public Opinion Polls 1947–1966* (Allensbach and Bonn: Verlag für Demoskopie). On polls and the representation of national unity in Australia, see M. Goot and T. Rowse (2007), *Divided Nation? Indigenous Affairs and the Imagined Public* (Carlton: Melbourne University Press). On polls in the context of a European public sphere, see A. Kruke (2006), 'Mit Umfragen zur europäischen Öffentlichkeit? Meinungsforschung, Parteien und Öffentlichkeit in Europa nach 1945', in J. Mittag (ed.), *Politische Parteien und europäische Integration. Entwicklung und Perspektiven transnationaler Parteienkooperation in Europa* (Essen: Klartext), pp. 405–32.
57. P. Wagner (2001), 'The Mythical Promise of Societal Renewal. Social Science and Reform Coalitions', in idem, *A History and Theory of the Social Sciences* (London: Sage), pp. 54–72.
58. On the early German interest in market research, as exemplified by the *Gesellschaft für Konsumforschung*, founded in 1934, see C. Conrad (2004), 'Observer les consommateurs. Études de marché et histoire des la consommation en Allemagne, des annèes 1930 aux annèes 1960', *Le Mouvement Social* 206, pp. 17–39, here 28–31.
59. Kruke (2007a), *Demoskopie*, pp. 49, 488–9.
60. See N. Luhmann (2000), *The Reality of the Mass Media* (Oxford: Polity), pp. 63–70.
61. M. Foucault (1991), 'Governmentality', in G. Burchell, C. Gordon, and P. Miller (eds), *The Foucault Effect. Studies in Governmentality* (Chicago: University of Chicago Press), pp. 87–104.
62. For an example from polls on religion, see B. Ziemann (2006b), 'Öffentlichkeit in der Kirche. Medien und Partizipation in der katholischen Kirche der Bundesrepublik 1965–1972', in F. Bösch and N. Frei (eds), *Medialisierung und Demokratie im 20. Jahrhundert* (Göttingen: Wallstein), pp. 179–206, here 191–3.
63. J. Link (1999), *Versuch über den Normalismus. Wie Normalität produziert wird*, second edition (Opladen and Wiesbaden: Westdeutscher Verlag).

13

Consumers, Markets, and Research: The Role of Political Rhetoric and the Social Sciences in the Engineering of British and American Consumer Society, 1920–1960

Stefan Schwarzkopf

The contested nature of the market as governance structure

This chapter addresses the role that the idiom of social-technical expertise and social science research methods have played in ensuring the ongoing hegemony of market- and choice-based approaches in contemporary social policy-making. I argue that far from being 'objective' and 'apolitical', European and American market and consumer researchers at times deliberately connected their social and scientific expertise to ideological questions about power structures within market capitalism. The chapter therefore studies in detail how specific rhetorical devices that 'mobilized'[1] and redefined people as consumers came to be used by market research companies and advertising agencies in order to project the marketplace as a democracy of goods in which consumer choices act as 'votes' that decide the fate of products and companies.

The ideal of the consumer as a sovereign and as independent 'voter', and the myth of the market as a form of 'democracy', have their origins in the interwar battles over the legitimacy of marketing as a social–economic technique and the legitimacy of open markets as a form of social and political governance.[2] Rather than merely produce qualitative and quantitative data about consumers, professional marketers and market researchers fostered an economic–political language, which brought marketing practice and social science methods much closer together than had been the case before the First World War. Through consumer interview techniques, focus groups, panel surveys, consumer juries, product testing panels, advertising attention studies, memory studies, readership surveys, and so on, marketing emerged during the interwar period as a much more 'scientific' endeavour, and this scientization helped legitimize the practices of markets and marketing.[3] With the production and popularization of graphs, tables, and statistics, advertising agencies and market research companies projected the marketplace as the *agora* of a consumer democracy and also promoted a vision of mass consumer opinion as its independent jury and sceptical arbiter. Of all the newly developed research tools, it was perhaps the consumer jury that symbolically aligned the voting and the verdicts of citizens most closely with the act of consumer choice. Just as citizen jurors in a fair and democratic judicial process delivered an unbiased verdict based on the evidence they saw, so the market research process allowed consumers to deliver a verdict over the products and services manufacturers offered on the market. The claim that consumer choices resembled the verdicts of an honest and fair jury consequently rebalanced the disparity in bargaining power between consumers and producers: as in a democracy, producers had to offer themselves up to the shrewd judgement of consumers, whose power of choice put them in the driving seat of a free market economy. Advertising agencies and marketing professionals engaging in market and consumer research could thus proffer themselves symbolically as true agents of a market democracy, which achieved a rational balance between the needs of consumers, workers, and producers, thus forestalling the socialist threat of reordering this power balance through government and collective action.

The historical and political origins of a marketing myth

As T. J. Jackson Lears, Kathleen Donohue, and Charles McGovern have argued, the origins of the consumer–citizen equation are to be found in the transatlantic transformation of liberal economic and political thought between 1890 and 1940, which increasingly began to stress the role of consumers and of consumption in the political economy of Western democracies.[4] The rise of new market research methods in Europe and the United States coincided with this transformation. These relatively new social science research methods sought to go beyond mere measurement of aggregate consumer expenditure in certain, often broadly defined, product categories, and instead allowed individual consumers to develop a voice and express opinions with regard to their daily choices. The coming together of new economic consumer theory in the wake of the marginal revolution in economic thought on the one hand, and increasing demand for consumer research on the other, sparked off a number of innovations in European–American marketing and advertising practices.[5] During the 1930s, for example, the European immigrants Paul Lazarsfeld, Ernest Dichter, and the son of immigrants Robert K. Merton developed the focus group method of consumer analysis.[6] In these groups, consumers answered open-ended questions, were guided through unstructured interviews, and discussed problems in an explorative way. Symbolically, these methods afforded consumers an authoritative voice and invited comparisons between the processes of democratic competition and deliberation (voting) on the one side and consumers' choice on the other.

Ironically, this equation between consumption and democratic political decision-making resulted in more power being conferred on corporations and governments than on individual consumers and citizens. As Lears has argued, this equation necessitated an entirely new terminology of consumer measurement and a new body of techniques for consumer and voter observation, which blurred the boundaries between the discourses of public and private:

> Survey research advanced under the banner of popular sovereignty. Through opinion polling and market research, government and business were supposed to become more responsive to their masters, the people. But what actually happened, as Tocqueville had predicted, was that the more responsive huge organizations seemed to be, the more they were able to set the boundaries of public discourse.[7]

Both the British and the American advertising industries proved enthusiastic in developing and exploiting this equation between mass consumers and a democratic electorate, or a jury. This symbolic equation allowed the industry to present itself as a force that made the market more responsive to consumer needs. Through the application of market research, markets could be promoted as mechanisms that gave consumers the power to keep manufacturers on their toes. For this equation between consumers and citizens to work and to permeate public discourse, however, political elites had to come in as well and lend their authority to the claim that markets were essentially the consumers' democracy.

As early as 1927, the then British Colonial Secretary Leo Amery proposed that advertising and market-driven consumerism were democratizing forces that helped spread the wealth of new products and technologies more evenly. At the opening ceremony of the 1927 London Advertising Exhibition, Amery described advertising as the art of helping the public know their needs and of 'fulfilling them in the best and cheapest manner'.[8] Amery saw in advertising an important 'social service': 'It enable[s] new ideas, new comforts, new reforms in the whole manner of living to be equally spread throughout the country. Our whole political life depend[s] on it.'[9] By the late 1930s, this idea of advertising and marketing as a social service for democracy had made its way into the intellectual debate around the legitimacy of commercial advertising. At the opening ceremony of the 1938 Advertising Convention in Glasgow, the president of the Advertising Association, Lord Southwood, declared that 'advertising's greatest value to society was that it proclaimed the customer as king'. Southwood portrayed advertising as an individualizing force, which enabled people to enjoy an ever greater choice of products, and as a democratizing force, which helped modern riches reach the most disadvantaged classes: 'We realise that the poorest of us to-day has indeed advantages denied even to kings and princes in the past.'[10]

During the 1930s, the Advertising Association in Britain spent tens of thousands of pounds on propagating the democratic ideals of advertising to a mass audience that still held many misgivings about product commercials and the 'ad-men' behind them. In 1939, the Vernon advertising agency in London was commissioned by the Advertising Association to devise a press and poster campaign that made 'the case for advertising'. In its campaign, Vernon's proposed that 'thanks to Advertising' people of the modern world were better off than medieval kings and queens, as the latest inventions of science and industry were now brought 'within the easy reach of modest means'. The agency's advertisements featured men and women consumers ('He is richer than Richard III' and 'She is richer than Cleopatra' – Figure 13.1) individually profiting from advertising.[11]

The Vernon campaign in 1939 also linked consumer affluence to issues of democracy and citizenship. A second set of advertisements devised by the agency featured photographs of a crowd. Over the photos were superimposed outlines of the heads of a king and of a judge. This design symbolized the quasi-royal, sovereign status of the citizen in a consumer democracy – the person who ultimately decides which product is to survive on the market. The advertisement using the shape of a king's head (Figure 13.2) claimed 'The customer is king', and explained that 'advertising is the democratic method of making the best readily available to the greatest number'.[12] The advertisement featuring the head of a judge over a photo of a crowd (Figure 13.3) made this point about advertising, consumers, and democracy even more strongly: 'The final arbiter of advertising is the common man. That is why advertising can fairly claim to represent the best methods of democracy.' In the view of the people at the Vernon agency, the supreme role of the common man in a consumer democracy necessitated the development of market research: 'That is why advertising is never tired of studying the needs, desires and fears of the ultimate consumers of advertised goods and services.'[13]

Similar campaigns based on the idea of the sovereign consumer as judge, jury, and king influenced marketing thought during the 1930s, a decade that witnessed a burgeoning of business propaganda connecting the advertising and the market research industries with the aims of democracy. In 1937, the Greenly advertising agency in London published an advertisement for its own services in numerous trade journals and the popular press, directly comparing consumers with a jury and advertising to a lawyer trying to persuade this jury to deliver a positive judgment. Ingeniously, the agency played with the term 'consumer trial'. Featuring a photograph of sworn jurors sitting in a courtroom and listening intently to the presented evidence (Figure 13.4), the advertisement's copy read: 'What is the verdict? You want people to "try" your goods. Remember – the public verdict is greatly influenced by the persuasive counsel of your advertising.'[14]

The American J. Walter Thompson advertising agency (JWT) outdid its competitors in symbolically leveraging the consumer–citizen equation for the purposes of legitimizing commercial advertising and its market orientation. While British pro-advertising propaganda was careful about naming its enemies and left a lot to the political imagination, its American counterpart went straight to the heart of the matter. 'What is the difference between state capitalism and private capitalism?' asked a JWT pamphlet in 1937 entitled 'A Primer of Capitalism'. The answer lay in the sovereign consumer, and once again the comparison of consumer with juror and judge was employed:

> In state capitalism the *politician* is the boss. He may call himself a king, a general, a dictator, a fuehrer or a super mass-man. But whatever he calls himself, he is a politician. He commands factories to open or close. He tells consumers what they can buy. ... Under private capitalism, the *Consumer*, the

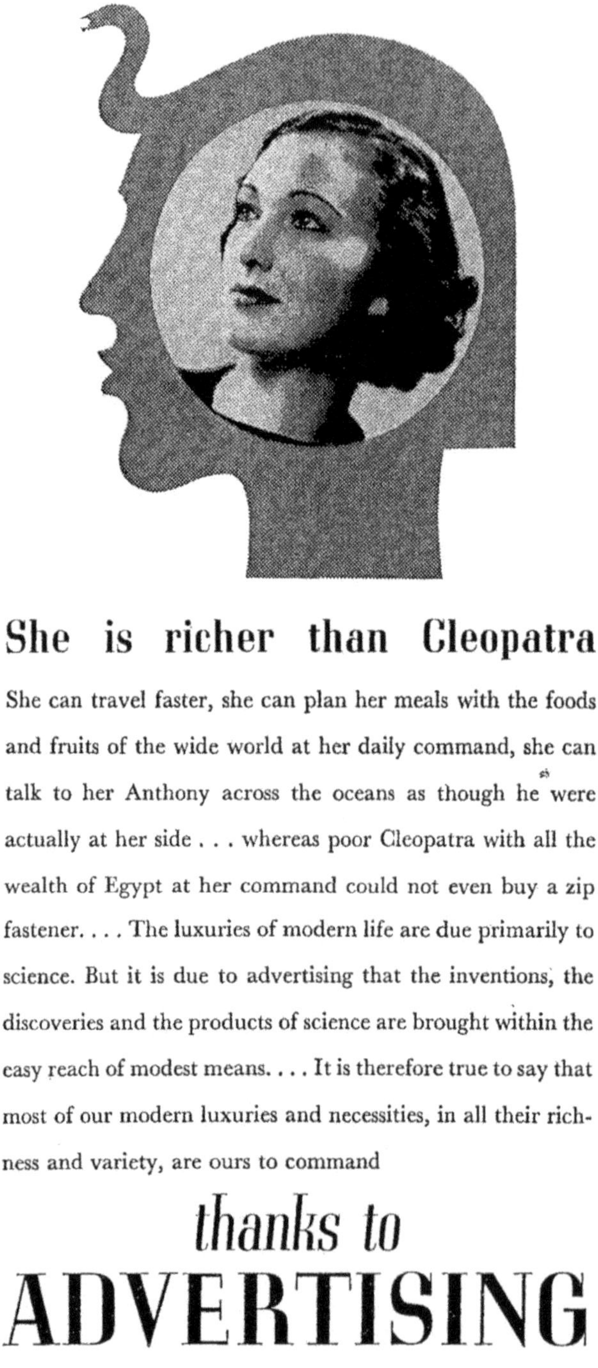

Figure 13.1 'She is richer than Cleopatra': C. Vernon & Sons' Campaign for Advertising, 1939 (© The History of Advertising Trust, Norwich)

Figure 13.2 'The Customer is King': C. Vernon & Sons' Campaign for Advertising, 1939 (© The History of Advertising Trust, Norwich)

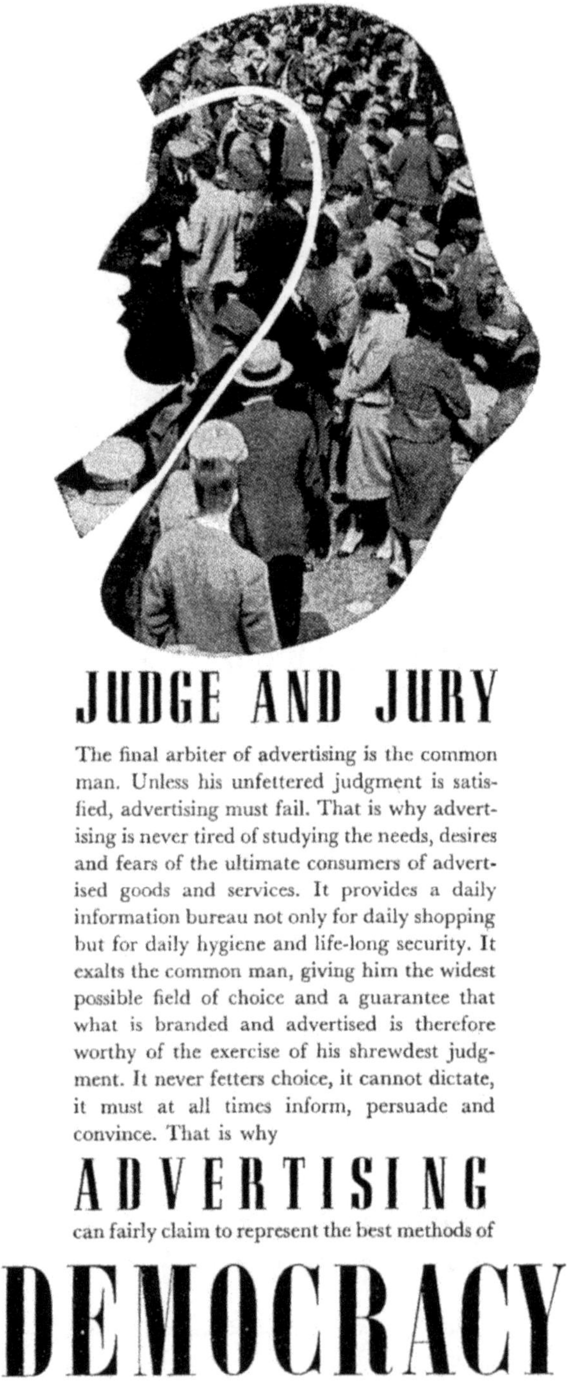

Figure 13.3 'Judge and Jury': C. Vernon & Sons' Campaign for Advertising, 1939 (© The History of Advertising Trust, Norwich)

Figure 13.4 'What is the Verdict?': Greenly's Ltd Trade Advertisement, 1937 (© The History of Advertising Trust, Norwich)

> *Citizen*, is boss. The consumer is the voter, the juror, the judge and the executioner. And he doesn't have to wait for election day to vote. He needn't wait for Court to convene before he hands down his verdict. The consumer 'votes' each time he buys one article and rejects another.[15]

In many ways, JWT's pamphlet summarizes the key features of 1930s market populism: its idiosyncratic equation of consumers and jurors/citizens; its equally idiosyncratic equation of consumer power with spending power, which denied the public the right to state-funded or civil society-based empowerment through protection of vulnerable consumers; and its cultural equation of abundant choice provided by large corporations with the American way of life: 'In all history there has been nothing remotely like modern American business as a sensitive index to popular likes and dislikes. It is democracy plus.'[16] The triptych of consumption, citizenship, and democracy painted by the JWT agency, however, would not be complete without a depiction of the fallen angel that still haunted the earthly paradise of Mr and Mrs Consumer. This demon was the politician, the little dictator who had set out to put the fulfilment of state directives above the needs of ordinary workers and housewives. The entire narrative of secular salvation hidden in the consumer–citizen/juror equation received its unique purgatorial power from the demonization of the politician – JWT took a swipe at the popular Franklin D. Roosevelt when putting 'Führer' and 'Super Mass-Man' side–by side – and the delegitimization of the state as a macro-institution.

Image campaigns on behalf of advertising had appeared in Britain, Canada, and the United States from about 1910, yet they always focused on the information value and the quality/price assurance provided by advertised goods. From about 1935, advertising was presented as a necessary part of a free consumer democracy that needed to defend the most precious of its values: people's freedom of choice. By creating equal and free citizens, advertising as a business force served an economy of plenitude and a consumer's democracy. Rhetorically, the terms 'freedom' and 'democracy' became the saviours of the image of advertising in the late 1930s. This new language infuriated left-leaning intellectuals like George Orwell and F. R. Leavis. Denys Thompson, himself a member of the *Scrutiny* circle around Leavis, dissected and chastised advertising slogans that sold telephones as 'the democratic instrument of a democracy' or promoted beer by referring to the village inn as 'the very school of democracy'.[17]

The burgeoning culture of left-wing intellectualism and the rise of totalitarian regimes on the European continent played into the hands of both the American and the British advertising industries and their attempts at portraying themselves as a vital part of the mechanism of democracy.[18] Even during the war, the Advertising Association in Britain commissioned press and poster advertising that defended the role of commercial advertising in the public sphere. Once again, the campaign deployed the political language of democracy, choice, and citizenship. Some of the Advertising Association's advertisements during 1940 went very far in their political and moral defence of advertising:

> This war is a war between two sorts of economy, the force economy and the free economy. The Nazis believe in regimentation, we in freedom of choice. We are confident that our system is better and will win, but let us not forget that freedom of choice requires leadership and guidance. Politically, the function of leadership devolves on the party system, which crystallizes and expounds the various political choices. In the economic sphere exactly the same function is performed by advertising. The nation-wide reputation of brand good-will enables the public to exercise its freedom of choice on a basis of knowledge instead of guesswork.[19]

Rhetoric and the authority of economic thought

Such slogans, published in trade magazines, popular newspapers, and on posters, were part of the rhetoric produced by an insecure and embattled advertising industry which, throughout the 1930s and 1940s, found itself exposed to unprecedented levels of public scrutiny. Following the writings of anthropologists like Franz Boas and Gregory Bateson and the linguists Edward Sapir and Benjamin Whorf, I argue that the language used to make sense of consumers, of consumption, and of consumer society and the range of epistemological tools available to create the notion of consumption as a field of investigation are closely interconnected. In other words, the intellectual construction of terms like 'free choice' and 'consumer sovereignty' and the rhetorical and symbolic defence of marketing communication in the public sphere went hand in hand with innovations in actual marketing and market research practice.

On both sides of the Atlantic, the 1930s were the key decade, which saw the coupling of a new language of consumer representation with new practices and techniques of consumer investigation. An essential group that was at the heart of this coupling has so far received little attention by historians of the consumer society in Britain and America. Liberal economists, especially the members of the Austrian School of Economics and those closely connected with it, were instrumental in a shift of attention away from the phenomena of production to the political economy of consumer decision-making. This new group of economists, evicted from their home country Austria, proclaimed that the consumer was the sovereign driver of the economy in the same way as the citizen-voter was sovereign in a political democracy. Joseph Schumpeter and Friedrich von Hayek argued pervasively that the institutional structure of democracies made them very much like markets.[20] Schumpeter advocated a 'minimalist' model of democracy, which stressed the 'discovery' function of democratic competition. This model presented democracy as a mechanism that enabled competition between political leaders and between the policy programmes of their parties, much as in a market structure. Periodical votes by the electorate represented the necessary symbolic affirmation of a government's legitimacy and kept it accountable.[21]

The idea of the powerful consumer whose aggregate choices ultimately moved the wheels of the economy originated in the Scottish Enlightenment, particularly in Adam Smith's *The Wealth of Nations* (1776), which had pronounced the consumer and consumption as the true defining powers in market exchange, their demand alone determining what was produced. The new generation of economists from Austria, however, made one crucial change to classical liberalism: those who followed Adam Smith viewed consumer sovereignty as an ideal, while the Austrian School saw consumers in 'free' markets as the *actual* sovereigns. Any imperfections in the market that might impede the actions of consumers in their role as sovereigns were explained away by non-market forces like welfare regulations, tax burdens, trade barriers, and trade union attempts to wage class war. Adam Smith had warned about the innate drive in entrepreneurs to seek inflated rents, to deceive and to conspire against the public, but no such scepticism was allowed in the new economic theories propagated by the Viennese. For Hayek, Mises, and Schumpeter it was no longer just an ideal that markets could be perfect and that the bargaining power between consumers and producers could be levelled: by allowing free choice, markets actually *were* perfect democracies and consumers *were* the sovereigns – so long as the state kept its hands out of the game.

Advertising and marketing practitioners were eager to use such political–philosophical constructs in order to legitimize their role in economic and social life. As early as 1928, Paul T. Cherington, a marketing pioneer at Harvard Business School, Director of Research at the J. Walter Thompson advertising agency and business associate of the opinion pollster Elmo Roper, endorsed the idea that 'every day is election day. The buyer casts his vote wherever he goes. With his Dollars the buyer votes for his representatives in industry.'[22] Invoking a similar democratic imagery, the advertising agent and later congressman Bruce Barton declared in his agency's house magazine, *The Wedge*, that 'advertising seems to have been accepted by the American people, voted for and elected as the medium through which they choose to learn of new things to eat, to wear, to protect themselves, and to add to their pleasure'.[23]

While this would have been dismissed as industry propaganda before the late 1920s, the later Nobel Prize winner Hayek provided a much-needed intellectual authority to narrow claims like this. In his widely read 1944 classic *The Road to Serfdom*, Hayek stated categorically: 'It is often said that democracy will not tolerate capitalism. If capitalism means a competitive system based on free disposal over private property, it is far more important to realise that only within this system is democracy possible.'[24] After the war, Hayek's long-term colleague and rival, Ludwig von Mises, continued to produce political economic theory that directly attacked all attempts at regulating market activities. By arguing that 'what has been called the democracy of the market manifests itself in the fact that the profit-seeking business is unconditionally subject to the supremacy of the buying public', von Mises inadvertently sang from the same hymn sheet that the advertising industry had introduced during the late 1920s.[25]

Hayek and Mises found ardent readers during the 1930s at the London School of Economics, among them the young British economist William Hutt, who, in an unpublished essay in 1931, introduced the term 'consumer sovereignty', which denoted the alleged rule or 'sovereignty' of consumers over the production of goods.[26] Since consumers had the choice between accepting or rejecting a market offering (product, service, employment and so on) and could always take their spending power elsewhere, the market was a system that reflected people's preferences quite adequately. In this perspective, the autonomy and preferences of individual consumers had a normative status. The idea of consumer sovereignty was meant to provide an answer to the question of who 'ruled' the market place and differentiated market economies. While the Historical School of Economics pointed at governments and various sets of institutions, and Marxists pointed at the bourgeoisie and the owners of the means of production, the 'new' Liberals declared the consumer to be the ultimate sovereign.

As stated earlier, the idea of sovereign consumers, whose choices put them at the steering wheel of the economy, originated with Adam Smith as early as 1776. The notion was reintroduced and confirmed in late nineteenth-century marketing and economics textbooks.[27] In 1911 the American economist Frank Fetter proposed that 'the market is a democracy where every penny gives a right of vote'.[28] After Hayek's admirer, Hutt, had written of market democracies and sovereign consumers in his essays on the structures of competition in free markets, Hayek himself drew on these ideas in 1935, and the German ordoliberal thinker Wilhelm Röpke in the same year spoke about capitalist market economies as a 'democracy of consumers'.[29]

Market research practice

It is no mere coincidence that during the 1930s the political–philosophical idea of the sovereign consumer as 'juror' and 'voter' became, for the first time, directly translated into market research practice. This translation hinged on the introduction of new survey methods that, in a quasi-experimental setting, allowed actual consumers to deliver a verdict on a given product or commercial message. The consumer jury and the consumer panel were such methods, and they made the idea of the powerful, informed, and discerning consumer visible in a social setting that brought consumers, researchers, and marketers together in ways not experienced before.

Consumer juries, carefully selected groups of consumers given the task of selecting their favourite make of product, most appealing advertisement design and so on, had been used by the fast-moving consumer goods company General Foods in the United States since 1926.[30] From 1931, the British chocolate- and sweets-manufacturer Rowntree used its advertising agency J. Walter Thompson (JWT) and the National Institute of Industrial Psychology (NIIP) to test advertisements, packaging appeal, and chocolate flavours with juries of consumers.[31] But it was only with Neil Borden's (Harvard Business School) study of the consumer jury method in ranking advertisements in 1935 that a proper methodological discussion of the research uses of small consumer groups began.[32] Borden and his collaborators had chosen a number of series of advertisements, the relative effectiveness of which had been established from circumstantial evidence such as direct sales records (coupon method) and enquiries to companies' sales forces and distributors. Borden submitted these advertisements to various consumer juries (controlled trial) in order to determine the relationship between the preferences of these juries and the known relative effectiveness of the advertisements in the various series. Borden's study concluded that this method of ranking advertisements according to their perceived effectiveness was generally a valid methodological device for judging differences in advertisements' layout and design.[33]

What from today's standpoint sounds trivial was indeed a revelation in the mid-1930s: it was possible to consult consumers in order to establish the effectiveness of advertisements, packaging or other forms of marketing communication; and consumers, it had been shown, could be trusted when they stated their preferences in market and consumer surveys. Rather than relying on a frustrating process of trial and error when launching new advertising campaigns or going through expensive test campaigns comparing the take-up of a product in a range of localities where different advertisement and poster designs had been in circulation, a company could put a jury of consumers together and optimize a sales message before the campaign launch and the introduction of a product. In other words: the new sovereign consumer had a voice, and this voice could be trusted as it guided the manufacturer and the advertising agency towards product innovations and towards product messages to which the market would actually respond.

Given these advantages, advertising agencies, their clients, and various social research organizations in the United States and Britain developed the consumer jury method further by turning ad hoc juries into standing, continuous panels of consumers – panels of individual consumers and 'household' panels as well. The behaviour of consumer panel members could be studied over long periods of time to appraise people's expenditure in various product categories, their behaviour in switching brands and sales outlets, and also their perception of social issues and of political candidates.[34] This development of continuous consumer panels followed the introduction of the first retail panel (retail audit) by the US market research organization A. C. Nielsen in 1933.

It is important not to underestimate the importance of these techniques of testing and 'polling' the consumer electorate when accounting for the very survival of consumer goods manufacturers in competitive markets after the economic depression of 1930–1. In his study of how Rowntree's adopted a marketing orientation – as opposed to a product or production orientation – the business historian Robert Fitzgerald has summed this up:

> Before 1930, Rowntree had in practice adhered to the principle that the proven quality of products, as determined by the company and without reference to consumers, was the best method of securing sales. There was an underlying belief that there would always be a market for the 'best' goods that could be made. Rowntree had been a production or quality-oriented business, but, as products were marketed merely on the hunches and personal opinions of directors and management, product development failures had been significant. ... The application of systematic market research enabled production-oriented companies to transform themselves into marketing-oriented businesses. It supplied them with the consumer data required to make all their activities cater and respond to markets rather than to the preconceptions of entrepreneurs, managers and production experts. ... Becoming a marketing-led firm was essential to [Rowntree's] survival and subsequent growth, and was symptomatic of twentieth-century industry.[35]

While, in the 1930s, market researchers made significant progress in applying and refining large-scale sampling methods to quantitative market research, the 'close-up' study of individual consumers through focus groups, depth interviews, consumer panel surveys and other means was still largely undiscovered terrain. During the 1940s and early 1950s, marketing organizations refined the use of consumer panels, consumer focus groups, depth interviews, and consumer juries.[36] Active organizations included the advertising agencies J. Walter Thompson (USA) and the London Press Exchange (UK), Ernest Dichter's Institute for Motivational Research in Croton-on-Hudson (USA), and Frank Stanton's research unit at the Columbia Broadcasting Station (USA); social research organizations like Mass Observation (UK), the BBC's Audience Research Department (UK), and Paul Lazarsfeld's Bureau for Applied Social Research (USA); and also government research units under Rensis Likert and Samuel Stouffer in the United States, and Mark Abrams in the United Kingdom.[37]

Some of these market research practitioners were united only by deep rifts of mutual suspicion. Both Lazarsfeld and Abrams, for instance, harboured strong professional misgivings about what they saw as amateurish charlatans at work at Mass Observation and at the Dichter Institute. Yet they also shared the belief that consumers deserved closer and more intensive analysis in a modern market economy. Again, advertising agencies symbolically exploited the existence of the new research methods and turned these innovations into a public propaganda advantage. In May 1959, the JWT agency published a trade advertisement in *Fortune* magazine, *Advertising Age*, and the *Wall Street Journal*, which extolled the usefulness of its 5000-strong panel of housewives. Under the headline 'A remarkably accurate marketing tool', JWT described how it had refined the consumer panel method since 1939 and developed it into a research tool that delivered a 'continuing, day-by-day picture of how the American people are actually buying by region, market size, age, education, income, race'.[38] As over 5000 housewives reported their families' day-to-day purchases in over 30 product categories, marketers learned which brands housewives bought, at what price and amount, in which store, and when, and from what brand they switched. The consumer panel was thus a method that allowed these marketers to poll the opinion of 'King Consumer' on a daily basis, and the advertising agency appeared in a favourable light too. Rather than overselling or manipulating consumers, advertising agencies and their scientifically trained market research personnel enabled the democratic voice of the real sovereign of America – the hard-working housewife, the honest shopper – to be heard as loudly and as clearly as possible.

This JWT advertisement, and in particular its headline celebrating the 'remarkable' accuracy of the new panel method, should make us suspicious. Although, at that time, the JWT agency ran one of the world's largest market research organizations, it nevertheless felt obliged to press home the reliability and accuracy of this relatively new consumer research tool. This alerts us to the fact that managers were still sceptical about investing time and money in tracking consumers' past behaviour in order to predict their future choices. Their scepticism was not unfounded, given the failure of *Reader's Digest* and the Gallup organization – in 1936 and in 1948 respectively – to forecast the American election outcome correctly. Similarly, albeit on a smaller scale, market research blunders occurred throughout the postwar era and gave rise to a considerable academic cottage industry testing and verifying various consumer research tools, including the consumer panel.[39]

As a young socialist in Vienna, Paul Lazarsfeld had been drawn to new modes of social research investigation because he noticed the '*methodological* equivalence of socialist voting and the buying of soap'.[40] This equivalence, of course, was established by the behavioural similarity of a consumer and a voter 'choosing' between different offerings. Yet this phenomenological identity did not necessarily have to be translated into a noumenal identity of citizen/voter and consumer/shopper, and those who did make the translation opened themselves up to a range of methodological pitfalls. Lazarsfeld and other market and opinion researchers made a deeply political – not scientific – decision to equate consumer choices and political choices as equivalent or even identical.[41] At the heart of this engineered transformation of voting (once an expression of political struggle) into mere 'choice' was the consumer/voter panel and the survey questionnaire, which facilitated the 'convergence of consumer choice and political knowing, of voting and soap-buying in the lives of citizens' and sidelined alternative methodologies that could have enquired 'into the origins of this convergence instead of taking it methodologically for granted'.[42]

Despite the epistemological and methodological uncertainties surrounding the validity and reliability of many of the new consumer research and opinion-polling techniques, advocates of a liberal market economy in Britain and the United States continued to exploit the symbolic and merely theoretical similarity between 'consumer' and 'voter' or 'jury member' to legitimize marketing and market research practices. When launching his market research company MarPlan in London in 1959, its director Harry Henry claimed that market research enabled a 'new consumer democracy':

> In these fascinating market developments, I see the research company as a bridge between the consumer and the manufacturer, working for the benefit of each. We can offer the manufacturer control of his market by giving him the facts and understanding to serve it effectively. In doing this we give the housewife and the ordinary people of the country control of the manufacturer too, because we enable them to express their preferences in ways that immediately change the trend of manufacture.[43]

In the same year, the vanguard of the New Right at the Institute of Economic Affairs in London declared that 'in a dynamic society where consumer choice is allowed free play' modern manufacturers must use advertising to talk to their scattered customers and thus should 'employ market research to let the public "talk back"'.[44] The same rhetoric was employed across the Atlantic in the United States, where Corning Glass Works, one of the largest makers of household goods, and the National Association of Manufacturers produced promotional films that hailed market and consumer research as social scientific research tools that turned average housewives into 'research partners' and average citizens into the 'most important politicians' of the country.[45]

The discourse of the choosing and 'voting consumer had by now escaped the compound of methodological debates within the social sciences and had entered the battleground of the cultural Cold War in which the subjects of marketing and consumer research took a prominent part'.[46] Throughout the late 1950s and 1960s, the standard-bearers of the Right celebrated the idea that 'the market offers you one-man, one-vote every day. Politics offers you one-man, one-vote every thousand days. Purchasing power is to own a vote.'[47] By 1969, the conservative British politician Enoch Powell claimed that the 'free enterprise economy is the true counterpart of democracy'. Because free markets allowed everybody to become a choosing consumer, there was a 'great and continuous general election' under way in which 'nobody, not even the poorest, is disenfranchised: we are all voting all the time'.[48]

Ideas, social practice, and political structure

Today, we have come to accept the equation between democracy, voting, and consumer power to such an extent that there is a danger of forgetting the ideological past of the concept. The history of the consumer–voter equation outlined above calls not so much for factual studies of the history of market and consumer research as for a historical account of the symbolic and linguistic representations of certain market research tools and their sociopolitical meanings. Consumer research techniques like the jury test or the panel method need to be stripped of their veil – their pretence of being 'objective' scientific techniques, which were embedded by their proponents in specific visions of markets and democracies. As the science historian Paul Feyerabend reminded us some 30 years ago, there is no science, no experimenting, no measurement technique, no laboratory regime outside the ideological framework that a society, at a given time, sets up for the researcher.[49]

From this vantage point, free market economists and market researchers like Harry Henry and the JWT advertising agency are what Lenin called a 'revolutionary class': a group of people who want to change a part of society, like science, or society as a whole. According to Lenin, this class has to understand and finally govern all sections and all sides of social life without any exception.[50] While this might sound like an unreasonably ideological verdict, it chimes remarkably well with the visions that the Austrian School economists and other free market liberals developed of themselves. In his liberal manifesto *The Good Society*, the American sociologist Walter Lippmann prophesied that market-based liberalism would bring about 'nothing else but a revolution in the world's way of life', and that this revolution was the 'unfinished mission' of a reformed liberalism.[51] For advertising practitioners and market researchers, the success of this mission relied inevitably on a certain interpretation of society: a system constituted of a consuming public and ultimately structured by choice.

The advance of consumer and market research methodologies and the growth in public influence of those who stood to benefit from this advance was based on the illusion that more market research meant more democracy. As Feyerabend has shown, the advancement of science and of scientific methods is based on bluffs, deceptive strategies and opportunistic changes in argumentation and methodology. These strategies are part and parcel of scientific development, and the rise of consumer research as an applied social science is no different in this respect. At a time when social surveys and questionnaires were unknown, mistrusted and had a sinister 'left-wing' feel, market research pioneers like Cherington, Lazarsfeld, Gallup, and Dichter resorted to the idea of consumers casting a vote in the shop in order to legitimize their own scientific work. As for the rise of consumer and market research, the invention of the consumer jury, the survey panel, and the focus group were perhaps this struggling science's most ingenious 'political manoeuvre'.[52]

But, as in real politics, such political manoeuvres and revolutions mainly exist in order to benefit one set of rulers at the expense of another. Since the rhetorical equation of consumers and voters/citizens is the keystone of the political economy surrounding modern market and consumer research, this set of new rulers is easy to identify. Knowing and testing consumers and equating them with voters is 'big business' for those who benefit from this illusion. In Britain, market research services are worth over £1 billion a year and in the US, just under $7 billion is spent on market research annually. There can be no mistake: the idea that commercial advertising and market/consumer research are a necessary part of modern democracy is mainly propagated by those who benefit financially from testing, surveying, panelling, tracking, observing, asking, and measuring consumers. Or, in the words of Sir Robert Worcester, founder of the market research and opinion company MORI, one can 'make a mint while making market research fun'.[53]

It is intriguing to observe how political liberalism, economic thought, and marketing practice in the United States and in Britain largely followed similar trends in constructing a new mythology around markets and consumers and, by doing so, reconstructed the power base on which advertising agencies and the growing market research industry could rely. During most of the interwar period, the debates in the US were ideologically more nuanced than those in Britain. By the 1940s, however, the language of 'freedom of choice' became equally emotionalized and emphatic on both sides of the Atlantic. One of the key factors for understanding the emergence of this language is to recognize the consumer–voter equation as an elite discourse which, all along, needed political elites and intellectuals to lend authority to claims about the ontological identity of markets, marketing, and political democracy. Without the support of key figures within the state and within academia in both Britain and the US, claims about the freedom of the 'dollar vote' would not have been taken seriously outside the realm of marketing. The British and American states had various reasons for pursuing this idea of the citizen–consumer on the basis of the market–democracy equation. As shown by Thomas Stapleford, Meg Jacobs and others, both the left-liberal welfare state and the more conservative 'small' state of economic opportunity became interested in economic and social research methods.[54] These methods, with the new scientific elites in industry and academia behind them, were vital for long-term social planning and for the fulfilment of the increasing managerial functions that statecraft in both Europe and America committed itself to, especially after the First World War.

Through the interaction of these players, a new terminology and a new set of political myths came to be constructed. Using the methodological spectrum of the history of ideas, one can deconstruct the determinants that shaped the formation of this new language as it made its way from marketing practice into economic thought and back. My interpretation of this conceptual journey relies on assumptions about the close interrelationship between language, knowledge, and power, and it is thus closely related to Foucault's idea of knowledge-power. An entirely Foucauldian interpretation, however, would perhaps aim at governmentality as the major outcome of historical processes that bring together scientific expertise, political power, and human subjectivities – in this case those of consumers.

Followers of Foucault in a stricter sense would therefore look at the intellectual discourses around market research much more with a view on surveillance. Jackson Lears, quoted earlier, exemplified this approach when he summarized the situation with these words: 'Through market research, advertisers pioneered the statistical surveillance of private life, a practice that would become central to the maintenance of managerial cultural hegemony.'[55] The approach chosen in my chapter searches for evidence of governmentality much less; instead I have attempted to unpick the way language and symbols were used to reframe consumers and citizens. This method of intellectual critique was once championed by the likes of George Orwell and Raymond Williams. The advantage of this way of thinking is that it allows us to examine the concept of the 'sovereign consumer' and the 'citizen consumer' without necessarily getting embroiled in debates about the veracity and value of statistical research and social science research as such.

Societies live by social, political, and economic concepts, which have sometimes more, sometimes less heavy ideological baggage attached to them. In 1922, the German sociologist Ferdinand Tönnies presented the complexity of 'public opinion' as the total construct of such concepts, and 'the public' as an intellectual community that agreed on their validity, coherence, and consistency.[56] For Tönnies, public opinion therefore had the equivalent social functions in societies that religion had in traditional communities. The ideas that markets are like democracies, that consumers' choice should be 'free' in order to bring about the best results for everyone affected by such choices, and that political democracies have the task of defending and widening people's choices, remain intensely vital quasi-mythologies that are not restricted to providing the glue holding modern society together. These political mythologies also ensure that some concepts and ideas are seen as 'obviously' true while social–political alternatives, such as the principles of mutuality as practised by co-operatives and exemplified by traditional rural communities, are regarded as 'obviously' impractical for our age.

Notes

1. P. Miller and N. Rose (1997), 'Mobilizing the Consumer. Assembling the Subject of Consumption', *Theory, Culture and Society* 14(1), pp. 1–36.
2. See S. Schwarzkopf (2011), 'The Consumer as "Voter", "Judge" and "Jury". Historical Origins and Political Consequences of a Marketing Myth', *Journal of Macromarketing* 31(1), pp. 8–18.
3. For the emergence of market and consumer research as a social science, see P. White (1927), *Scientific Marketing Management. Its Principles and Methods* (New York: Harper and Brothers); F. Wheeler (1937), *The Technique of Marketing Research* (New York: McGraw-Hill), pp. vi–viii, 3–10; L. Brown (1937), *Market Research and Analysis* (New York: Ronald Press), pp. 58–83; L. Lockley (1950), 'Notes on the History of Marketing Research', *Journal of Marketing* 14(5), pp. 733–6. See also S. Strasser (1989), *Satisfaction Guaranteed. The Making of the American Mass Market* (New York: Pantheon), pp. 148–59, and D. Ward (2009), 'Capitalism, Early Market Research, and the Creation of the American Consumer', *Journal of Historical Research in Marketing* 1(2), pp. 200–23.
4. T. J. J. Lears (1994), *Fables of Abundance. A Cultural History of Advertising in America* (New York: Basic Books); K. Donohue (2003), *Freedom from Want. American Liberalism and the Idea of the Consumer* (Baltimore, MD: Johns Hopkins University Press); C. McGovern

(2006), *Sold American. Consumption and Citizenship, 1890–1945* (Chapel Hill: University of North Carolina Press).

5. S. Patten (1889), *The Consumption of Wealth* (Philadelphia: Johnson); idem (1907), *The New Basis of Civilization* (New York: Macmillan); T. Veblen (1899), *The Theory of the Leisure Class. An Economic Study in the Evolution of Institutions* (New York: Macmillan).
6. P. Lazarsfeld (1937), 'The Use of Detailed Interviews in Market Research', *Journal of Marketing* 2(1), pp. 3–8; P. Lazarsfeld and M. Fiske (1938), 'The "Panel" as a New Tool for Measuring Opinion', *Public Opinion Quarterly* 2, pp. 596–612; R. Merton, M. Fiske, and P. Kendall (1956), *The Focussed Interview. A Manual of Problems and Procedures* (Glencoe, IL: Free Press); P. Merton (1987), 'The Focussed Interview and Focus Groups. Continuities and Discontinuities', *Public Opinion Quarterly* 51, pp. 550–66; J. M. Converse (1987), *Survey Research in the United States. Roots and Emergence, 1890–1960* (Berkeley: University of California Press), pp. 133–44, 267–304; H. Kasserjian (1994), 'Scholarly Traditions and European Roots of American Consumer Research', in G. Laurent, G. Lilien, and B. Pras (eds), *Research Traditions in Marketing* (Boston, MA: Kluwer), pp. 265–79; Stefan Schwarzkopf and Rainer Gries (eds) (2010), *Ernest Dichter and Motivation Research. New Perspectives on the Making of Postwar Consumer Culture* (London: Palgrave Macmillan).
7. Lears (1994), *Fables*, pp. 244–5.
8. *The Times*, 18 July 1927, p. 15; *The Times*, 19 July 1927, p. 11.
9. *The Times*, 20 July 1927, p. 11.
10. *The Times*, 28 June 1938, p. 11.
11. *Advertiser's Weekly*, 17 August 1939, p. 180, and History of Advertising Trust Archive (HAT), Advertising Association Collection, AA 13/1/5.
12. HAT, AA 13/1/5.
13. Ibid.
14. *Daily Mail*, 14 June 1937, Brighton Advertising Convention Supplement.
15. 'A Primer of Capitalism (1937)'. J. Walter Thompson Archives, John W. Hartman Center for Sales, Advertising and Marketing History, Duke University, Durham, NC. Italics in original. An excellent interpretation of this material is in S. H. Walker and P. Sklar (1938), 'Business Finds its Voice', *Harper's Magazine*, January, pp. 113–23, and McGovern (2006), *Sold American*, p. 285.
16. 'A Primer of Capitalism (1937)'.
17. D. Thompson (1943), *Voice of Civilisation. An Inquiry into Advertising* (London: Frederick Muller), p. 116.
18. R. Griffith (1983), 'The Selling of America. The Advertising Council and American Politics, 1942–1960', *Business History Review* 57(3), pp. 388–412.
19. HAT AA 13/1/6.
20. J. A. Schumpeter (2008 [1942]), *Capitalism, Socialism and Democracy* (New York: HarperCollins); F. A. von Hayek (2001 [1944]), *The Road to Serfdom* (London: Routledge); K. Hoover (2003), *Economics as Ideology. Keynes, Laski, Hayek, and the Creation of Contemporary Politics* (Lanham, MD: Rowman and Littlefield), pp. 69–80.
21. Schumpeter (2008 [1942]), *Capitalism*, pp. 269–302.
22. P. Cherington (1928), *The Consumer Looks at Advertising* (New York: Harper), p. 58.
23. Quoted and discussed in R. Marchand (1985), *Advertising the American Dream. Making Way for Modernity, 1920–1940* (Berkeley: University of California Press), pp. 30–1, and in McGovern (2006), *Sold American*, pp. 67–77.
24. Hayek (2001 [1944]), *The Road*, p. 73.
25. L. von Mises (1952), *Planning for Freedom* (South Holland, IN: Libertarian Press), p. 112.
26. W. Hutt (1934), 'Economic Method and the Concept of Competition', *The South African Journal of Economics*, 2(1), pp. 3–23; idem (1936), *Economists and the Public. A Study of Competition and Opinion* (London: Jonathan Cape), pp. 257–72; idem (1940), 'The Concept of Consumers' Sovereignty', *The Economic Journal* 197, pp. 66–77.

27. D. Dixon (1992), 'Consumer Sovereignty, Democracy, and the Marketing Concept. A Macromarketing Perspective', *Canadian Journal of Administrative Science* 9(2), pp. 116–25; D. Shaw, T. Newholm, and R. Dickinson (2006), 'Consumption as Voting. An Exploration of Consumer Empowerment', *European Journal of Marketing* 40(9/10), pp. 1049–67.
28. F. A. Fetter (1911), *The Principles of Economics* (New York: Century), p. 394.
29. F. A. von Hayek (1935), *Collectivist Economic Planning* (London: Routledge), p. 214; W. Röpke (1935), 'Fascist Economics', *Economica* 2, pp. 85–100, quote on 93.
30. S. Levy (1999), 'Qualitative Research', in D. Rook (ed.), *Brands, Consumers, Symbols and Research. Sidney J. Levy on Marketing* (Thousand Oaks, CA: Sage), p. 455.
31. R. Fitzgerald (1995), *Rowntree and the Marketing Revolution, 1862–1969* (Cambridge: Cambridge University Press), pp. 305–45.
32. N. Borden and O. Lovekin (1935), *A Test of the Consumer Jury Method of Ranking Advertisements* (Cambridge, MA: Harvard Business School); A. Blankenship (1942), 'Psychological Difficulties in Measuring Consumer Preference', *Journal of Marketing* 6(4), pp. 66–75.
33. L. Lockley (1936), 'Review of *A Test of the Consumer Jury Method of Ranking Advertisements*', *Journal of the American Statistical Association* 194, pp. 449–50; C. H. Sandage (1948), *Advertising Theory and Practice* (Chicago: Business Publications), pp. 458–74; James Cross (1952), 'The Consumer Jury Method of Measuring Advertising Effectiveness – A Case Study', *Journal of Marketing* 16(4), pp. 435–9.
34. F. Cawl (1943), 'The Continuing Panel Technique', *Journal of Marketing* 8(1), pp. 45–50.
35. Fitzgerald, *Rowntree*, pp. 343–4.
36. S. Goldish (1954–55), 'Proceedings of the American Association for Public Opinion Research', *The Public Opinion Quarterly* 18(4), pp. 431–64; L. D. H. Weld et al. (1942), 'Significant Current Trends in Marketing', *Journal of Marketing* 6(4), pp. 7–21; D. J. Robinson (1999), *The Measure of Democracy. Polling, Market Research and Public Life, 1930–1945* (Toronto: University of Toronto Press), pp. 121–5; S. Igo (2007), *The Averaged American. Surveys, Citizens and the Making of a Mass Public* (Cambridge, MA: Harvard University Press), pp. 180–6.
37. Converse (1987), *Survey Research in the United States*; S. Herbst (2003), 'Polling in Politics and Industry', in T. M. Porter and D. Ross (eds), *The Cambridge History of Science*, vol. 7: *The Modern Social Sciences* (Cambridge: Cambridge University Press), pp. 577–90.
38. *Advertising Age*, 18 May 1959.
39. S. Sudman (1962), 'On the Accuracy of Recording of Consumer Panels', Ph.D. thesis, Graduate School of Business, University of Chicago.
40. P. Lazarsfeld (1969), 'An Episode in the History of Social Research. A Memoir', in D. Fleming and B. Bailyn (eds), *The Intellectual Migration. Europe and America, 1930–1960* (Cambridge, MA: Harvard University Press), p. 279. Italics are mine.
41. P. Lazarsfeld, B. Berelson, and H. Gaudet (1948), *The People's Choice. How the Voter Makes Up his Mind in a Presidential Campaign* (New York: Columbia University Press); E. Katz and P. Lazarsfeld (1955), *Personal Influence. The Part Played by People in the Flow of Mass Communications* (Glencoe, IL: Free Press).
42. T. Gitlin (1978), 'Media Sociology. The Dominant Paradigm', *Theory and Society* 6(2), pp. 205–53, quotation on 239. Equally critical is L. Glickman (2006), 'The Consumer and the Citizen in "Personal Influence"', *The Annals of the American Academy of Political and Social Science* 608, pp. 205–12.
43. H. Henry (1971), *Perspectives in Management, Marketing and Research* (London: Crosby Lockwood), p. 143.
44. R. Harris and A. Seldon (1959), *Advertising in a Free Society* (London: Institute of Economic Affairs), pp. 195, 193. On Harris, Seldon, and the role of the IEA, see R. Cockett (1994), *Thinking the Unthinkable. Think-Tanks and the Economic Counter-Revolution, 1931–1983* (London: HarperCollins), pp. 133–8.

45. Corning Glass Works (1960), 'American Women. Partners in Research' (13 minutes); National Association of Manufacturers (1964), 'The Most Important Politician in the World: You!' (18 minutes). Prelinger Archives at the Motion Picture, Broadcasting and Recorded Sound Division, Library of Congress, Washington, DC.
46. S. Schwarzkopf (2005), 'They Do it with Mirrors. Advertising and British Cold War Consumer Politics', *Contemporary British History* 19(2), pp. 133–50.
47. Arthur Seldon, cited in P. Hennessey (1990), 'Market "the measure of democracy"', *The Times*, 24 December.
48. E. Powell (1969), *Freedom and Reality* (London: Batsford), p. 33. A contemporary debate of this claim and an outline of a Galbraithian refutation of the consumer-voter is in P. Donaldson (1973), *Economics of the Real World* (London: Penguin), pp. 148–55.
49. P. Feyerabend (1993 [1975]), *Against Method. Outline of an Anarchistic Theory of Knowledge*, third edition (London: Verso), pp. 9–11.
50. W. I. Lenin (1967), *Left-Wing Communism – An Infantile Disorder*. Vol. 3 of *Selected Works* (Moscow: Progress Publishers), Section 10 (Conclusions).
51. W. Lippmann (1937), *An Inquiry into the Principles of the Good Society* (Boston, MA: Little, Brown), pp. 194–5.
52. Feyerabend (1993), *Against Method*, p. 16.
53. R. Worcester (2007), 'A Way of Life. Making a Mint Building MORI into a Household Name while Making Market Research Fun', *Journal of Public Affairs* 7(4), pp. 383–92.
54. T. Stapleford (2007), 'Market Visions. Expenditure Surveys, Market Research, and Economic Planning in the New Deal', *Journal of American History* 94(2), pp. 418–44, Meg Jacobs (2007), *Pocketbook Politics. Economic Citizenship in Twentieth-Century America* (Princeton: Princeton University Press).
55. Lears (1994), *Fables of Abundance*, p. 138. Similarly A. Arvidsson (2004), 'On the "pre-history of the panoptic sort". Mobility in Market Research', *Surveillance and Society* 1(4), electronic journal: www.surveillance-and-society.org; N. Rose (1990), *Governing the Soul. The Shaping of the Private Self* (London: Routledge).
56. F. Tönnies (1922), *Kritik der öffentlichen Meinung* (Berlin: Springer).

14

Business Organizations, Foundations, and the State as Promoters of Applied Social Sciences in the United States and Switzerland, 1890–1960

Emil Walter-Busch

Social scientists who have tried to apply their expertise in fields of professional practice have left – and are still leaving – a rich legacy behind them of published and unpublished texts. Sometimes they have left other artefacts too, psychotechnical test devices, for example. Despite this, textbooks on the history of the social sciences often ignore what really happened in the many practice-orientated sub-disciplines of the sciences whose history they recount. Surely, most of these textbooks do not practise 'disciplinary Whiggism' any more:[1] in other words, they do not construe the history of a social science as the unfolding of cumulatively more powerful descriptions and explanations, the origin of which may be discovered in the embryonic political science, economics, sociology, anthropology, and psychology of classical philosophers like Plato and Aristotle. Instead, the development of the modern social sciences is increasingly perceived as a 'continuous struggle by multiple participants to occupy and define a sharply contested, but never clearly demarcated, discursive and practical field'.[2] The resulting contents and borders of a discipline are 'the product as much of national cultures, local circumstances, and accidental opportunities as of intellectual logic'.[3]

Several historical overviews show clearly how the practical requirements of the legislative and judicial system, and of the economy and institutions like the schools, hospitals, and prisons of a nation influenced the development of its social scientific disciplines.[4] According to most experts, the prehistory of the modern social sciences ended around 1880 or 1890. Since then, the contemporary system of these disciplines has unfolded. Interestingly enough, this did not originally happen in Germany, the country that until the 1920s was generally acknowledged as having the best research universities in the world and very comprehensive 'sciences of state'.[5] The project of institutionalizing the core disciplines of the modern social sciences was most successfully carried out in the United States. Here, the American Social Science Association, founded in 1865, gave way, at first nationally, then internationally, to highly successful associations representing all the major social scientific disciplines:[6] the American Economic Association was founded in 1885, the American Psychological Association in 1892, the American Anthropological Association in 1902, the American Political

Science Association in 1903, and, finally, the American Sociological Society (later renamed 'Association') in 1905.

After the Second World War, the United States became definitively the country that fostered most of the internationally leading university departments of social scientific disciplines or, as they were sometimes called, the social, economic, and behavioural sciences. In the 1960s and 1970s, economics, the social science discipline that was apparently the most scientific, started a process of globalization (or – seen from the periphery – Americanization) that progressively reduced the initially quite considerable differences between nationally diverse styles of doing economic research.[7] Today, the number of articles published and cited by leading anglophone journals counts as the principal indicator of the academic standing that Economics departments and their individual researchers can achieve. The same now applies even more for other social scientific disciplines, especially psychology and political science.

Respecting the time limits set by this simple concept of the institutionalization of social scientific core disciplines after 1880 and their globalization or 'transnationalization' since the 1960s, the following case study describes the development of three sub-disciplines of the applied social sciences in the United States and in Switzerland between about 1900 and 1960. The three sub-disciplines selected – industrial psychology, industrial relations research, and market and public opinion research – cannot yet be easily located on maps drawn by textbooks treating the history of the social sciences. The index of *The Cambridge History of the Modern Social Sciences*, for example, does not contain any entries for the first two of the sub-disciplines referred to. Furthermore, in its chapter on 'polling', market research is almost entirely passed over.[8] The strongly psychology-centred *Norton History of the Human Sciences*, on the other hand, gives some information about occupational and industrial psychology, but says almost nothing about the other two fields of applied social research I have selected.[9]

The following case study presupposes that it still makes sense to speak of *applied* social sciences. Even 'fundamental' social scientific research is usually the result of struggles by participants 'to occupy a ... discursive *and practical field*' (as cited above).[10] Compared to the *independently* authoritative knowledge of 'pure' or 'fundamental' sciences, however, the findings of applied social sciences depend on the judgement of non-scientific clients concerning their practical usefulness. It is therefore, in the sense of Lindblom and Cohen, only *dependently* authoritative.[11] As the case study will show, anti-theoretical and anti-elitist opinions influenced the development of the social sciences, especially in Switzerland.

Industrial psychology, industrial relations research, and market and political opinion research in the United States and Switzerland, 1900–60

As the advancement of industrial psychology, of industrial relations research and of market and political opinion research in the United States is quite well known already, the following section dealing with this subject is much more superficial than the second one on the early history of the same sub-disciplines in Switzerland.

Development of the three sub-disciplines in the United States, 1900–60

Hugo Münsterberg, together with Walter D. Scott, James McKeen Cattell, and Walter Bingham, was one of the pioneers of industrial psychology in the United States, and in 1913 he wrote a book, *Psychology and Industrial Efficiency*. In it he assumed that many business establishments at the time were increasingly feeling a need 'to appoint professionally trained psychologists'. Such 'psychological engineers' or 'consulting psychologists' might, for example, 'devote themselves to the problems of vocational selection and appointment' or address 'questions of advertisement and display and propaganda'. They might investigate 'the psychological demands for the arrangement of the machines' and 'problems of fatigue, efficiency, and recreation'. Doing this would confront them, almost daily, with new challenges.[12]

Since Münsterberg's times, 'professional consulting psychologists' have indeed occupied the sometimes sharply contested, but in the long run fairly clearly demarcated, discursive, and practical fields just mentioned. Münsterberg anticipated the future of industrial psychology quite accurately. The essential fields of specialization for 'consulting [industrial] psychologists' are still vocational counselling, the selection, compensation, training and motivation of personnel, research on the well-being and efficiency of working men and women, market research, and ergonomics. Münsterberg was also right in assuming that in these fields, new problems would come up incessantly, waiting to be solved with well-known or new guiding ideas and methods.

After the United States had declared war on Germany in 1917, American psychologists served their nation primarily as experts on the mental testing of recruits and the selection of officers, aircraft pilots and other specialists of the military, while industrial psychologists in Britain mainly investigated working conditions and the 'fatigue' of munition workers. In 1919, Walter D. Scott founded a psychological consultancy firm. James McKeen Cattell's 'Psychological Corporation', established in 1921, was especially successful. Contemporary observers of these developments did not suppose, however, that the United States was already leading the international movement for the institutionalization of industrial psychology. For example, the progressive American entrepreneur Henry Dennison said, in one of his talks to British audiences in the early 1920s, that while his country had the best management schools of the world, Britain and Germany had gone further than America 'in using the psychologist in organization management'. According to Dennison, the 'ethical and psychological aspects' of management were nowhere else in the world studied as well as through Britain's National Institute of Industrial Psychology.[13]

This situation changed decisively with the breakthrough that Elton Mayo's Hawthorne approach to Human Relations achieved – at first, from the late 1930s, within the United States; then, after the Second World War, internationally.[14] Countries with militant socialist movements, especially France and Italy, tended to reject 'Harvard's Hawthorne gospel' (as some members of Mayo's team used to call their approach). Other politically less polarized countries such as Sweden, the Netherlands, and Switzerland, accepted the new 'paradigm' of Human Relations more readily. Starting in the late 1940s, the so-called behavioural sciences began to flourish in the United States. American industrial psychology profited from this movement. Ever since, it has kept and even reinforced the internationally leading position it gained in the 1930s. The research of many of the members of the American Society for Industrial and Organizational Psychology – 290 members in 1950 and 3600 in the year 2000 – set standards that could not be ignored elsewhere.

Until the 1960s the United States remained the country that nurtured a particularly strong tradition of research in the field of industrial relations. In 1920, at the University of Wisconsin, the economist and labour investigator John R. Commons started systematic studies on the relations between employers and workers. According to Bruce E. Kaufman, the foremost historian of this branch of socioeconomic studies, the new sub-discipline included two components. While the institutional labour economics school of Commons and other economists contributed the principal component, personnel management experts explored the social and psychological aspects of labour relations. During the first two decades of its history, industrial relations research depended greatly on the financial support of John D. Rockefeller Jr. At first Bruce Kaufman tended to underrate Rockefeller's importance.[15] In later works, he came to the conclusion (correctly, I think): 'The actual birth of the field took place in the United States of America, and its "midwives" were the unlikely duo of industrialist John D. Rockefeller, Jr., and university economist and labour investigator John R. Commons.'[16]

Paradoxically, the International Industrial Research Association, founded in 1966, institutionalized industrial relations research internationally only after the American 'golden age', which lasted from about 1945 to 1960, had ended.[17] Several countries, among them Switzerland, never practised this branch of social scientific studies. As in the case of industrial psychology after about 1930 and of industrial relations research from its beginnings, the United States played a leading role in the history of market and political opinion research. Two narratives exist, recounting its origins differently. According to one of them, the straw votes already organized by American journalists in the nineteenth century were the precursors, and a soil was formed on which political opinion research began to flourish in the 1930s. The other account emphasizes that market research techniques invented in the 1920s were a more important determinant for the breakthrough achieved by the survey research institutes of George Gallup, Elmo Roper, and Archibald Crossley in 1936 when they successfully predicted the re-election of President Franklin D. Roosevelt.[18]

In the long run, the academic success of three leading American research centres for applied social research was at least as important as the publicly acknowledged breakthrough of these commercial survey research institutes. Paul F. Lazarsfeld's Bureau of Applied Social Research (founded in 1938/44), Harry Field's National Opinion Research Center (1941), and Rensis Likert's Survey Research Center

(1946) considerably improved traditional methods of attitude measurement and the multivariate analysis of social data.[19] Their technologies for measuring 'opinions' as 'social attitudes' had, according to Kurt Danziger, an effect similar to the intelligence tests applied during and after the First World War. Like 'intelligence' defined by these tests, attitudes were to become 'whatever it was that attitude scales measured'. The attitude concept turned into a seemingly solid vehicle 'for psychological claims to have something to say about social issues' within the framework established by the new attitude measurement methods.[20]

Development of the three sub-disciplines in Switzerland, 1910–60

Both the first and the ninth congress of the International Association of Psychotechnics (called the International Association of Applied Psychology, IAAP, after 1955) took place in Switzerland. Edouard Claparède, the co-founder of the Institut Jean-Jacques Rousseau and its Cabinet d'Orientation professionelle at the University of Geneva, organized and chaired the first congress there in 1920. In 1949, the ninth congress convened again in Switzerland, now in its capital, Bern. It was formally organized by the French psychologist Henri Piéron, whom Claparède had befriended. It was Franziska Baumgarten, then one of the leaders in the field of industrial and occupational psychology in Switzerland, who really initiated and organized the conference, the first one held by the IAAP after the Second World War.

Apparently, the international movement for applying results and techniques of experimental psychology to problems of counselling, selecting, training, compensating, and motivating working men and women did not bypass Switzerland. Swiss entrepreneurs and practice-orientated academics recognized Frederick Taylor's Scientific Management and the psycho-technical movement relatively early, seeing them as possibly interesting challenges. In 1913, the influential entrepreneur Iwan Bally and three of his employees visited Frederick Taylor and Hugo Münsterberg in the United States. As a consequence of this tour, Jules Suter, an assistant in the Psychology department at the University of Zurich, was put in charge of experiments with workers in Bally's shoe factories from about 1913 to 1917.[21] From 1921 to 1923, Suter directed the Psychotechnical Bureau of the canton of Zurich. In 1923, he transformed this institution into a privately run Psychotechnical Institute. Its objective was to satisfy the growing demands both of the state and of private firms for psycho-technical services, mainly in the fields of vocational counselling and the selection of employees. This goal was met with great success after Alfred Carrard, an engineer converted to the ideals and techniques of psycho-technics, joined Suter's Institute in 1925. Even more than Suter, Carrard had difficulties in getting recognition from Germany's scientifically minded professors of Applied Psychology – men like William Stern and Otto Lipmann. Both Suter and Carrard were promoted rather belatedly to associate professorships at the University of Zurich and the Federal Institute of Technology in Zurich (in 1938 and 1945 respectively). Carrard nevertheless developed some successful programmes, at first teaching training methods for workers, then also leadership theories and practices.[22]

Another centre for the advancement and application of psycho-technical knowledge in Switzerland was Claparède's Institut Jean-Jacques Rousseau in Geneva. Claparède founded this institute together with Pierre Bovet in 1912. With Claparède's successor, Jean Piaget, as its director, it would become well known. In 1916, a Cabinet d'Orientation professionelle, to be directed by Léon Walther, was added to it. Walther and some other experts in industrial psychology, called '*technopsychologie*' in Geneva, applied their knowledge to several firms in the French-speaking part of Switzerland. In doing this, they were at first as successful as Suter's team in Zurich.[23] On Carrard's initiative, however, a Schweizerische Stiftung für Psychotechnik was founded in 1927. Extending its activities from the eastern regions of the country (St Gallen) to the west (Lausanne and Geneva), this institute, dominated by Carrard's Zurich group, succeeded in displacing the experts of Walther's team in Geneva. Carrard could even export his method of training workers, called the 'méthode Carrard', to France.[24]

Franziska Baumgarten, in many respects the most interesting personality among the pioneers of industrial psychology in Switzerland, promoted a third approach, which had its centre in Bern.[25] Baumgarten published a comprehensive survey on psycho-technical selection tests in 1928. She then carefully prepared her career as a *Privatdozent*, professor, and prospective director of an institute of Applied Psychology at the University of Bern.[26] The setback that the deep crisis of the 1930s inflicted on the rationalization movement, however, thwarted most of her plans. Baumgarten reacted to this by publishing countless articles and popular books on psychological problems, among them a pioneering treatise on the psychology of leadership. Some of these publications, although mostly forgotten today, were translated into several languages. A series of articles originally published in Swiss newspapers on *Democracy and Character* (1944) is still remarkable. In these articles, Baumgarten drafted a model of moral development from the level (I) of egocentrism to the highest level (III), which, according to her, was an indispensable precondition for democracy.[27] At roughly the same time, the authors of the *Authoritarian Personality* and Jean Piaget had similar ideas. As Baumgarten expressed them simply in commonsense terms, however, only the other social scientists, who gave them an academically acceptable form, achieved an assured place in the collective memory of psychology.

Baumgarten's career in Switzerland also suffered from her continual conflict with Carrard's practice-orientated school of '*Betriebs-*' or '*Industriepsychologie*', as it came to be called from the early 1950s. At the conference of the IAAP in Bern, Baumgarten was quite blunt in describing the tensions between the two schools of 'empirical' and 'scientific' psycho-technicians in Switzerland. According to her, only the latter group applied a sound kind of 'scientific psychology'. The numerically dominant group of 'empiricists' had no background in psychology, and was therefore incapable of doing scientific research.[28] Baumgarten's problem was that her own research projects, conducted at the University of Bern since the 1930s, did not fulfil what some of her earlier works published in the 1920s had promised. All in all, Switzerland's psycho-technicians solved practical problems fairly successfully, not only locally but, in the case of the 'méthode Carrard', internationally as well. Generally, however, Swiss theoretical or empirical contributions to industrial psychology were not impressive.[29]

After the Second World War, Swiss businessmen, as well as leaders of the trade unions, reacted quite positively to the human relations ideas of Elton Mayo's industrial research team at the Harvard Business School. Christian Gasser, temporarily a professor at the St Gall School of Business Administration and Economics ('*Handelshochschule*'), was an enthusiastic defender of this new movement in industrial psychology. Shortly before and after 1950, Gasser spread 'Harvard's Hawthorne gospel' not only in Switzerland, but also in Germany – and quite effectively.[30] The very success of the human relations movement in Switzerland, however, reduced the chances that the sub-discipline of industrial relations research, which in the United States had just entered its golden age, would also find Swiss supporters. In 1950, an article on American 'Institutes of Industrial Relations' and the 'new ways of American economic research' practised by them, was published in one of Switzerland's two leading management periodicals.[31] Yet there never emerged a Swiss initiative for institutionalizing this field of social scientific studies. While the institutional labour economics and the personnel management branches of industrial relations research were acknowledged as possibly interesting areas of expertise, no need to combine them under the umbrella of an autonomous sub-discipline was ever felt.

Arthur Lisowsky, who taught at the St Gall School of Business Administration and Economics from 1932 to 1952, was Switzerland's first full professor of Business Administration to focus his teaching and research on problems of marketing and market research. As a convinced critic of quantitative methods, Lisowsky preferred to advise publicity experts on the artistic quality and the unconscious effects of their advertisements. He thought that the essential truths of the art of marketing had to be intuited ('*ganzheitlich erfühlt*'). To him, market research was '*Gestaltforschung*'.[32] He therefore could not continue what Paul F. Lazarfeld's Vienna-based Wirtschaftspsychologische Forschungsstelle had initiated in Switzerland – the execution of at least three pioneering market research studies for the department store Globus and for the shoe stores Arola and Bally in Zurich (probably 1932–3). These promising beginnings of a methodologically sound tradition of quantitative market research were interrupted after the forced emigration of Lazarsfeld and other members of his Viennese research institute to Britain and the United States in the 1930s.[33]

The breakthrough of modern market and political opinion research in the United States, though, did not remain unnoticed in Switzerland. In 1942, Peter Kaufmann, a pioneer of Swiss market research, published a brochure on Gallup's Institute. In this booklet, he described quite accurately how the research institutes of George Gallup, Elmo Roper, and Archibald Crossley had successfully predicted President Roosevelt's re-election in 1936 and in 1940.[34] One year earlier, in May 1941, Kaufmann and seven other interested practitioners had founded the Gesellschaft für Marktforschung (GfM). The market research knowledge developed by this association and its offspring, the GfM-Befragungsdienst and the Schweizerisches Institut zur Erforschung der öffentlichen Meinung, however, lagged far behind the

American models they tried to imitate, until about the 1960s. The first big survey using an allegedly representative sample of Swiss citizens – 2160 living in the German-speaking part of Switzerland, 600 living in its French-speaking part and 180 in its Italian-speaking part – had serious methodological deficiencies.[35] This survey was not well suited to disperse the doubts that, until about the late 1960s, many Swiss citizens, and even leading politicians, commonly held about public opinion surveys and their dubious claims to scientific objectivity.[36]

Contextual factors for the different development of the three sub-disciplines

When we assess similarities and differences between the advances of the three sub-disciplines, two questions may be asked. First, what were the primary push factors explaining the strong development of the three fields of applied social science in the United States? And, secondly, what specific circumstances could explain the neglect of industrial relations research in Switzerland, on the one hand, and, on the other, the deficiencies of industrial psychology deplored by Franziska Baumgarten and the difficult beginnings there of market, and especially of public opinion, research?

Apart from the obvious factor, country size, the following clusters of contextual variables might contribute to a better understanding of the different paths taken by these social scientific sub-disciplines in the two countries: certain features of managerial capitalism, as conceptualized by Alfred Chandler; the role of private foundations and the state as promoters of social scientific research; and, finally, a political culture of democratic federalism and populist anti-elitism. Because of space limitations, only a small fraction of the rich historical data supporting the suggested conjectures can be described here. Hopefully, however, the few empirical illustrations given are ones that cannot be easily contradicted by counter-examples.

The power of competitive managerial capitalism and private foundations in the United States

According to Alfred Chandler, by the second half of the nineteenth century, the United States was already the country *par excellence* of managerial capitalism. A revolution in transportation and communication, led by its great railroad companies of the 1850s and 1860s, made possible mass production and mass distribution of a magnitude that soon surpassed that achieved in Britain, the home country of the industrial revolution. The 'visible hand of management' reigned very effectively over the managerial hierarchies of the United States' rapidly expanding big enterprises – its Standard Oil trust, US Steel, AT&T, General Electric, Sears & Roebuck, General Motors, the Ford Company and so on. The United States became the 'seed-bed of managerial capitalism' not least because the necessary investments in managerial capabilities took place there in good time. Britain, relying instead on more traditional ways of training its business elite, missed the chance of founding exemplary institutions like the Wharton School of

Finance or Harvard's Graduate School of Business Administration. According to Chandler, therefore, the management style prevailing in Britain in the first half of the twentieth century was not properly 'managerial', but just a style of 'personal capitalism'.[37]

Countries like Germany or Switzerland, too, invested enough of their resources into the production facilities, the distribution networks, and the managerial capabilities of their business enterprise system. Since about 1900, therefore, the biggest German or Swiss firms, as for example Krupp, BASF, Bayer, CIBA, Nestlé, Siemens, AEG, BBC, Sulzer or Bally, were also practising a relatively progressive form of managerial capitalism. In contrast to the competitive style of managerial capitalism in the United States, however, that of Germany and Switzerland was cooperative. In 1890, overwhelming majorities in the US House of Congress and Senate confirmed 'fundamental American values' by voting for the Sherman Antitrust Act.[38] This act prohibited formal cartels, any other agreements to exclude competition, and all attempts to monopolize commerce in the United States. The fundamental values of Germany and of Switzerland, however, were then quite the opposite of those that had led to the Sherman Antitrust Act. In the same year, 1897, 'when the United States Supreme Court upheld the constitutionality of the Sherman Act', the German *Reichsgericht* decided that 'contractual agreements as to price, output, and allocated markets were enforceable in courts of law'.[39]

As a consequence of this striking difference between the economic regimes of the United States on the one hand, and those of Germany and Switzerland on the other, large American firms used to compete 'aggressively for market share and profits', while major firms in Germany or in Switzerland 'preferred to cooperate', as their courts even endorsed cartels and other inter-firm agreements that tried to prevent 'destructive competition'.[40] As late as 1958, a popular initiative aimed at making cartels illegal in Switzerland was rejected by a national vote of its male citizens, following the unanimous recommendation of the federal government, employers' organizations, and trade unions.

According to Chandler, the American way of competitive managerial capitalism was particularly dynamic. The big private foundations of Russell Sage, Andrew Carnegie, John D. Rockefeller senior, Henry Ford and other American entrepreneurs were in a certain sense both a product and an important component of this dynamism. Progressively managed firms and private foundations – most of them also managed very competently – opened up great opportunities for practice-orientated social scientists like Walter Bingham, George Gallup, Paul Lazarsfeld, Kurt Lewin, Rensis Likert, Elton Mayo, Hugo Münsterberg, and Walter D. Scott. It is a well-established fact that without the support of rich foundations, the American social sciences would not have surpassed those in Europe as swiftly as they did in the 1930s and 1940s. The professionals running the foundation units in charge of social scientific research sometimes seemed to know better than anyone else what was going on in the social sciences, nationally as well as internationally. The knowledge base that the heads of the Rockefeller foundations had at their disposal by funding – most generously – the Social Science Research Council was particularly fabulous.[41]

In any case, it is extremely unlikely that an academic outsider like Elton Mayo would have found in any other country a job opportunity similar to the one offered him in 1925–6 by the Dean of Harvard's Graduate School of Business Administration, Wallace Donham. Mayo's research professorship at the Harvard Business School was financed with a generous grant by the Laura Spellman Rockefeller Memorial. Initially free of any obligation to teach, this job allowed him to do 'industrial research' during the next five years with the sole aim of creating that language, soon to be called the language of Harvard's human relations approach, that Dean Donham, while lecturing some years before on problems of business policy, had still sadly lacked.[42] The professional experiences of Kurt Lewin, the German immigrant, and the Austrian Paul Lazarsfeld were quite similar to those of the Australian Mayo who, after having entered the United States in 1922, was soon penniless. Without the salary that John D. Rockefeller Jr at first paid personally, Mayo's American adventure would most probably have soon ended.

In the first half of the twentieth century, then, the United States' dynamic version of managerial capitalism and one of its consequences, American foundations with their immense resources, made it possible for social scientists to establish themselves quite comfortably as academically accredited experts in the fields of industrial psychology, of market and public opinion research, and – so long at least as this speciality was not swallowed up by the longer-living sub-disciplines of labour economics and industrial and organizational psychology – industrial relations research.

Switzerland's social sciences under the influence of cooperative managerial capitalism, democratic federalism, and populist anti-elitism

Originally, neither the federal state of the USA nor that of Switzerland supported social scientific research systematically. This began to change in the United States in the late 1930s, and in Switzerland about two decades later. Principles of democratic federalism characterize the political culture and structure of both countries. Each of the current 50 American states and 26 Swiss cantons has its distinct identity, defended eagerly against the prerogatives of the central state. Both countries think highly of the principle that nothing should be delegated to upper political levels that could as well or better be accomplished by self-organizing citizens themselves. Citizens have been granted extensive political rights, especially in Switzerland. In the 1860s, a democratic movement established the right of popular initiative and, in many cantons, of the referendum for legislation. The constitution of 1874 granted to all citizens the right of referendum against laws voted by the bicameral federal assembly. In 1891 the right of initiative for revisions of the federal constitution was added. Swiss citizens were then empowered to decide on a constitutional revision that was proposed by at least 50,000 of them. Concerned observers suspected that this right might be misused by demagogues. Americans interested in international politics discovered in the 1890s that the tiny Swiss 'sister republic' far away in old Europe had implemented some of their own ideals of democratic republicanism quite radically.[43] As the votes on first initiatives for amendments to the Swiss constitution proved to be more traditionalist than revolutionary, the fears expressed by sceptical observers vanished. Among advocates of democratic federalism, the Swiss counted then, perhaps rather prematurely, as ready for direct democracy.

How did these political circumstances and the regime of cooperative managerial capitalism impact on the course of the social sciences in Switzerland? Reviewing relevant results from historical studies, the following answer to this question might make sense: while in the United States competitive managerial capitalism and its important side-effect, the existence of private foundations, forcefully promoted the social sciences, the decision-makers who followed the rules of Switzerland's political system and cooperative managerial capitalism supported only social scientific (sub-) disciplines that appeared academically proven already or that seemed for certain practically useful. Anti-elitist and anti-intellectual attitudes were popular even among leading Swiss businessmen and politicians. In the first half of the twentieth century, therefore, social scientific disciplines they considered to be too theoretical, academic or even useless in practice had slim chances of being established at a Swiss university.

These assumptions can be further elaborated, first by pointing at two typically Swiss equivalents to exploring public opinion and the relations between employers and employees scientifically, namely popular votes (point 1), and delegating the governmental task of observing economic and social developments to high-level committees of practically experienced experts (point 2). Finally, a few examples will be discussed of anti-intellectual commonplaces that circulated among Swiss decision-makers and that tended to hamper the seemingly useless social sciences (point 3).

1 *Direct democracy and social scientific polling techniques*

According to two comparative studies, 382, or more than half of all nationwide referenda ever arranged by a state up to 1990 took place in Switzerland. If these worldwide comparisons had also included referenda at the communal or, in Switzerland, at the cantonal level, the results would probably have been even more skewed.[44]

In her richly documented history of how the American public at large responded to the findings of some new types of applied social research, Sarah Igo has argued that political elections and referenda are both robust 'litmus tests' for public opinion research and excellent opportunities for validating common suppositions about political movements and trends.[45] As Switzerland's direct democracy every year offers dozens of such opportunities, Swiss public opinion researchers found it particularly difficult to convince politicians and citizens of the usefulness of their job. Taking into consideration the numerous referenda carried out each year at the level of Swiss communities, cantons, and the federal state, what could results of public opinion research actually teach observers accustomed to compare their guesses about political developments regularly with the results of referenda? In Switzerland, Elmo Roper's bold statement in 1944 that polling techniques were the 'greatest contribution to democracy since the introduction of the secret ballot' would have sounded rather strange.[46]

In fact, Swiss public opinion researchers initially supplemented customary ways of observing the political situation less by predicting the results of referenda or elections than by trying to assess periodically people's moods and Switzerland's 'psychological climate'. Surveys trying to predict or analyse after the fact the results of national referenda or elections came into being only in the late 1970s.

2 Extra-parliamentary committees entrusted with the task of observing economic and social developments

As long as the cantons kept Switzerland's federal administration comparatively slim, as they did until about the 1960s, the country's network of extra-parliamentary committees was a crucial component of the legislative, as well as the executive work of the central government. Most members of these committees were recruited from cantonal or communal administrative offices, from professional or other interest associations, from private firms, and from universities. The federal administration could thus profit from the experience of the leaders of many significant organizations. From the 1870s, it gave the principal interest groups financial assistance when they fulfilled useful tasks. Even the office of the workers' movement (the 'Arbeitersekretariat') received such assistance after it had satisfied the condition that socialist and Christian unions should collect data needed for federal projects of social policy cooperatively. In Switzerland's 'consociational' or 'consensus democracy' – a more or less unintended consequence of its federalism and direct democracy – extra-parliamentary committees fulfilled indispensable functions. They were, and still are to an extent, 'the place where the main compromises of Swiss consociational politics are forged'.[47]

In 1930, 46 extra-parliamentary committees worked permanently or temporarily for the federal administration.[48] The 'stabilization policy committee' (Kommission für Konjunkturbeobachtung, later called Kommission für Konjunkturfragen) was founded in 1932. Entrusted with the task of publishing quarterly reports on recent economic trends and analysing problems of Switzerland's stabilization policy, it was one of the most important commissions of the federal government in the 1930s and during the Second World War.

The director of the federal department of commerce, Lorenz Stucki, opened the committee's first session by saying that if its outstanding members felt 'the right spirit of cooperation', they would be able to do the work of a whole research institute. Some of the committee's members who had been assigned the laborious task of collecting sensitive data, however, felt that in Switzerland, generally, the scientific knowledge of economists was neither appreciated nor applied sufficiently well. Other countries, the Netherlands for example, were apparently far more generous in this respect. Jacob Lorenz, the founder of the committee and its first chairman, responded to this critique by defending once more the Swiss 'principle of militia' (*Milizprinzip*). Not every country needed to support an economic research institute. According to Lorenz, 'Research', written with a big R, tended to become bureaucratic and sterile, while 'our work is connected to life'.

Moreover, the committee's reports were at least as good as those of other countries that were equipped with economic research institutes.[49]

Eugen Böhler, a professor of Economics at the Federal Institute of Technology in Zurich, was a regular member of the committee from 1937 and, up to 1965, its fourth chairman. In 1937–8, a group of influential Swiss businessmen accepted his project of establishing a private association (the Erfagruppe für Konjunkturbeobachtung) and, connected to this 'Erfagruppe', a research institute, the Forschungsstelle für Konjunkturbeobachtung. Böhler's 'Forschungsstelle' was assigned the task of supplementing the studies of the stabilization policy committee by observing economic trends in foreign countries. The studies of the two committees and other ideas were regularly discussed at roughly a hundred meetings arranged by the 'Erfagruppe' in the first five years of its existence.

Reading the records kept by the stabilization policy committee and the 'Erfagruppe' in the 1930s and 1940s conveys fascinating insights into the thinking and some of the activities of Switzerland's federal administration and parts of its business elite in a politically most difficult period. The range of themes discussed by the members of the 'Erfagruppe' was understandably wider than that of the federal commission. It also included debates on the living conditions of Swiss workers and employees, their moods and protest potential, and how firms could satisfy, improve, and prepare their personnel for the highly uncertain time after the war.[50] Members were generally satisfied with the quality of the knowledge exchanged at meetings of their 'Erfagruppe'. They therefore felt no need for other types of scientization than that offered by Böhler's practice-orientated research institute.

3 *Swiss varieties of anti-elitism*

Both in the United States and in Switzerland, citizens by tradition feel free to articulate their anti-intellectual and other populist opinions openly.[51] Taking into account the small size of the country, anti-elitism and anti-intellectualism seem to have influenced the development of Switzerland's cantonal universities until about the 1960s more powerfully than in the United States. In the United States, leading universities usually operated on levels that depended only partially on local or regional interest groups. Switzerland's federalist and anti-elitist culture, on the other hand, prevented the country from concentrating substantial resources on centres of excellence, such as, for example, a national university.[52] Typically, the idea of establishing a centre for advanced studies, where highly promising or famous scholars would be allowed to work out their ideas freely, was never seriously considered in Switzerland.

In the first half of the twentieth century, Switzerland's practice-orientated social scientists were quite proud of their practical achievements. Hans Biäsch, a member of Suter's and Carrard's Psychotechnical Institute, described its chief principle as the 'indirect method' of 'implicit psychology'. Working with clients, psychologists should, according to Biäsch, 'never speak of psychology, but always think psychologically'.[53] Biäsch knew, on the other hand, the dilemma of relevance versus rigour that is a crucial problem in every applied social science. He recognized the theoretical and empirical shortcomings of his 'implicit psychology'. Even Alfred Carrard complained that working exclusively 'out of practice for the sake of practice' had the disadvantage of not leaving enough time for carrying out academically sound research projects.[54]

From the 1890s until the 1950s, most Swiss social scientists described themselves, or were described as being, strongly practice-orientated, non-theoretical academics. In the two decades just before and after 1900, almost every professor of Economics teaching at a Swiss university was a German, Austrian, Frenchman or Italian. Two German professors of Economics who had taught for many years in Switzerland, August Oncken and Heinrich Herkner, agreed on the fact that the Swiss usually approached the economic, social, and political problems of their country competently. The knowledge accumulated by practice-orientated academics like Emil Richard, the secretary of the Kaufmännische Gesellschaft Zürich, or Naum Reichesberg, the professor of Political Economy, Statistics, and Social Policy at the University of Bern from 1898 to 1928, is in fact highly impressive even today.[55] While Switzerland, according to Oncken and Herkner, had by tradition 'excellent practitioners of economic and social policy', almost every professorship of theoretical Economics was filled by a foreign academic.[56] Eugen Böhler belonged to the first generation of Swiss economists to show some interest in problems of economic theory. But as late as 1940, he said at a meeting of his 'Erfagruppe' that most Swiss economists and administrators were not yet capable of thinking in terms of theoretical models.[57]

Until the 1950s, not many attempts were made to establish those branches of the traditional 'sciences of state' that focused on the political or the societal sector of modern societies as autonomous disciplines. Such attempts were sometimes fiercely, even nastily, resisted. According to the memoirs of a member of the board of politicians supervising the University of Zurich, for example, he and some of his colleagues 'liked to laugh at the intrigues of academic would-be celebrities and their over-ambitious projects that looked ... like an academic mirage'.[58]

The author of this memoir did not reveal the identity of the pseudo-scientific disciplines at which he used to laugh. It is obvious, however, that one of them must have been sociology – at that time (around 1950) a particularly suspect discipline in the eyes of many citizens of the German-speaking part of Switzerland.[59] In 1952–3, a pioneering community study of a suburban quarter of Zurich produced a terrible wave of public resentment against allegedly arrogant sociologists ready to invade the private lives of honourable citizens.[60] In this particular case at least, Zurich's response to one of the first Swiss community studies was the exact opposite of how the inhabitants of Muncie rewarded Helen and Robert Lynd's famous Middletown study of their prototypical community: in 1941 the principal street of one of Muncie's urban extension projects was to be called 'Lynd Avenue'.[61] Imagining that Switzerland's town planners or politicians would ever propose giving a new street the name of one of its social scientists still seems bizarre. In the United States, though, the golden age of the social sciences is also gone. Even there, it is now unlikely that a proposal like that made in Muncie in the early 1940s will be someday repeated.

Table 14.1 Structural and cultural preconditions of the growth of applied social sciences in the United States and Switzerland, 1900–60

Economic and political conditions:	**USA, 1900–60**				**Switzerland, 1900–60**			
	Condition relevant		**Effect on growth of applied social sciences**		**Condition relevant**		**Effect on growth of applied social sciences**	
	Yes	**No**	**Positive**	**Negative**	**Yes**	**No**	**Positive**	**Negative**
Competitive managerial capitalism	X		X			X	--	--
Private foundations	X		X			X	--	--
Cooperative managerial capitalism		X	--	--	X		(X)[a]	
Educational and research policy of the state at the federal level	(X)[b]		(X)[b]		(X)[c]		(X)[c]	
Educational and research policy of the state at the level of states/cantons	(X)		(X)		(X)		(X)[a]	
Anti-elitist and populist tendencies of the political culture	X		---	---	X			X

Notes: Condition/effect: X clearly existing/strong or clearly positive, negative; (X) partially relevant/not so strong or weak; -- irrelevant; --- no effect. [a] Favouring only practically useful or academically proved (sub-) disciplines. [b] Relevant/effective only since the late 1930s. [c] For the social sciences relevant/effective only since the 1950s.

Conclusion

The following four propositions are offered as summarizing the main results of the preceding case study (cf. Table 14.1).

1. Two forces – competitive managerial capitalism and, as one of its side-effects in the United States, rich private foundations – explain best why since the 1920s American social sciences expanded more dynamically than the same disciplines in Europe. Without the help of these two growth factors, Europe's social scientists were seriously disadvantaged in comparison to American colleagues.

2. Under the regime of the cooperative managerial capitalism that reigned in Switzerland in the first half of the twentieth century, the political and business elites of Swiss cantons preferred to promote practically useful or academically already proven (sub-) disciplines of the traditional sciences of state and the novel social sciences. At the federal level, the Swiss state embarked on supporting social scientific research only in the 1950s.
3. Switzerland's political structures and culture, since the nineteenth century, had much more in common with those of the United States than with the political regimes of its nearest neighbours – in the first instance, Germany. As both cause and consequence of its direct democracy, anti-elitism and populist anti-intellectualism have been, and still are, politically powerful forces in Switzerland. During the whole period under scrutiny, these forces tended to prevent the establishment of social scientific (sub-) disciplines that from an anti-elitist standpoint were academically or practically useless. In the United States, on the other hand, quite similar attitudes and movements influenced the expansive course of the social sciences not at all or at best only insignificantly.
4. In the first half of the twentieth century, Switzerland's political elite and its leading businessmen had apparently good reasons for being generally satisfied with how the dense Swiss network of private interest associations and public institutions kept them informed. They therefore confined themselves to putting academics in charge of developing and exploiting locally available, practice-orientated knowledge resources. The task of promoting fundamental research in the new social sciences was left to bigger, in this respect less frugal, countries.

Critical observers of the development of the social sciences in Switzerland have repeatedly criticized these attitudes of Swiss decision-makers. In 2007, a team of researchers led by Claudia Honegger and Hans-Ulrich Jost deplored once again the chronically underdeveloped and politically dependent status of Switzerland's social sciences.[62] In the United States, of course, it was also, and still is, possible to look critically at social scientists as being 'servants of power'.[63] Occasional laments about their provincial status there, however, soon gave way to generally quite optimistic attitudes about the great potential of social scientific research.

Since the 1970s, a growing number of Switzerland's social scientists, especially those of its university departments of Economics, Business Administration, Psychology, and Political Science, have been following the American model of doing research.[64] It seems that the system of the social sciences developed by the United States embodies the *telos* towards which all other national varieties of social scientific activities tend to converge. There is no longer much room left today for the two kinds of American and Swiss exceptionalism described in the present case study. If we accept this assumption, a crucial question arises: can scientifically minded experts of a social scientific domain still learn something from the early history of their discipline?

In Lutz Raphael's broad overview of the scientization of the social since the nineteenth century, the fourth and, for the time being, the last phase of this process is described as the stage of 'secondary scientization' in Ulrich Beck's sense. At this stage, social scientists must increasingly deal with positive and negative consequences of their own activities.[65] In a world where almost every field of human practice is influenced by the language, and sometimes also the techniques, of a social science, these disciplines have (so to speak) lost the innocence of their youth. Fritz Giese, a pioneer of industrial psychology in Germany, promised in 1925 that among one hundred personnel selection decisions made by psychological experts, only about ten decisions would be erroneous.[66] This prediction, of course, proved to be too optimistic. Generally, however, the establishment and further development of social scientific disciplines seems more or less independent of whether such promises can be kept. Most social scientific innovators like to exaggerate the theoretical or practical benefits of their discoveries. The rhetoric of technical breakthroughs or theoretical paradigm changes is in a certain sense an indispensible vehicle of social scientific discourse. It is often very difficult, if not impossible, to separate unequivocally the rhetorical façade from the positively or negatively effective consequences of a social scientific innovation. The forces at work for and against accepting, then institutionalizing new ways of describing, explaining or changing human behaviour tend to be almost hopelessly over-complex. In this case study, only a tiny fraction of them could be brought to the light. Much more needs to be done. In the case of the sometimes exotic history of Switzerland's social sciences, for example, we need more research to generate richer descriptions of their prehistory and the fields of practice connected with them.

Describing how concerned actors typically react to the results of the investments they have made in social scientific activities is of itself a demanding task. Even more challenging would be the job of reviewing critically the value added by social scientific knowledge to what the enlightened practitioner and layperson endowed with experience and common sense 'always already knows'.[67] For both types of studies, a useful guide could be the following version of Odo Marquard's ironical reversal of the famous thesis of Karl Marx: social scientists have until now only tried to *change* the world; what matters however is *to understand it*.[68]

Notes

1. T. M. Porter (2003), 'Genres and Objects of Social Inquiry, from the Enlightenment to 1890', in T. M. Porter and D. Ross (eds) (2003), *The Cambridge History of Science*, vol. 7: *The Modern Social Sciences* (Cambridge: Cambridge University Press), pp. 13–39, at 13.
2. M. Ash (2003), 'Psychology', in Porter and Ross, *The Modern Social Sciences*, pp. 251–74, at 252; D. Ross (2003), 'Changing Contours of the Social Science Disciplines', in ibid., pp. 205–36 at 206.
3. Ross, 'Changing Contours', p. 206.
4. H. Cravens (1985), 'History of the Social Sciences', *Osiris* II 1, pp. 183–207; Porter and Ross, *The Modern Social Sciences*; L. Raphael (1996), 'Die Verwissenschaftlichung des Sozialen als methodische und konzeptionelle Herausforderung für eine Sozialgeschichte des 20. Jahrhunderts', *Geschichte und Gesellschaft* 22, pp. 165–93; R. Smith (1997), *The Norton History of the Human Sciences* (New York, London: Norton); P. Wagner (1990), *Sozialwissenschaften und Staat. Frankreich, Italien, Deutschland 1870–1980* (Frankfurt and

New York: Campus); idem (2001), *The History and Theory of the Social Sciences* (London, Thousand Oaks, and New Delhi: Sage); P. Wagner, C. Hirschon Weiss, B. Wittrock, and H. Wollmann (eds) (1991), *Social Sciences and Modern States. National Experiences and Theoretical Crossroads* (Cambridge: Cambridge University Press); P. Wagner, B. Wittrock, and R. Whitley (eds) (1991), *Discourses on Society. The Shaping of the Social Science Disciplines* (Dordrecht, Boston, MA, and London: Kluwer Academic Publishers).

5. D. Lindenfeld (1997), *The Practical Imagination. The German Sciences of State in the Nineteenth Century* (Chicago and London: University of Chicago Press).
6. T. Haskell (1977), *The Emergence of Professional Social Science. The American Social Science Association and the Nineteenth-Century Crisis of Authority* (Urbana, Chicago, and London: University of Illinois Press).
7. M. Fourcade-Gourinchas (2006), 'The Construction of a Global Profession. The Transnationalisation of Economics', *American Journal of Sociology* 112, pp. 145–94.
8. S. Herbst (2003), 'Polling in Politics and Industry', in Porter and Ross, *The Modern Social Sciences*, pp. 577–90.
9. Smith, *Human Sciences*, pp. 606–16.
10. Ash, *Psychology*, p. 252 (my emphasis); cf. also T. M. Porter (1995) *Trust in Numbers. The Pursuit of Objectivity in Science and Public Life* (Princeton: Princeton University Press).
11. C. E. Lindblom and D. K. Cohen (1979), *Usable Knowledge. Social Science and Social Problem Solving* (New Haven and London: Yale University Press), p. 44.
12. H. Münsterberg (1913), *Psychology and Industrial Efficiency* (Boston and New York: Houghton Mifflin), pp. 307–8.
13. Unpublished lecture, cited by E. Walter-Busch (2006), *Faktor Mensch. Formen angewandter Sozialforschung der Wirtschaft in Europa und den USA, 1890–1950* (Konstanz: UVK), pp. 126–7.
14. L. Baritz (1977), *The Servants of Power. A History of the Use of Social Science in American Industry*, third edition (Westport, CT: Greenwood Press); E. Walter-Busch (1989), *Das Auge der Firma. Mayos Hawthorne-Experimente und die Harvard Business School, 1900–1960* (Stuttgart: F. Enke); R. Gillespie (1991), *Manufacturing Knowledge. A History of the Hawthorne Experiments* (Cambridge and New York: Cambridge University Press); J. Jansz and P. v. Drunen (2004), *A Social History of Psychology* (Malden, MA, Oxford, and Carlton: Blackwell).
15. B. Kaufman (1993), *The Origins and Evolution of the Field of Industrial Relations in the United States* (Ithaca, NY: ILR Press), pp. 46, 57.
16. B. Kaufman (2004), *The Global Evolution of Industrial Relations. Events, Ideas and the IIRA* (Geneva: International Labour Office), p. 9.
17. Ibid., pp. 301–80.
18. J. Converse (1987), *Survey Research in the United States. Roots and Emergence 1890–1960* (Berkeley, Los Angeles and London: University of California Press), pp. 87–8, 441, n.5; S. Igo (2007), *The Averaged American. Surveys, Citizens, and the Making of a Mass Public* (Cambridge, MA and London: Harvard University Press), Chapter 3.
19. Converse, *Survey Research*, Chapters 9–11.
20. K. Danziger (1997), *Naming the Mind. How Psychology Found Its Language* (London, Thousand Oaks, and New Delhi: Sage), pp. 152, 146.
21. R. Jaun (1986), *Management und Arbeiterschaft. Verwissenschaftlichung, Amerikanisierung und Rationalisierung der Arbeitsverhältnisse in der Schweiz 1873–1959* (Zürich: Chronos); R. Rüegsegger (1986), *Die Geschichte der Angewandten Psychologie 1900–1940* (Bern, Stuttgart, and Toronto: Huber).
22. Walter-Busch, *Faktor Mensch*, pp. 150–2, 301–3.
23. F. Baumgarten (1961), *Zur Geschichte der angewandten Psychologie in der Schweiz* (Münsingen: Buchdruckerei B. Fischer); M. Leimgruber (2001), *Taylorisme et Management en Suisse Romande, 1917–1950* (Lausanne: Éditions Antipodes).
24. Rüegsegger, *Angewandte Psychologie*, pp. 244–5.

25. E. Daub (1996), *Franziska Baumgarten. Eine Frau zwischen akademischer und praktischer Psychologie* (Frankfurt: Peter Lang); Walter-Busch, *Faktor Mensch*, pp. 292–308.
26. F. Baumgarten (1930), 'Die Psychologie im Wirtschaftsleben', *Schweizerische Zeitschrift für Betriebswirtschaft und Arbeitsgestaltung* I(36), pp. 257–71.
27. F. Baumgarten (1944), *Demokratie und Charakter* (Zurich: Rascher).
28. F. Baumgarten (ed.) (1949), *Progrès de la Psychotechnique 1939–1945* (Bern: Francke), p. 293; cf. also idem, *Geschichte der angewandten Psychologie*.
29. Rüegsegger, *Angewandte Psychologie*, p. 292.
30. Walter-Busch, *Auge*, p. 154; idem, *Faktor Mensch*, pp. 165–75.
31. H. U. Baumgartner (1950), 'Neue Wege amerikanischer Wirtschaftsforschung. Das "Institute of Industrial Relations"', *Die Unternehmung* IV(5), pp. 146–52.
32. A. Lisowsky (1947), *Vom Denken in Zusammenhängen in der Marktforschung* (n. p., Report 35 of the Gesellschaft für Marktforschung); idem (1950), 'Käuferschichtung als absatzwirtschaftliches Problem', *Zeitschrift für handelswissenschaftliche Forschung*, N.F. 2, pp. 527–41.
33. Walter-Busch, *Faktor Mensch*, pp. 389–99.
34. P. Kaufmann (1942), *Wie das Gallup-Institut die öffentliche Meinung ermittelt. Mit 37 Erfahrungen und Anregungen, die dem Politiker und dem Marktforscher nützlich sein können* (Thalwil: Oesch Verlag).
35. NHG (eds) (1948), *Die Schweiz hält durch. Buch der Volksumfrage unter dem Patronat der Neuen Helvetischen Gesellschaft* (Zürich and Wädenswil: Neue Helvetische Gesellschaft), pp. 140–80.
36. F. Keller (2001), 'Ein Tag im Nirgendwo. "Gulliver" an der Expo 64 oder: Wer hat Angst vor der Meinungsforschung?' (unpublished paper).
37. A. D. Chandler (1990), *Scale and Scope. The Dynamics of Industrial Capitalism* (Cambridge, MA, and London: Harvard University Press). Alfred D. Chandler, the leading American business historian of his generation, seems to me to have used the term 'managerial capitalism' as one of those bridging concepts that, according to Jürgen Kocka, might invite 'economic historians to incorporate social, cultural, and legal dimensions, while encouraging social, cultural, and other historians to take the economy more seriously'. J. Kocka (2010), 'Writing the History of Capitalism', *Bulletin of the German Historical Institute, Washington DC* 47, pp. 7–24, at 11. It is justified therefore to add Chandler's name to Kocka's list of the rather small number of non-Marxist historians who have succeeded in applying the concept of capitalism quite effectively.
38. Chandler, *Scale and Scope*, p. 72.
39. Ibid., p. 423.
40. Ibid., p. 395.
41. Walter-Busch, *Faktor Mensch*, pp. 313–21; cf. with regard to American ways of thinking, which generally favoured the growth of the social sciences, R. Dahrendorf (1963), *Die angewandte Aufklärung. Gesellschaft und Soziologie in Amerika* (Munich: Piper); cf. also E. C. Lagemann (1992), *The Politics of Knowledge. The Carnegie Corporation, Philanthropy, and Public Policy* (Chicago and London: University of Chicago Press).
42. Walter-Busch, *Auge*, pp. 49–54.
43. E. P. Oberholtzer (1891), 'Law-Making by Popular Vote. Or, the American Referendum', *Annals of the American Academy of Political and Social Science*, 2, pp. 36–56; L. Wuarin (1895), 'Recent Political Experiments in the Swiss Democracy', *Annals of the American Academy of Political and Social Science* 6, pp. 1–20.
44. W. Linder (1994), *Swiss Democracy. Possible Solutions to Conflict in Multicultural Societies* (Basingstoke and New York: Palgrave Macmillan), pp. 138–9.
45. Igo, *Averaged American*, pp. 135, 185; cf. also her chapter in this volume.
46. Cited Igo, *Averaged American*, p. 121.
47. R. Germann (1991), 'Arenas of Interaction. Social Science and Public Policy in Switzerland', in Wagner, Wittrock, and Whitley (eds), *Social Sciences*, pp. 191–206, at 195.

48. K. Arnold (1969), *Verwaltungs- und Regierungstätigkeit durch eidgenössische Kommissionen* (Winterthur: Schellenberg), pp. 3, 25–6.
49. *Protokolle* (Minutes, 1932–1943), no. 1–106 of the Kommission für Konjunkturbeobachtung, vols I–IV (kept by the Archives of the Swiss National Bank); no. 1, 6 February 1932, pp. 4–5, and no. 47, 20 April 1937, p. 11; parenthetically, the cited opinion of Lorenz is still a popular *topos* among Swiss civil servants.
50. Walter-Busch, *Faktor Mensch*, pp. 156–65.
51. R. Hofstadter (1963), *Anti-Intellectualism in American Life* (New York: Vintage).
52. However, this has also been an unfulfilled 'enduring dream' in the United States; cf. D. Madsen (1966), *The National University. Enduring Dream of the USA* (Detroit, MI: Wayne State University Press).
53. Cf. Walter-Busch, *Faktor Mensch*, pp. 303–4.
54. Carrard, in Baumgarten, *Progrès*, p. 302.
55. Cf. especially the lengthy handbooks by E. Richard (1924), *Kaufmännische Gesellschaft Zürich und Züricher Handelskammer 1873–1923*, vols I and II (Zurich: Züricher Handelskammer), and N. Reichesberg (ed.) (1903–11), *Handwörterbuch der Schweizerischen Volkswirtschaft, Sozialpolitik und Verwaltung*, vols I, II, III(1), and III(2) (Bern: Verlag der Encyklopädie). Unfortunately, most social scientists and historians exploring the development of the social sciences in Switzerland have, until now, ignored these treasures of knowledge, which are rooted in the practice-oriented 'sciences of state' of the nineteenth century.
56. The best-known among them today are Léon Walras and Vilfredo Pareto; cf. H. Herkner (1929), 'Begrüssungsrede', in F. Boese (ed.), *Verhandlungen des Vereins für Sozialpolitik in Zürich 1928* (Munich and Leipzig: Duncker and Humblot), pp. 3–12, at 4–5; cf. also A. Oncken (1895), 'Political Economy in Switzerland', *The Economic Journal* 5(17), pp. 133–7.
57. *Protokolle* (Minutes, 1937–1943), nr 1–100, of the Erfahrungsaustauschgruppe für Konjunkturbeobachtung, vols I and II (kept by the Archives of the Swiss National Bank), no. 63, 31 October 1940, p. 7.
58. P. Schmid-Ammann (1978), *Unterwegs von der politischen zur sozialen Demokratie. Lebenserinnerungen* (Zurich: Morgarten), pp. 253–4.
59. M. Zürcher (1995), *Unterbrochene Tradition. Die Anfänge der Soziologie in der Schweiz* (Zurich: Chronos), pp. 239–85.
60. Ibid., pp. 278–85.
61. Igo, *Averaged American*, p. 93.
62. C. Honegger, H-U. Jost, S. Burren, and P. Jurt (2007), *Konkurrierende Deutungen des Sozialen. Geschichts-, Sozial- und Wirtschaftswissenschaften im Spannnungsfeld von Politik und Wissenschaft* (Zurich: Chronos).
63. Cf. Baritz, *Servants of Power*.
64. Honegger, Jost, Burren, and Jurt, *Deutungen des Sozialen*, pp. 241–50; cf. also S. Burren (2010), *Die Wissenskultur der Betriebswirtschaftslehre. Aufstieg und Dilemma einer hybriden Disziplin* (Bielefeld: Transcript).
65. Raphael, *Verwissenschaftlichung*, pp. 166, 178.
66. F. Giese (1925), 'Auswahl und Verteilung der Arbeitskräfte', in J. Riedel (ed.), *Arbeitskunde* (Berlin and Leipzig: Teubner), pp. 283–308, at 304.
67. Cf. the outline of a sceptical theory of social scientific reasoning in Walter-Busch, *Faktor Mensch*, pp. 431–50.
68. Cf. O. Marquard (1973), *Schwierigkeiten mit der Geschichtsphilosophie* (Frankfurt: Suhrkamp), p. 13: 'The philosophers of history have only changed the world differently; what matters is taking care of it.'

15
Catholic Church Reform and Organizations Research in the Netherlands and Germany, 1945–1980

Benjamin Ziemann and Chris Dols

At first glance, the Catholic Church is a highly unlikely candidate for the practical application of social scientific knowledge and techniques such as opinion polling or organizations research. In the years after 1945, official ecclesiology reaffirmed the traditional idea of the Church as the 'mystical body of Christ', as outlined in the encyclical *Mystici Corporis*, which was issued by pope Pius XII in June 1943. According to this notion, the Church was not only a strictly hierarchical body. Even more, as a mystical body, the Church had to be understood not in terms of a worldly, man-made bureaucratic structure with changeable elements. Quite to the contrary, it was instituted by God and designed by him as a closely knit community, and as an organism where the individual elements had to fulfil a given and fixed role.[1] As the Church was conceptualized as a metaphysical entity with 'ascribed' roles and elements (to use the terminology of Talcott Parsons), no entry point for social scientific ideas on organizational change seemed to exist. The human relations movement had coined the phrase 'overcoming resistance to change' as a motto for the pivotal aim of this type of organizations research.[2] Seen as a mystical body, the Church was predicated precisely on the rejection of institutional change not only as a practical option, but also as an inherent possibility.

Even in the Catholic Church, however, a discrepancy existed between the official edifice of dogmatic constitutions and promulgations on the one hand, and the practical requirements of pastoral care on the other. After 1945, this discrepancy between the façade of ecclesiological statements and the daily encounter with problems and deficiencies evolved into a yawning gap when the Church had to observe the effects of functional differentiation in society or, in other words, a process that can be reconstructed as 'secularization'.[3] Historians are well advised to use the concept of secularization with great diligence, and to avoid portraying the Church as a passive victim of an anonymous and unstoppable process of societal change. Analyzing the encounters between Catholics and organizations research in the period from 1945 to 1980 can help to do precisely this: to understand how laymen and Church officials at various levels of the hierarchy actively engaged with new forms of social science knowledge in order to reinvigorate and adapt the teaching of the gospel in a rapidly changing social environment, and thus to respond to 'secularization'.

Such a search for new institutional forms coincided with and was supported by a major reorientation with regard to the relation between Church and the secular world. The documents of the Second Vatican Council, which had gathered in Rome from 1962 to 1965, left the hierarchical structure of the Church basically intact, at least in institutional terms. At the same time, the council promoted a new understanding of the Church as the 'people of God', a notion that promoted – metaphorically speaking – the laity to being a crucial constituent part of the Church and its task, which was to articulate the message of the gospel in a dialogue with the secular world.[4] A dialogue, however, required a common language and a more specified understanding of the dynamics of secular society. For this reason, the pastoral constitution of the council, *Gaudium et Spes*, explicitly demanded a substantial study and reception of the secular social sciences, including sociology and psychology, at all levels of the Church. Only the systematic incorporation of an 'external perspective', as provided by the concepts and insights of the modern social sciences, would allow proper implementation of the substantial reform agenda of the council.[5]

The application and implementation of social scientific approaches in the Catholic Church, from statistics, role sociology, and opinion polling to organizations research, was thus part and parcel of a post-Vatican II drive towards reform and an *aggiornamento* of the Church with changing social conditions. But the practical application of methods from the social sciences had already begun before the council legitimized and supported such an endeavour. These early attempts to use the social sciences emerged in a variety of specific and to some extent contingent contexts. In the following, we will analyze these contexts and the forms and practical implications of social science knowledge in the Catholic Church in a comparison between two national cases: first is the Netherlands, which from the mid-1960s was often cited as an example of a rebellious church that gave the laity unprecedented leverage in their pursuit of progressive and liberalizing ideas. The Dutch case is juxtaposed with the Federal Republic of Germany, where the practical usage of social science knowledge began later and in a less coherent fashion than in the Netherlands. In both cases, however, the application of organizations research and related methods was bound up with the increasingly heated struggle between different currents in the Church over the proper implementation of the reform agenda of the council. In a final step, we discuss both comparative aspects and some conceptual consequences of our empirical study.

'Pastoral pedagogy': Social scientists and the dynamic flywheel of reform in the Catholic Church in the Netherlands

Before the Second World War, the applied social sciences were practically non-existent in the Catholic Church in the Netherlands. Catholics understood and approached the world primarily through theological and socio-philosophical lenses.[6] After 1945, the tide changed, and between 1945 and 1960 practitioners of the social sciences gained much territory.

Early protagonists of the social sciences couched their activities in discourses of evangelization and of moral and religious decline. How and where did such decline originate, and which circumstances influenced it? What did social scientific research results mean for the organization of pastoral care? The experts, who included churchgoers, put together voluminous blueprints, and wrote numerous reports, bringing these and similar questions to the fore.[7] George Zeegers noted in 1948:

> Everywhere in the world, the Church of Jesus Christ is in great danger. ... Compared to other countries, the Netherlands has been struck less hard. But also in our country, there are evident signs of decay. By means of a complete understanding of the actual situation, and with the help of God, we hope to find the methods to bring that decay to an end.[8]

Catholics who acted as main protagonists of the social sciences – both academically trained men and 'amateurs'– integrated well-chosen references and metaphors that almost everyone could understand into their missionary discourses. In 1956, the Latin words 'in hoc signo vinces' ('by this sign thou shalt conquer') explicitly referred to the sign of the cross, and implicitly to the famous battle between Constantine the Great and his opponent Maxentius in AD 312.[9] An abundant use of medical terms such as 'anamnesis', 'disease', 'diagnosis', 'prognosis', 'thermometer', 'suffering', 'cure', and 'therapy' is also meaningful. Apart from semantic narratives, visual discourses were constructed and communicated. The depiction of Saint George on the cover of a brochure, killing a dragon, indicates that it was the fight against heathendom that practitioners of the social sciences in the Catholic Church in the Netherlands were engaging in. As Lutz Raphael argues, references and metaphors linked different discourses together, in this case medical and social scientific narratives to languages with a religious and moral nature.[10]

Three Catholics who greatly stimulated the ascent of the applied social sciences were the Redemptorist Leonardus Buys, the Franciscan Montanus Versteeg, and the aforementioned Zeegers, who was a layman. In 1946, they established the Katholiek Sociaal-Kerkelijk Instituut (KASKI, the Catholic Institute for Social-Ecclesiastical Research) as a small study group.[11] This study group was soon transformed into a full-fledged institute (1947). At that time, KASKI was the only institute in Europe that was devoted to the practical application of the social sciences in the Catholic Church.[12] Its foundation has to be set against the background of the staging and scripting of parish missions in the Netherlands, often in so-called diaspora areas (see Figure 15.1). A more general technocratic understanding of planned change must also have influenced Buys, Versteeg, and Zeegers. The latter was KASKI's general manager. In order to stimulate and facilitate local and regional research, the institute founded offices in the different dioceses, small and modest in size. Its headquarters were located in The Hague.

KASKI's formative years can be divided into the two periods 1946–51 and 1951–7. The first was characterized by a serious degree of friction and uncertainty. As a pioneering institute, one should note, KASKI was not founded because there was

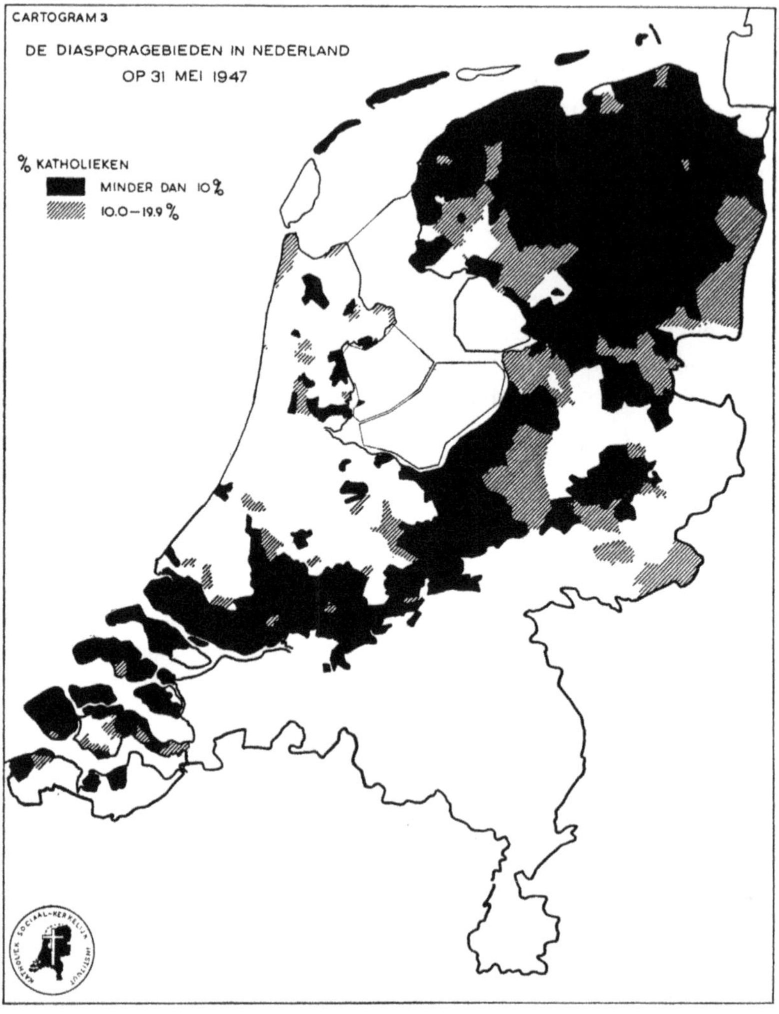

Figure 15.1 A KASKI map, showing so-called diaspora areas in the Netherlands, coloured in black. Visual representations like this were the trademark of KASKI social research among Catholic sociologists across Europe
Source: KDC – KliB Nijmegen.

a huge demand for social scientific knowledge, and even the inexhaustible Zeegers could not, at this time, fall back on large sums of money. From 1951 onwards, although the institutional and financial basis remained fragile, the institute had survived its teething troubles. KASKI became well known in Europe, and enjoyed a period of expansion. The number of regional offices (also abroad), employees, and research assignments increased substantially. Between 1946 and 1951, KASKI published 26 research reports (an average of 6.5 per year), a number that increased to no less than 215 reports for the period between 1950 and 1958 (an average of

30.7 per year).[13] During the 1950s, KASKI – and this can be regarded as one of its most important contributions to the scientization of the social in the Catholic Church – provided an intellectual infrastructure, through networks and channels that circulated concepts, ideas, and people, and by facilitating a number of Ph.D. projects in the sociology of religion.

Catholics who consumed early social scientific knowledge came from many different backgrounds. Vicars, priests, father and mother superiors, media bosses, politicians, and church councillors were all confronted and sometimes taken by surprise by transformative societal developments, and hence found their way to KASKI. Statistics and sociography were the dominant technologies and approaches in the Catholic Church in the Netherlands during the 1940s and 1950s.[14] Ecclesiastical statistics were used in a more sophisticated way by the French sociologist Gabriël le Bras in the interwar period, before they made their way to the Netherlands.[15] With the help of statistics, KASKI and its clients strove for social accounting, that is, they wanted, for example, to establish long-term observations and comparisons of churchgoers. It was important, as the staff of the institute underlined, to discover trends and signals for societal change, so that Church leaders could act as quickly and effectively as possible. The sociographical school of thought, launched in the 1920s in Amsterdam by Professor Sebald Steinmetz, combined historiography, demography, geography, and ethnology.[16] Sociography was often used in cases of territorial planning. This kind of research has metaphorically and methodologically opened the doors of the Church. Where statistics had addressed its interior, sociography focused on the relationships between society and Church.

Zeegers paid special attention to the Dutch bishops. They had to be convinced that theological forms of knowledge and thinking – which were still dominant during the 1950s – had to be confronted with and sometimes replaced by social scientific discourses and research results. From the very beginning, it was one of his primary aims to bring about a profound translation, a 'turnover in the bishop's way of thinking'.[17] He wanted nothing less than to shunt these Church leaders on the right track, and technological-sounding reports that featured abundant attractive maps and graphs had to contribute to this purpose.

Although the scientization of the social was anything but a straightforward success, Zeegers and his colleagues did manage to convince the Dutch bishops that the social sciences were academically trustworthy, and legitimate, helpful, even essential ingredients for decision-making processes.[18] The Church leaders increasingly turned to these sciences to help them understand and react to societal phenomena. They commissioned several research projects during the 1950s, and Walter Goddijn, a Franciscan and social scientist, was hired as Pieter Nierman's personal advisor. Nierman was the bishop of Groningen, where Goddijn ran a KASKI office. In 1952, Bernard Alfrink – archbishop-coadjutor of Utrecht from 1951, archbishop from 1955, and cardinal from 1960 – stated that the Lord certainly approved the use of the social sciences in the Church. He did not only give Catholics his mercy and grace, Alfrink argued, but He also gave them a mind and the power of reason, and He wanted them to use these gifts.[19]

Figure 15.2 Photograph taken on 3 April 1952, showing how Church officials and social scientists with a Catholic background literally gathered around the visual representations of the social sciences. Archbishop-coadjutor Bernard Alfrink (centre) is taking a closer look at a KASKI map, which is explained by the Franciscan and social scientist Walter Goddijn
Source: KDC – KliB Nijmegen.

In the years from 1958 to 1963, two important transformations took place.[20] On the one hand, Alfrink moved from a rather conservative position to the liberal left.[21] Someone like Gerard de Vet, in addition, the bishop of Breda (appointed in 1962), was in many respects even more progressive. He stood out as relatively modern, and willing to adapt to the rapidly changing society. Instead of adopting a more repressive stance, the Dutch bishops opted to stimulate a coherent process of reform – not only in order to prevent a breakdown of the societal order, but also because they truly believed that a transformation of Catholicism was beneficial.[22] Convinced that reform was unavoidable and necessary, they became keen on social scientific knowledge and expertise.

On the other hand, as Zeegers left KASKI at the beginning of 1958, and his successor – not quite a visionary and charismatic man, and not a social scientist – failed to live up to high expectations, a power vacuum emerged. Moreover, it became evident that the work of KASKI, and the statistical and sociographical methods more generally, had serious limitations.[23] The practical application of social science methods by KASKI had a regional and fragmented character. It was thus difficult to develop an overarching, national plan of pastoral care. Subsequently, the institute had to cope with the fact that clergymen in key positions often considered the institute, guided by laymen who did not always speak the 'ecclesiastical language', to be much too keen on commercial success. In addition, statistics and sociography were not always able to generate the desired knowledge and lead to the right prognoses and diagnoses. Lastly, because of its small financial and personal base, KASKI could hardly stimulate bold experiments, and keep up with scientific developments.

These two larger developments fuelled the rise of an institutional framework in which two new institutes set the tone: PINK and ITS. PINK (Pastoraal Instituut van de Nederlandse Kerkprovincie, the Pastoral Institute of the Dutch Church Province) was founded by the bishops in 1962, with Goddijn as its manager. PINK was established to stimulate scientific reflection on Catholicism, coordinate already existing and new initiatives of study and reform, guide pastoral experiments, and give practical and scientific advice to the bishops and father and mother superiors.[24] 'You partly have to act like a plumber,' Goddijn was told by Bishop De Vet, who hereby referred to the metaphor of the social scientist as a social engineer. Several buckets, De Vet continued, were heavily 'leaking'.[25] These leaks had to be repaired, and time was running out. ITS (Instituut voor Toegepaste Sociologie, the Institute for Applied Sociology) was established in 1964. The Flemish social scientist Edward Leemans, who was a professor at the Catholic University of Nijmegen, was one of the founders. As a strong advocate of the practical application of the social sciences, he wished to expand Nijmegen-based sociology. Contrary to PINK and KASKI, which continued to exist, ITS did not mainly focus on Catholicism. Nonetheless, ITS and its protagonists became important players with regard to the investigation of vocations, liturgy, celibacy, birth control, and the Catholic education system.

In 1965, the reform of Catholicism in the Netherlands accelerated. It was De Vet's plan to implement the ideas, decisions, and spirit of the Second Vatican Council into the local Catholic Church by organizing a Dutch council, which was to function as a follow-up to Vatican II.[26] The other bishops agreed. A 'renovation' was 'highly necessary', the Church leaders would write to all Catholics in 1968.[27] In their view, there was no other option than the choice between 'renewal' or 'fossilization'. From 1966 to 1970, the Pastoral Council was held. The bishops invited laypeople to engage in an active dialogue. During six sessions (1968–70), bishops, priests, and laypeople discussed with each other subjects such as authority, missionary work, marriage, and the priesthood. In the Netherlands, the flywheel of reform was spinning at full speed.

These years were the heyday of the applied social sciences in the Catholic Church, the years that saw planning euphoria, and an increasing application of opinion polling, organizations research, group counselling, and role sociology. During the second half of the 1960s, theology was indeed, as the historian Louis Rogier has observed, caught up with if not sometimes surpassed by the social sciences.[28] It has to be said, however, that the application of the social sciences remained a disputed and controversial phenomenon.

The Pastoral Council can be described and analyzed as a culmination in the reform of Catholicism, and, not surprisingly, also as a culmination of the application of the social sciences and the concomitant 'narratives of the new'. Social scientists acted as key agents of change, that is, as leading architects of a national *aggiornamento*.[29] Many of them were progressive – some in a more radical way than others. They vigorously promoted discourses of change, and, contrary to the early social scientists, were very critical about the pope and his encyclicals. Asked about the things that should be done in favour of reform, the Franciscan Osmund Schreuder, who was a professor in the sociology of religion at the Catholic University of Nijmegen, stated that he no longer believed in a top-down reform. Rather, he preferred bottom-up reform, and this reform, he said, was 'only possible through conflicts, say, through a *revolution* in the Church'.[30]

During the second half of the 1960s, social scientists were widely acknowledged as experts in the Catholic Church. Sometimes, they were even regarded as a 'rescuer' or as a 'Messiah of Renewal'.[31] Social scientists, as a consequence, were members of study groups that PINK established for the preparation of the Pastoral Council. Schreuder, for example, was the chairman of the committee on the priesthood, and, to a considerable degree, laid semantic and conceptual foundations for the discussions on this crucial matter.[32] At the macro-level it was Goddijn who staged and scripted the process of reform. In 1965, the bishops gave him a wide-ranging mandate and appointed him secretary-general of the Pastoral Council. This function, combined with the directorship of PINK, made him one of the most influential and powerful but also one of the most controversial men in the local Catholic Church. 'Pope of Holland' – Goddijn's well-known nickname – reflected both respect and disdain.

From 1964 to 1970, PINK commissioned several large-scale and ambitious social scientific interventions. Convinced that the Catholic Church was in definite need of social scientific schemes of thought and a national 'pastoral pedagogy', and mandated to organize the Pastoral Council and to coordinate research and experiments, Goddijn tried to maximize and monopolize social scientific research on religious matters.[33] He and his colleagues did not assign a single extra theological project in preparation of the Pastoral Council – only an impressive range of social scientific research projects. Together, these projects cost around 450,000 Dutch Guilders, that is, around 204,000 Euros in current terms.[34]

Social scientists such as Goddijn and Schreuder couched their interventions in discourses that differed from those employed by Zeegers and other protagonists during the 1950s. Instead of emphasizing evangelization and the need to counter moral and religious decline, the younger generation stressed 'renewal' and 'adaptation' to a changing society. By means of the social sciences, as Goddijn and his brother Hans Goddijn – who also was a prominent Catholic and social scientist – wrote in 1964, the Church had to determine what was out of date and therefore could be abolished, and what was reparable and provided new points of departure.[35] From a distance, it had to achieve self-observation, self-knowledge, and self-criticism. Church leaders could not any longer allow themselves 'to turn a blind eye on the facts of society'.[36] They had to face reality, and leave unprofessional, passive, and defensive leadership behind, before it was too late:

> A traditional and hierarchal Church structure, the feudalistic system of orders and congregations, and a paternalistic pastoral care do simply not fit any

> more in a modern, industrial and pluralistic society of fully-grown, free, and enlightened people.[37]

Schreuder remarked that mankind could not live any more without permanent reflection, a 'Dauerreflexion', as he called it, citing the German social scientist Helmut Schelsky.[38]

Social scientific discourses of 'crisis' and 'renewal' contributed significantly to a dynamic and religiously explosive climate that put an enormous and probably unforeseen pressure on the Dutch bishops to reform. It has to be said, however, that they were deeply interested in social scientific research results themselves. It is known that the Church leaders, by steering a so-called middle course, adopted a rather progressive strategy. This meant that they, with regard to burning questions such as celibacy, liturgy, and birth control, pursued a policy that could rest on the support of around 70 per cent of Dutch Catholics.[39] Social data were something to hold on to, they wrote in 1968. These were both 'route indicators' and 'warning signs'.[40]

After the Pastoral Council had come to an end, in 1970, social scientists quickly lost influence. Goddijn, PINK, and the council in particular, and the Catholic Church in the Netherlands in general, were being identified by many (also abroad) with ultra-progressivism, and regarded as too disturbing and rebellious.[41] Moreover, it became evident that social scientists did not have an answer to all the societal problems that had arisen. It also became clear that the large-scale, progressive, and ambitious social plans and interventions could hardly be implemented in the context of an overarching, national Church policy.

At the beginning of the 1970s, the Holy See intervened by appointing two conservative bishops. In its opinion, the reform had run out of control and the dynamic flywheel had to be slowed down This policy resulted in a backlash against the scientization of the social, which was strikingly reflected by Walter Goddijn's dismissal and the liquidation of PINK in 1972. ITS no longer received research assignments on religious matters, but it repositioned itself in the market, and still exists today. KASKI also managed to survive the backlash, but never recovered former levels of activity and influence.

The pitfalls of participation: Sociological approaches and the failure of reform in the German Catholic Church

It was only in the late 1950s that a substantial reception and discussion of sociological organizations research in the Catholic Church in West Germany began. To be sure, Church statistics, that is, the aggregation of data about the number of Catholics who attended Sunday services and received communion at Easter, had been pioneered in Germany. After lengthy debates, an official statistical office of the German Catholic Church started to collect and publish these data in 1915.[42] The next step in the scientization of the Church, the introduction of sociographical methods, was heavily depending on the conceptual ideas and methodological premises that pastoral theologians and church sociologists in Belgium, France, and the Netherlands had developed since the early 1940s. For the application of sociography, the German Church also relied on external help to develop institutional resources. In 1951, KASKI established a branch in Königstein near Frankfurt/Main, one of the many branches founded in other European countries. Its initial focus was on the consequences on pastoral services of the massive influx of refugees from the former German territories east of the Oder–Neisse line. Together with similar institutes in Münster, Essen, and Landstuhl (Palatinate), the focus soon shifted to empirical surveys of the pastoral situation in local settings. Most of these surveys were conducted in urban neighbourhoods, in an attempt to gauge the decline of practised piety among working-class Catholics. The results were rather disappointing, indicating that children, women, and middle-class men provided the bulk of the active 'flock'. Attempts to reformulate pastoral strategies according to these data, however, were incoherent, and the sociographical approach received only lukewarm support from many West German bishops.[43]

However, sociological forms of knowledge moved closer to the institutional core of the Church as an organization when the activities of the parish priests were considered in terms of a professional role. Initially, attention towards the priesthood was driven by concerns about a quantitative lack of younger candidates, following on from Jan Dellepoort's comparative survey on priests in Europe, published in 1959.[44] Early responses to these data discussed the overburdening of parish priests as a result of overwork, and social scientists contemplated ways of streamlining daily routines in order to make efficiency savings. Beginning in the early 1960s, the focus then shifted towards a qualitative idea of the priest as someone who performed a professional role. In 1958, Ralf Dahrendorf had published *Homo Sociologicus*, a short introduction into the concepts of role sociology, a book that familiarized not only many Catholics with the core ideas of this strand of research.[45] Henceforth, difficulties of and with the clergy were described as a growing divergence between different expectations into their professional roles by the laity. Part and parcel of these reflections was the insight that traditional metaphors for pastoral care could no longer be sustained amid rapid social change. To describe the priest as a 'good shepherd', as pastoral discourse had done for a long time, reflected a paternalistic attitude that was increasingly deemed to be problematic, not only among the laity, but also among clergy and bishops. A more open and less authoritarian role definition was required, and the ability to balance contradictory expectations at the intersection between laity and Church hierarchy.[46]

During the 1960s, empirical surveys about the practical tasks of the parish clergy and their 'role-set' were rare. They encountered resistance from diocesan authorities, particularly when problems with celibacy and the authoritarian structures within the Church were expected to come to the fore, as it was inevitably the case in the late 1960s. Role sociology was less relevant in practical terms, as no substantial changes to the education or professional employment of priests were implemented until the early 1970s. But sociological terminology did inform and drive forward a seismic shift in the semantics used to describe the clergy and its position vis-à-vis the laity. A vivid expression of these changes is a memorandum that Karl Forster, secretary-general of the German Bishops Conference, prepared for the meeting of this group in 1975. Drawing on data from the various surveys and opinion polls among the West German clergy, which had been conducted since 1970, Forster discussed options to alleviate the pressure from an 'overload of roles' that characterized the situation of parish priests. Increasing the responsibility of active members of the laity was one option. Others included attempts to develop the ability to work well in a team.[47] Together with an increased specialization on specific functional aspects of pastoral care (among the youth, for married couples, etc.), teamwork was meant to take away some responsibilities from the priest and thus to reduce the complexity of his 'role-set'. As Forster acknowledged at least implicitly, however, this option was bound to produce increased complexity, as the delicate capability to work in a team was added to the many role expectations.[48] Describing the work of a priest in terms of role sociology, it turned out, led to further ambivalences, where many bishops had expected it to produce clear-cut results. One result, however, was certain. A hierarchically defined notion of priesthood, which situated the priest in the charisma of the Church, was giving way to a horizontal one, which defined the priest in functional terms and with regard to his 'performance'. Already in 1970 Joseph Ratzinger (later Pope Benedict XVI), then a professor in Regensburg, was alarmed about the dangers of turning the priest into a mere 'social functionary' (*Sozialfunktionär*). As a functionally defined specialist, he warned, the priest could be too easily replaced by psychotherapists and other experts from the human sciences.[49]

Within 15 years, from 1960 to the mid-1970s, the reception of role sociology fundamentally altered the terms of the debate on the work and livelihood of Catholic priests, even though a discussion of practical reforms with regard to celibacy remained off limits. In the late 1960s, another dimension was added to this sociological understanding of the Church as an organization. Beginning in 1967/8, many of the 21 Catholic dioceses in the Federal Republic began to develop blueprints and practical proposals for a substantial reform of their institutional structures. Pastoral planning was the keyword for these endeavours. There were three main reasons for the fact that various initiatives were more or less simultaneously started across the country. First, an imminent shortage of candidates for priesthood had become apparent since the mid-1960s, and made urgent action mandatory in order to stabilize the provision of pastoral services. Secondly, the Second Vatican Council had encouraged a coordination of pastoral activities beyond the level of the individual parish and to break up parochial attitudes. Thirdly, optimistic expectations for the possibility of planning pervaded the political system, particularly after the Social Democrats had gained power in 1969. Using models from organizations research and cybernetics, it seemed possible to combine planning and increased participation.[50]

As they discussed the broad outlines of pastoral planning and organizational reform, theologians, bishops, and Catholic social scientists knew that they were about to fundamentally alter the premises on which the Church was based. Historically, the Church had been predicated on tradition. Now, decisions were taken in the knowledge that they were contingent and could produce contradictory results. Whereas bishops and priests had in the past relied on the frameworks provided by dogmas, they were now about to engage in a process of 'permanent reflection'. It was, by the way, a Protestant sociologist, Helmut Schelsky, who had introduced this notion into West German debates on the sociology of religion, indicating the extent to which the practical application of sociological ideas transcended the divide between the confessions. All in all, pastoral planning opened up a rather ambivalent horizon and was thus part and parcel of the broader societal trend of 'planning for the future'.[51]

Moving from abstract debates to detailed analysis and concrete institutional change, most West German dioceses implemented organizational reforms in the years after 1967. In one way or another, most of these plans intended to merge local parishes and to supplement them by pastoral services provided for the whole town or city, thus also changing the allocation of priests to different tasks and enforcing cooperation. In most, but not all cases these changes at the level of the parish community were accompanied by a reshuffling of the departments in the diocesan administration or *Generalvikariat*, aiming to organize the departments according to more clearly defined functional responsibilities.[52]

Aims and ambivalences of organizational change in the Catholic Church can be analyzed in detail by taking the *Strukturplan* (structural plan) in the diocese of Münster in North-West Germany as an example. Basically, the *Strukturplan* was a booklet of 50 pages with the title *Überlegungen und Vorschläge zur Struktur der Seelsorge im Bistum Münster* (Thoughts and Suggestions for the Structure of Pastoral Care in the Bishopric of Münster). Starting in 1967, the text had been drawn up by a planning committee headed by Wilhelm Stammkötter, a member of the cathedral chapter. The key proposal of the plan was to replace the traditional territorial parish with a *Großpfarrei*, a greater parish. Comprising between 20,000 and 100,000 Catholics and thus in its size resembling the traditional deanery, the *Großpfarrei* was meant to offer four core functions of the Church that the authors had identified: service to the world, pronouncement of the gospel, *Diakonia* or social service, and liturgy. Within the *Großpfarrei*, tasks should be divided between pastoral specialists, both priests and pastoral assistants from the laity. The territorial parish was to be downgraded to a mere 'substructure' of the larger entity. In addition, the plan envisaged a reorganization of the councils for lay participation that had been established after Vatican II, henceforth equally equipped with subcommittees according to the four core functions. Finally, the *Generalvikariat* should be reorganized in a more cooperative fashion.[53]

Sociological knowledge informed this proposal in various ways.[54] First, at a personal level, Egon Golomb was invited to the meetings of the planning committee. Golomb was the head of the Institute for Pastoral Sociology in Essen, which had been founded in 1958 when the new diocese of Essen was created, with an assignment to study the structural underpinnings of pastoral care in an industrial region. Apparently, Golomb helped to formulate the four core functions of the *Strukturplan*, which had a striking resemblance with the AGIL scheme of Talcott Parsons's general theory of action systems (adaption, goal attainment, integration, and latent pattern maintenance). Secondly, the idea of the *Großpfarrei* was taken from a recent book in French pastoral sociology, which argued that the unity of work, leisure, and religious activities in a local community had dissipated, and that the more mobile life patterns of city dwellers should be matched by a variety of functionally defined pastoral services. Thirdly, the department for pastoral care in the *Generalvikariat* hired two professional sociologists in spring 1970, after Bishop Heinrich Tenhumberg had distributed 24,000 copies of the plan across the diocese and invited clergy and laity to embark on a substantial discussion of the overall pastoral concept and the detailed proposals. The sociologists' assignment was to arrange and evaluate the incoming suggestions and criticisms, arriving in their hundreds from parish councils and lay groups across the diocese, and to offer pertinent ideas for an appropriate implementation of the plan.

The older one of the two sociologists who were hired in 1970 had extensive experience in Church service. After completing a Ph.D. in the sociology of industrial relations, Philipp von Wambolt, born in 1918, had taught courses for Catholic high school students in the academy Franz Hitze-Haus in Münster. From 1967 to 1970, von Wambolt had taught at the Catholic University in Valparaíso in Chile, where he experienced both the perennial problems of underdevelopment and capitalist exploitation in Latin America, and the student uprising of 1968, which led to sit-ins and to a campus occupation. Highly politicized in a leftist and critical fashion, von Wambolt returned to Münster and embarked on his new professional assignment. In various memoranda and letters to his line manager, Hermann-Josef Spital, the head of the department for pastoral care, von Wambolt developed a fundamental critique of the *Strukturplan* and its agenda for a reorganization of pastoral services. Starting from the premise of a fixed goal of the Church as an organization, the plan would try to implement changes that would facilitate a smooth running of routine operations and the attainment of this goal. Thus, the plan conceptualized organization along the lines of the 'human relations' approach pioneered by Elton Mayo, rather than highlighting the significance and the legitimate role of conflicts in organizations, which could foster a permanent reform of internal procedures.[55]

But von Wambolt did not only spell out conceptual shortcomings in the plan in the light of more recent developments in organizations research. He also reflected on his own role as a professional sociologist, and arrived at conclusions that implied nothing less than a fundamental critique of the functionalist underpinnings of the *Strukturplan*. Drawing on C. Wright Mills and his critique of Parsons, but also on psychoanalytical authors such as Erich Fromm and Karen Horney, von Wambolt rejected the notion of a set of stable functions that had to be fulfilled. This line of reasoning would gloss over the key problem in the contemporary Catholic Church, the lack of legitimate space for the articulation of divergent and pluralistic subjectivities. In a memorandum with the telling title *Thirst for a Human Church*, von Wambolt explained why the 'betrayal of freedom' within the Church was the key problem for him, and why it was necessary to create space for the articulation of personal 'experiences of hardship'.[56]

To be sure, von Wambolt was not the only one who criticized the plan. The vicar general of the archdiocese of Freiburg, Robert Schlund, was sure that pastoral planning, which he supported in principle, should not postulate rationalist principles in a deductive fashion. Wholesale blueprints for reform like the structural plan were for Schlund nothing more than 'progressive sociologism', that is, an attempt to replace theological core notions with left-leaning sociological abstractions.[57]

However, not only high-ranking Church officials criticized the plan. Equally widespread, and in the end more decisive for the fate of the plan, was simmering discontent from below. When Bishop Tenhumberg invited parish committees to discuss the plan and voice their concerns, this was meant to be an exercise in participatory planning, where the laity could contribute to a better understanding of the need for reform. And indeed, many of the written comments by the active Catholics who sat in the parish committees started from the premise that the local parish community was the 'base' and hence the 'foundation' of the Church.[58] The pastoral needs and organizational structures outlined in these submissions were, however, diametrically opposed to the sociological ideas of the 'structural plan'. Laypeople at the local level did not only resent the abstract, functionalist terminology of the plan and the very notion of rapid organizational change. At the heart of their objections was the plan to reorganize pastoral care, and the roles of the priests in particular, along functionalist criteria. While further functional differentiation was the core of the plan, the laity clung on to the notion that their parish priest had to be versatile in all aspects of the Christian faith, and had to attend to their personal problems in the first instance. The projected loss of the territorial parish as a place where believers could meet and communicate in the closed setting of personal encounters also caused the wrath of the laity. Couched in the conservative semantics of 'community' (*Gemeinschaft*), the parish committees, with the implicit or explicit support of the local clergy, rejected the functionalist sociology of the *Strukturplan* out of hand.[59]

When Bishop Heinrich Tenhumberg announced details about the implementation of the plan to the media in April 1971, the widespread resentment among laity and parish priests was surely a major factor that had forced his hand. The deans had also voiced their concerns, since the deanery as an intermediary structure between parish and diocese would have been replaced by the *Großpfarrei*. With Tenhumberg's announcement, the far-ranging functionalist ideas of the plan were shelved. Some of its ideas, however, were implemented, albeit in much less coherent form. Subsequently, between three and five individual parishes should cooperate in a confederation (*Pfarrverband*), which entailed a pooling of resources and a coordination of pastoral initiatives. In line with attempts to improve coordination within the diocese, structures for regional cooperation were built up. Finally, the structure and responsibility of departments within the diocesan administration were adapted.

These were elements of a piecemeal strategy for organizational reform that was not only practised in Münster, but widely adopted across the Federal Republic during the 1970s. Working within the parameters of a *Pfarrverband* actually comprised some of the core reform elements the *Strukturplan* had envisaged. In Münster, all laypersons and priests who worked in pastoral care were to be trained by a team of experts in pastoral sociology, pastoral counselling, and social pedagogy. Identifying pastoral problems, surveying opinions among the laity through self-organized opinion polls, establishing benchmarks for planning and cooperation among the clergy, training staff in the soft skills that were required for a more cooperative work-pattern and, last but not least, periodic supervision by therapists in order to review the successes and failures of these professional engagements: these were core elements of the agenda set for practitioners in the *Pfarrverbände*. Reflecting the general trend towards a more flexible approach in pastoral planning, this range of activities was complemented by the 'Communio-plan', a new blueprint for pastoral reform that had been discussed since 1971, and which was finally promulgated in 1980. Rather than outlining core functions, this revised agenda conceptualized pastoral services in the context of a highly individualized society, where the Church had to address the diversity of the spiritual needs of individuals.[60]

The incremental implementation of new pastoral strategies and practices was by no means the only form of sociological intervention into the Catholic Church as an organization during the 1970s. At a more general level, professional sociologists discussed the need to conceptualize the Church using the theory of the most recent approaches to the study of formal organizations. The most prominent and widely heard voice in these debates belonged to Franz-Xaver Kaufmann, professor of sociology at the University of Bielefeld. His theoretical and highly critical interventions into controversies about the ongoing *aggiornamento* created widespread resonance in relevant circles. Kaufmann advocated the use of sociological thinking explicitly in order to rid the Church of the 'ballast' of dated normative ideas derived from natural law.[61] One particular concern for Kaufmann was the issue of a 'non-ecclesiastic religiosity', which emerged because a growing number of baptized Christians no longer accepted the institutional Church as the core reference point for their highly individualized piety. Reflecting these and other developments in their relevance for pastoral decisions, Kaufmann argued, made a substantial incorporation of sociological knowledge paramount in order to make the Catholic faith 'sustainable in the future' (*zukunftsfähig*).[62]

This concept in itself reveals how much the West German Church had been transformed, in the three decades from 1950 to 1980, under the impact of sociological knowledge: from an institution that was predicated on the continuation of a long-established tradition to an organization that was, at least in the view of some prominent Catholic academics, predicated on the notion of constant change. Sweeping and ambitious blueprints for organizational reform, as the *Strukturplan* in Münster, had hardly any chance of being implemented. But the semantics being used to describe the Church had changed substantially. Ultimately, decisive steps towards a structural reform of pastoral services were finally taken across the Federal Republic from the late 1990s, starting in the dioceses of Mainz and Essen. In a stark contrast to the early 1970s, however, these initiatives were not based on an intensive dialogue with sociologists. Rather, the bishops employed the services of consultancy firm McKinsey in order to drive down costs and make efficiency savings. It thus appears that finance was the hidden curriculum of Church reform in the Federal Republic. As long as the bishops could rely on a steady rise of income from Church tax – in fact a fourfold increase in nominal terms from 1963 to 1978 – they were not pressed hard enough in order to embark on the uncertainties of a full-fledged reform of pastoral services.[63]

Comparisons and conclusions

A comparison between the application of sociological knowledge in the Dutch and West German Catholic Church reveals both similarities and differences with regard to the timing, context, and limits of these initiatives. Starting in the immediate postwar period in the Netherlands, and with some delay in the Federal Republic, specialized sociological institutes such as KASKI with its Dutch and German branches began to offer statistical and sociographical knowledge for the reform of pastoral services. In a somewhat mechanical fashion, three different roles of the Catholic sociologists working in these institutes can be distinguished: 'consultant', 'practitioner', and 'researcher'.[64] In the Netherlands, in the 1960s, KASKI basically worked within the parameters of the first two roles, whereas ITS and PINK were able to develop a substantial analytical research agenda, which also extended the basic parameters of pastoral sociology. Similar initiatives in West Germany had to rely on the expertise offered by KASKI, or were situated in contingent circumstances, such as the Institute for Pastoral Sociology in Essen. Institutes for pastoral sociology in Germany, usually smaller than their Dutch counterparts, struggled to be embedded in crucial networks within the Church, and to make their voice heard at the top level of decision-making. Conducting substantial research was beyond their reach, and a number of Catholic sociologists soon took up university posts, as, for instance, Philipp von Wambolt, who quit his job in Münster in 1975, frustrated by the apparent lack of appetite for progressive reform.

These remarks highlight the different trajectories of pastoral sociology in the two countries, in terms of institutional resources and of interconnections within the national Church. There was, however, one important commonality, and that was the impact of Vatican II. The documents of the council not only legitimized concerted efforts to reform the Church with the help of the human and social sciences, including sociology. In addition, the council suggested the foundation of national bishops' conferences, and by implication mandated these bodies with the paramount task of implementing the *aggiornamento* of the council in a coordinated process and with concrete outcomes. This happened in the Dutch Pastoral Council from 1965 to 1970, and with the 'Joint Synod' of the West German Dioceses, planned since 1969 and gathering in Würzburg from 1971 to 1975. Both occasions provided Catholic sociologists with a crucial opportunity to shape the ongoing process of reform with their own ideas and insights. This was most obvious in the Netherlands, where Walter Goddijn served as secretary-general of the council. While individual sociologists did not play a central role in the preparation of the German synod, social scientific expertise was crucial for its preparation, too, as the bishops commissioned a comprehensive set of opinion polls that were meant to gauge the attitudes and preferences of the laity.[65]

While the council was a crucial turning-point for the scientization of the Catholic Church, it also marks the difference to the Protestant Churches in both countries. Since the late 1940s, Reformed Protestants in the Netherlands and Lutherans in West Germany had practised sociography, opinion polling, and organizations research in different pastoral contexts. In the Netherlands, social scientists in the Neo-Calvinist Churches often faced an uphill struggle when they tried to engage theologians and pastors with social scientific insights and perspectives. Gerard Dekker, the former manager of the Gereformeerd Sociologisch Instituut (GSI, the Sociological Institute of the Neo-Calvinist Churches), established in 1954, described his work within the Church as a 'hard fight'.[66] Additionally, the institutional basis for applied social research was weaker. The GSI, for instance, had already been disbanded by 1966.[67] In Germany, the reception of social science approaches among Protestants was initially equally lukewarm. But the 1960s saw a change of attitudes, and the Evangelische Kirche in Deutschland (EKD) continued to make substantial use of opinion polls and organizations research throughout the 1970s and 1980s.[68] In both countries, accordingly, the scientization of the Protestant churches was more incremental, and the backlash after 1970 was less significant. It should not be forgotten, though, that both Catholics and Protestants, often in an ecumenical fashion, have continued to use psychotherapy pastorally until the present.[69]

Both in the Dutch and in the German context, the pinnacle of a practical application of social scientific knowledge in the Catholic Church during the council and its immediate aftermath revealed that experts were not simply flag-bearers of objectivity. Even though the sociologists claimed to be rationalizing the controversial reform debates and to be underpinning them with empirical data, their interventions contributed towards a substantial politicization of the ongoing struggles about the implementation of Vatican II. The politicization of social scientific expertise was part and parcel of the scientizing of the social in the Catholic Church. It was thus a logical corollary of the increasing backlash against the ideas of the council since the mid-1970s that sociological approaches for a reform of the Church came under pressure. Influential theologians attacked sociological thinking within the Church in their attacks on Latin American liberation theology with its Marxist underpinnings. But they were also critical of what they perceived as a tendency of sociology to valorize 'society' as an almost god-like term.

A good example is a dissertation, published by German theologian Reinhard Marx in 1990 under the title 'Is the Church different?', that is, was the Church an institution that cannot be properly conceptualized in sociological terms. The answer for Marx was a resolute yes, and he minced no words in his suggestion that theologians should resist any further dissemination of sociological worldviews in the Church.[70] This conviction has certainly not hindered Marx's steep elevation to high-ranking office under Popes John Paul II and Benedict XVI. First appointed as archbishop for Paderborn in 1996 and then for Munich and Freising in 2007, Marx was accepted in the College of Cardinals in November 2010.

The growing backlash against sociological approaches to understanding and reforming the Church is also relevant in conceptual terms, as it underscores the fact that the scientizing of the social is no secular, inevitable, or irreversible process. As the example of the Catholic Church in the Netherlands and in West Germany demonstrates, any advancement of organizations research and related approaches was dependent on a reform coalition between social scientists, theologians, and progressive bishops, which emerged in a contingent situation during the run-up to and immediate aftermath of the Second Vatican Council.

Notes

1. For the text of the encyclical: H. Denzinger (1991), *Kompendium der Glaubensbekenntnisse und kirchlichen Lehrentscheidungen*, 37th edition (ed.) P. Hünermann (Freiburg: Herder), pp. 1046–58.
2. Cited in A. Etzioni (1964), 'Industrial Sociology. The Study of Economic Organizations', in idem (ed.), *Complex Organizations. A Sociological Reader* (New York: Holt, Rinehart and Winston), p. 130.
3. For such a concept of secularization, see B. Ziemann (2007a),'The Theory of Functional Differentiation and the History of Modern Society. Reflections on the Reception of Systems Theory in Recent Historiography', *Soziale Systeme* 13, pp. 220–9.
4. J.W. O'Malley (2008), *What Happened at Vatican II* (Cambridge, MA: Belknap Press), p. 50.
5. E. Klinger (1996), 'Das Aggiornamento der Pastoralkonstitution', in F-X. Kaufmann and A. Zingerle (eds), *Vaticanum II und Modernisierung. Historische, theologische und soziologische Perspektiven* (Paderborn: Schöningh), pp. 171–87, quote on 184.
6. L. Winkeler (2010), 'Sociografie en pastoraal beleid, 1946–1957. Franciscanen en de opkomst van de godsdienstsociologie in Nederland', in J. van Gennip and M. A. Willemsen (eds), *Het geloof dat inzicht zoekt. Religieuzen en de wetenschap* (Hilversum: Verloren), pp. 178–92.
7. Katholiek Documentatie Centrum, Nijmegen (KDC), KASKI: 2670.
8. Circular letter, 3 May 1948: Sociaal Historisch Centrum voor Limburg, Maastricht (SHCL), EAN 64: 366.
9. W. Goddijn and C. M. Thoen (1956), *Het Katholiek Sociaal-Kerkelijk Instituut. Research Centrum en Planbureau, 1946–1956. Ontstaan, organisatie en werkwijze* (The Hague: KASKI), p. 28.
10. See his chapter in this volume.
11. C. Dols (2010), 'Een Limburgs wespennest. Het Bureau Maastricht van het Katholiek Sociaal-Kerkelijk Instituut (KASKI), 1948–1951', in W. A. den Boer, J. G. J. van Booma, A. P. J. Jacobs, and J. Y. H. A. Jacobs (eds), *Onder 't Kruys. Kerkelijk en religieus leven in het gebied van Maas en Nederrijn* (Gouda: VNK), pp. 191–208.
12. Report 'Expertise über die Tätigkeit des Katholiek Sociaal-Kerkelijk Instituut (KASKI) in Holland' by A. Holl, 1971, p. 3: KDC, KASKI: 3626.
13. G. Dierick, A. Maes, and J. Tettero (1977), *Veertig jaar KASKI-onderzoek, 1946–1986* (Nijmegen: Dekker and Van de Vegt), pp. 42–114.
14. For the use of these technologies and approaches in the German Catholic Church, see B. Ziemann (2007b), *Katholische Kirche und Sozialwissenschaften, 1945–1975* (Göttingen: Vandenhoeck and Ruprecht), pp. 27–130.
15. *Expertise über die Tätigkeit des KASKI*, 10–12: KDC, KASKI: 3626.
16. M. Gastelaars (1985), *Een geregeld leven. Sociologie en sociale politiek in Nederland, 1925–1968* (Amsterdam: SUA), pp. 66–94.
17. Letter by Zeegers to Goddijn, 7 August 1952: KDC, KASKI: 2568.
18. Dols, 'Een Limburgs wespennest'.
19. Report of a meeting, 23 January 1952: KDC, KASKI: 4287.

20. J. A. Coleman (1978), *The Evolution of Dutch Catholicism, 1958–1974* (Berkeley: University of California Press).
21. T. van Schaik (1997), *Alfrink. Een biografie* (Amsterdam: Anthos), pp. 239–367.
22. J. Kennedy (1995), *Nieuw Babylon in aanbouw. Nederland in de jaren zestig* (Amsterdam: Boom), pp. 82–116.
23. Report 'De structuur van het Katholiek Sociaal-Kerkelijk Instituut, in het bijzonder van de afdeling godsdienstsociologie' by Walter Goddijn, March 1962: KDC, KASKI: 631.
24. M. Derks and Chr. Dols (2010), 'Sprekende cijfers. Katholieke sociaalingenieurs en de enscenering van de celibaatcrisis, 1963–1972', *Tijdschrift voor Geschiedenis* 123, pp. 414–29.
25. W. Goddijn (1993), *De moed niet verliezen. Kroniek van een priester-socioloog* (Kampen: Kok), p. 62.
26. J. Y. H. A. Jacobs (1997), 'Verankerd of weggedreven? Het Pastoraal Concilie van de Nederlandse Kerkprovincie (1966–1970) in het spanningsveld tussen lokaliteit en universaliteit', *Trajecta* 6, pp. 187–203.
27. The Dutch Bishops (1968a), *Vernieuwing en Verwarring. Vastenbrief 1968* (Heemstede: Actie voor God).
28. L. J. Rogier (1974), *Vandaag en morgen* (Bilthoven: Amboboeken), p. 22.
29. Interviews by Chr. Dols with Bishop H. Ernst (2 March 2010) and Bishop J. Bluyssen (10 May 2010).
30. O. Schreuder and E. Simons (1969), *Revolution in der Kirche? Kritik der kirchlichen Amtsstruktur* (Düsseldorf: Patmos), p. 73. Our emphasis.
31. A. van Heijst, M. Derks, and M. Monteiro (2010), *Ex Caritate. Kloosterleven, apostolaat en nieuwe spirit van actieve vrouwelijke religieuzen in de 19e en 20e eeuw* (Hilversum: Verloren), p. 941.
32. O. Schreuder (1964), *Het professioneel karakter van het geestelijk ambt* (Nijmegen: Dekker and Van de Vegt), and PCNK (1969), *Ontwerp-rapport Over het Ambt* (Rotterdam: PCNK).
33. 'Walter Goddijn wil meer lijn brengen in Nederlandse zielzorg', *Brabants Dagblad*, 8 February 1964.
34. List of extra studies in preparation of the Pastoral Council, 1971: KDC, PINK: 698.
35. H. Goddijn and W. Goddijn (1964), 'Voor een dialoog tussen theoloog en socioloog', *De Maand. Algemeen tijdschrift voor culturele en sociale bezinning* 7, pp. 487–97.
36. H. Goddijn and W. Goddijn (1966), 'De tweede emancipatie', in H. Goddijn and W. Goddijn (eds), *De kerk van morgen. Een postconciliair toekomstbeeld van de katholieke kerk in Nederland* (Roermond: Romen), pp. 7–23.
37. Ibid., p. 13.
38. O. Schreuder (1969), *Gedaanteverandering van de kerk. Aanbevelingen voor vernieuwing* (Nijmegen: Dekker and Van de Vegt), p. 21; H. Schelsky (1957), 'Ist die Dauerreflexion institutionalisierbar? Zum Thema einer modernen Religionssoziologie', *Zeitschrift für evangelische Ethik* 1, pp. 153–74.
39. *Expertise über die Tätigkeit des KASKI*, p. 7: KDC, KASKI: 3626.
40. The Dutch Bishops (1968b), *Kerk worden in dienst van een veranderende wereld* (Heemstede: Actie Voor God).
41. G. Fittkau (1969), *Der Kampf um den Zölibat, insbesondere in Holland* (Essen: n. p.).
42. Ziemann (2007b), *Katholische Kirche*, pp. 46–59.
43. B. Ziemann (2003), 'Auf der Suche nach der Wirklichkeit. Soziographie und soziale Schichtung im deutschen Katholizismus 1945–1970', *Geschichte und Gesellschaft* 29, pp. 409–40.
44. Core results with regard to the German clergy were published in J. Dellepoort, N. Greinacher, and W. Menges (1961), *Die deutsche Priesterfrage. Eine soziologische Untersuchung über Klerus und Priesternachwuchs in Deutschland* (Mainz: Matthias Grünewald).

45. R. Dahrendorf (1958), *Homo Sociologicus. Ein Versuch zur Geschichte, Bedeutung und Kritik der Kategorie der sozialen Rolle* (Cologne and Opladen: Westdeutscher Verlag). By 1974, the book had already gone into 13 editions.
46. Ziemann (2007b), *Katholische Kirche*, pp. 206–10.
47. Karl Forster, 'Entscheidungen und pastorale Initiativen für die kirchlichen Berufe. Hinweise aus den Ergebnissen sozialwissenschaftlicher Studien der letzten Jahre' (n. d. [1975]): Bistumsarchiv Münster, GV NA, A-0-966.
48. Ziemann (2007b), *Katholische Kirche*, pp. 211–19.
49. Compare ibid., pp. 220–5, quote on 224.
50. M. Ruck (2000), 'Ein kurzer Sommer der konkreten Utopie – Zur westdeutschen Planungsgeschichte der langen 60er Jahre', in A. Schildt, D. Siegfried, and K. C. Lammers (eds), *Dynamische Zeiten. Die 60er Jahre in den beiden deutschen Gesellschaften* (Hamburg: Christians), pp. 362–401.
51. L. Roos (1971), 'Kann man den Heilsdienst der Kirche planen?', *Lebendige Seelsorge* 22, pp. 111–22, quotes on 112–13; compare Schelsky (1957), 'Dauerreflexion'.
52. J. Hofmeier (1971), 'Kirchliche Strukturplanung', *Stimmen der Zeit* 188, pp. 230–46.
53. *Überlegungen und Vorschläge zur Struktur der Seelsorge im Bistum Münster. Strukturplan*, n. d. [Münster 1969], on the *Großpfarrei*, pp. 11, 13.
54. B. Ziemann (2007c), 'Organisation und Planung in der katholischen Kirche um 1970. Das Beispiel der Diözese Münster', *Schweizerische Zeitschrift für Religions- und Kirchengeschichte* 101, pp. 185–206.
55. Ibid., pp. 191–4.
56. Cited ibid., pp. 195–7.
57. R. Schlund (1970), 'Pastoralplanung im Erzbistum Freiburg', *Informationen. Berichte, Kommentare, Anregungen der Presse- und Informationsstelle des Erzbistums Freiburg* 3, pp. 20–4, at 20.
58. Ziemann (2007c), 'Organisation', pp. 198–9.
59. Ibid., pp. 200–3.
60. Ziemann (2007b), *Katholische Kirche*, pp. 245–56.
61. F-X. Kaufmann (1973), *Theologie in soziologischer Sicht* (Freiburg, Basel, and Wien: Herder), quote p. 5; cf. idem (1979), *Kirche begreifen. Analysen und Thesen zur gesellschaftlichen Verfassung des Christentums* (Freiburg: Herder).
62. Kaufmann (1979), *Kirche begreifen*, pp. 111–46; idem (1989), *Religion und Modernität. Sozialwissenschaftliche Perspektiven* (Tübingen: Mohr), pp. 235–75.
63. B. Ziemann (2007d), 'Die Katholische Kirche als religiöse Organisation. Deutschland und die Niederlande, 1950–1975', in F. W. Graf and K. Große Kracht (eds), *Religion und Gesellschaft. Europa im 20. Jahrhundert* (Cologne: Böhlau), pp. 329–51, here 348–9.
64. A. Holl (1971), 'La recherche appliqué. Le cas du KASKI aux Pays-Bas', *Social Compass* 18, pp. 621–37, at 624.
65. B. Ziemann (2006a), 'Opinion Polls and the Dynamics of the Public Sphere. The Catholic Church in the Federal Republic after 1968', *German History* 24, pp. 562–86.
66. G. Dekker (2005), *Een moeizaam gevecht. Mijn geschiedenis met de kerk* (Hilversum: Verloren), pp. 32–3.
67. L. Laeyendecker (1967), 'The Development of Sociology of Religion in the Netherlands since 1960', *Social Compass* 14, pp. 58–66.
68. Pastoralsoziologisches Institut der Evangelischen Fachhochschule Hannover (ed.) (2001), *'Gesellschaft in die Kirche tragen' oder: 30 Jahre Pastoralsoziologie in der hannoverschen Landeskirche* (Hanover: Blumhardt).
69. B. Ziemann (2006b), 'The Gospel of Psychology. Therapeutic Concepts and the Scientification of Pastoral Care in the West German Catholic Church, 1945–1980', *Central European History* 39, pp. 79–106.
70. R. Marx (1990), *Ist Kirche anders? Möglichkeiten und Grenzen einer soziologischen Betrachtungsweise* (Paderborn: Schöningh), pp. 447–9.

Index

GPSR Compliance
The European Union's (EU) General Product Safety Regulation (GPSR) is a set of rules that requires consumer products to be safe and our obligations to ensure this.

If you have any concerns about our products, you can contact us on

ProductSafety@springernature.com

In case Publisher is established outside the EU, the EU authorized representative is:

Springer Nature Customer Service Center GmbH
Europaplatz 3
69115 Heidelberg, Germany

www.ingramcontent.com/pod-product-compliance
Ingram Content Group UK Ltd.
Pitfield, Milton Keynes, MK11 3LW, UK
UKHW021015050925
462611UK00012B/1269
9781349326808